49 Springer Series in Solid-State Sciences

Edited by Peter Fulde

Springer Series in Solid-State Sciences

Editors: M. Cardona P. Fulde H.-J. Queisser

Volume 40 **Semiconductor Physics** - An Introduction By K. Seeger

Volume 41 **The LMTO Method** Muffin-Tin Orbitals and Electronic Structure
By H.L. Skriver

Volume 42 **Crystal Optics with Spatial Dispersion and the Theory of Excitations**
By V.M. Agranovich and V.L. Ginzburg

Volume 43 **Resonant Nonlinear Interactions of Light with Matter**
By V.S. Butylkin, A.E. Kaplan, Yu.G. Khronopulo, and E.I. Yakubovich

Volume 44 **Elastic Media with Microstructure II** Three-Dimensional Models
By I.A. Kunin

Volume 45 **Electronic Properties of Doped Semiconductors**
By B. I. Shklovskii and A. L. Efros

Volume 46 **Topological Disorder in Condensed Matter**
Editors: F. Yonezawa and T. Ninomiya

Volume 47 **Statics and Dynamics of Nonlinear Systems**
Editors: G. Benedek, H. Bilz, and R. Zeyher

Volume 48 **Magnetic Phase Transitions**
Editors: M. Ausloos and R. J. Elliott

Volume 49 **Organic Molecular Aggregates,** Electronic Excitation and Interaction Processes
Editors: P. Reineker, H. Haken, and H. C. Wolf

Volume 50 **Multiple Diffraction of X-Rays in Crystals**
By Shih-Lin Chang

Volume 51 **Phonon Scattering in Condensed Matter**
Editor: W. Eisenmenger

Volumes 1 - 39 are listed on the back inside cover

Organic Molecular Aggregates

Electronic Excitation and Interaction Processes

Proceedings of the International Symposium on Organic Materials at Schloss Elmau, Bavaria
June 5–10, 1983

Editors: P. Reineker, H. Haken, and H. C. Wolf

With 113 Figures

Springer-Verlag
Berlin Heidelberg New York Tokyo 1983

Professor Dr. *Peter Reineker*
Universität Ulm, Abteilung für Theoretische Physik,
D-7900 Ulm, Fed. Rep. of Germany

Professor Dr. Dr. h. c. *Hermann Haken*
Institut für Theoretische Physik der Universität Stuttgart, Pfaffenwaldring 57,
D-7000 Stuttgart 80, Fed. Rep. of Germany

Professor Dr. *Hans Christoph Wolf*
Physikalisches Institut der Universität Stuttgart, Pfaffenwaldring 57,
D-7000 Stuttgart 80, Fed. Rep. of Germany

Series Editors:
Professor Dr. Manuel Cardona
Professor Dr. Peter Fulde
Professor Dr. Hans-Joachim Queisser
Max-Planck-Institut für Festkörperforschung, Heisenbergstrasse 1
D-7000 Stuttgart 80, Fed. Rep. of Germany

ISBN-13:978-3-642-82143-1 e-ISBN-13:978-3-642-82141-7
DOI: 10.1007/978-3-642-82141-7

Softcover reprint of the hardcover 1st edition 1983

2153/3130-543210

Preface

This volume contains the talks presented at the International Symposium on Electronic Excitations and Interaction Processes in Organic Molecular Aggregates which was held at Schloss Elmau, Bavaria, Germany from June 5 to June 10, 1983.

In the recent years the investigation of organic materials has developed rapidly and has led to the construction of materials with interesting properties. The discovery of highly conducting and superconducting organic materials is definitely one of the reasons for the enormous increase in research activity in this field. Interesting applications have been realized or seem possible, such as the application of organic materials in electrophotography. The conductivity of organic polymers may be varied within a large range by doping and such materials have been used to construct an organic battery. Some time ago it was suggested that organic glasses and polymers could be used as storage materials in computer technology. With the development of preparation techniques and modern methods of investigation, for example, high resolution and picosecond spectroscopy, a large amount of experimental data is available not only for the conducting materials but also for organic semiconductors and insulators.

With this rapid accumulation of experimental material the microscopic theoretical understanding could not keep pace. Although several new concepts have been developed and discussed in recent years, for example, those of charge and spin density waves or of the coupled coherent and incoherent exciton motion, a lot of questions are still open. In the context of exciton motion, for example, the nature of the moving particle is still unclear. Is the moving excitation a free or a clothed particle? Is there a transition from free to clothed excitations and what are the relaxation times?

Another open question is the description of energy transfer in mixed molecular crystals. Have methods to be used as in the description of Anderson localization or is percolation theory more appropriate? Further unsolved problems are under discussion in connection with the description of the interaction of electronic excitations with each other, with the transitions from Frenkel to charge-transfer excitons and finally to electron-hole pairs, which are responsible for photoconductivity, or with the investigation of guest molecules in organic glasses. Finally, another interesting aspect should be mentioned: similar questions arise in the investigation of electronic excitations in organic molecular aggregates and in the investigation of photosynthesis in biological systems.

In view of the discrepancy between the huge amount of available experimental information and its theoretical exploration, a symposium on electronic excitations in organic molecular aggregates seemed necessary with more weight on theoretical questions. Another point of view in organizing the symposium was to invite not only researchers working in the field of electronic excitations in organic solids but also scientists doing research on similar problems in other fields of condensed-matter physics, in chemistry and biophysics.

Taking into account these aspects, a program for the symposium arose where, in about half of the talks, the questions mentioned above are approached from the theoretical point of view, and in the other half, more from the experimental side. Two introductory talks review basic experimental and theoretical methods and results. The aim of the symposium was to exhibit concepts and ideas which were formed in the investigation of organic solids, to analyze models and their consequences and to discuss the procedures in the organic field as compared to those in other fields.

The organizing committee would like to use this opportunity to thank the Volkswagenwerk Foundation for its support. Finally, the editors thank K. Kassner, A. Scheuing, B. Schmid, and U. Schmid for their help in preparing these proceedings.

Stuttgart/Ulm
November 1983

H. Haken P. Reineker H.C. Wolf

Contents

Part I

Basic Concepts, Methods and Results

Excited State Interaction and Energy Transfer Between Molecules in Organic Crystals – Basic Experimental Methods and Results

H.C. Wolf

Physikalisches Institut Teil 3, Universität Stuttgart, Pfaffenwaldring 57
D-7000 Stuttgart 80, Fed. Rep. of Germany

1. Introduction

One of the most interesting aspects of the physics and chemistry of organic molecular crystals is their ability to transfer energy which is absorbed at one specific site of the crystal to some other place where this energy is used for inducing processes which need energy: sensitized luminescence, charge carrier release, induction of photochemical reactions and others. In the past, many different experimental methods have been used to study such energy transfer processes. In my introductory talk, I want to give a survey of these methods or at least several of them, discuss their possibilities and try to give some ideas for future work.

2. Early History

The work on excited state interaction in organic crystals was initiated as early as 1907 by BECQUEREL [1]. A more systematic study started in the early twenties with the attempts of PRINGSHEIM and KRONENBERGER [2] to measure low-temperature absorption spectra of organic crystals, especially of benzene and in the following years by several Russian physicists [3]. Soon it became clear that in these crystals excited state interactions are relatively weak with the result that crystal absorption spectra are very similar to those of the isolated molecules. But it also became clear that there are characteristic differences. It took many years until finally in 1948 DAVYDOV [4,5] gave the fundamental explanation for the appearance of excited state interactions in crystal spectra in terms of the model of Frenkel excitons. He showed that the so-called Davydov splitting is one direct spectroscopic way to a study of excited state interactions in these crystals.

Any given molecular energy level is shifted and split into one or more bands when building a crystal from the molecules, depending on the number of inequivalent molecules in the unit cell. For a crystal with two molecules in the unit cell like naphthalene, there are two exciton bands for each molecular level, for the case of one molecule per unit cell there is one band only. The splitting between the $k = 0$ states of these bands is known as Davydov splitting, it depends on the interaction between molecules which are translationally inequivalent. The shift (translational shift) depends on the interaction between equivalent molecules. The width of the different exciton bands depends on both types of interaction. Therefore, from measuring crystal spectra one gets information on the pairwise excited state interaction matrix elements

$$J_{12} = (\phi_1\phi_2^*|V|\phi_1^*\phi_2)$$

corresponding to the transfer of excitation energy between molecules 1 and 2, where the asterix designates an excited state and V the interaction energy. One does not get the individual transfer elements, but rather always a sum of transfer elements which are due to interaction with different neighbour positions. Since very often one or two of them are by far larger than the others, measuring of crystal spectra is a good first way to study the static transfer elements. These parameters determine not only the exciton band structure and, in conjunction with scattering processes, the rates of energy transfer - the dynamic transfer parameters - for which, on the other hand, there is an independent second approach.

The second early approach originated from measuring the fluorescence of mixed crystals. In 1934, WINTERSTEIN, SCHÖN and VETTER [6] were the first who identified the "chrysogene" which was responsible for the green emission of anthracene crystals as a very low concentration of tetracene molecules, as impurities in anthracene, and so they initiated a tremendous amount of papers dealing with sensitized fluorescence in organic molecular crystals. The first reviews were given as early as 1938 and 1949 by FRANCK, TELLER and LIVINGSTON [7,8], and already by this time the importance of energy transfer as studied in mixed crystals by sensitized fluorescence for biological problems, especially photosynthesis, became clear. "Trivial" transfer - this means by reabsorption - long-range transfer (Förster type) and transfer by exciton type interaction was discussed. At that time, the review authors believed that exciton transfer in the anthracene crystal can be excluded by claiming [8]: "the absorption spectrum of anthracene shows no indication of a strong coupling of the molecules"... and ..."the time every particular molecule stays excited must be much greater than the duration of one atomic oscillation".

In the 35 years following the review by Franck and Livingston, a huge amount of experimental work has been published dealing with many kinds of excited state interaction and energy transfer. In the following, I want to summarize the most important experimental approaches. The most recent review is the book by POPE and SWENBERG [9].

3. Present Situation: Static Interaction Parameters

In this section, I want to summarize the methods to measure the static interaction parameters which have been developed from different types of experiments within the last 30 years, Table 1. The specific types of information which one can get from diffe-

Table 1. Static Interaction Parameters

Observation	Information	Quantity which can be deduced from measurements	Singlet or Triplet	Technique	Ref.
Davydov splitting	Sum of interaction elements	ΣJ_{ik}	S, T	Absorption and emission spectroscopy	4, 5
Splitting of molecular levels by aggregation	Individual interaction matrix elements (mini-excitons)	J_{11}, J_{12}	S, T	Pair spectroscopy High resolution excitation spectroscopy, isotopic mixed crystals	15, 16
Additional ESR lines ("M lines")	same	J_{12}	T	ESR in isotopically mixed crystals	17
Shift of levels in relatively dilute mixed systems	Squares of interaction matrix elements: quasi-resonance	J^2/U_0	S, T	Optical spectroscopy	18, 19
Shift of near band guest states by the presence of the exciton band	Location of band edge, band structure: Rashba effect	Band properties	S, T	Optical spectroscopy	20, 21, 16
Band-to-band transitions, including vibrational bands	Density of states in the exciton band	Band properties	S, T	Optical spectroscopy	22, 23, 24
Zero field ESR or ODMR line shape	same	Band properties	T	ESR, ODMR	25
Pair and exciton splitting in optical spectra	Exciton dimensionality	J_{11}	T	Optical spectroscopy	24, 26

rent types of observations using various experimental techniques are given as a survey. The references are far from being complete and I also did not always try to mention the most original papers. I rather tried to list those references which may give a good access to the different entries, sometimes also the first and the most recent or advanced paper.

4. Present Situation: Dynamic Interaction Parameters

Methods to measure the dynamic interaction parameters which are directly connected with energy transfer processes are summarized in Table 2. Again the references are somewhat subjective since the table covers an extremely large number of experimental papers. The references are again a mixture of the very first, the most advanced and reviewing papers. From this table and the preceeding one it is clear how many different experimental techniques have been applied to study excited state interaction processes in molecular crystals.

We can summarize the present experimental situation as follows. The static parameters are well understood, both in singlet and triplet states. The dynamic parameters are not so clearly established by experiments, especially not in the singlet state. In the triplet state the situation is somewhat better due to the possibility of measuring correlation times by spin resonance.

5. The Importance of Defects

One can hardly overemphasize the important role which defects and impurities play in all types of energy transfer experiments in molecular crystals. The purification of crystals has been developed into a very advanced technique [10]. But even in very impurity free crystals there are many defects which can hardly be avoided. The defects most characteristic for organic crystals are the so-called X traps [11]. These are host molecules which are slightly disturbed both in orientation and energy levels by the neighborhood of impurities or crystallographic defects.

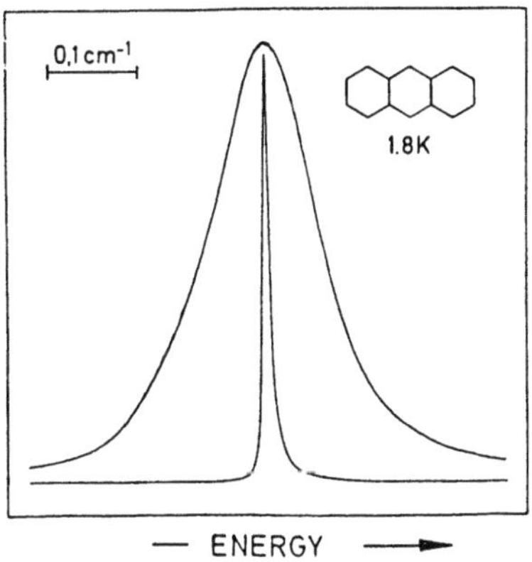

Fig. 1:

Excitation spectra ($T_1 \leftarrow S_0$) of the lower Davydov component at 14738 cm^{-1} in anthracene at 1.8 K for a typical Bridgman and sublimation crystal, respectively [16]

Table 2. Dynamic Interaction Parameters

Observation	Information	Quantity which can be deduced from measurements	Singlet or Triplet	Technique	Ref.
Sensitized fluorescence	Energy transfer constants	Energy transfer rate constant k_{HG}, diffusion coefficient D, capture rate	S, T	Host and guest fluorescence intensity in mixed crystals	27, 28
Sensitized delayed fluorescence	same	same	T	same, but using delayed fluorescence	29
Diffusion length	Diffusion coefficient of excitons	D	S	Direct measurement using thin crystals	30
	same	D	T, S	Ronchi ruling and transient grating technique	31, 32 33
Rise and decay function of sensitized emission	Exciton transfer and capture times	k_{HG}, capture rate	S	Time resolved (psec) spectroscopy	34
Line width and line shape of exciton ESR	Exciton correlation or hopping time	t_c, t_h, transfer rates P_{11}, P_{12}	T	ESR	35, 36
Influence of excitons on nuclear T_1	same	same	T	NMR	37

Table 2. continued

Observation	Information	Quantity which can be deduced from measurements	Singlet or Triplet	Technique	Ref.
Line shape and line-width of optical and ESR lines	Phonon scattering	scattering rate $P_{kk'}$	S, T	Optical spectroscopy, ESR, ODMR	38, 40 36, 39 25
Kinetics of delayed fluorescence	Exciton diffusion and annihilation	annihilation coefficient γ_1, D	T	Optical spectroscopy	41
Line shape and width, spectral structure of exciton ODMR and ESR	same	same	T	DF-ODMR, RYDMR	42, 43
Dependence of fluorescence intensity from excitation intensity	same	same	S	Optical spectroscopy at higher excitation densities	34, 44
Sensitized emission in isotopic mixed crystals at higher (>1%) concentration	Exciton transfer in "dilute" exciton systems	k_{HG}, D	S, T	Intensities of guest emission with different trap depth in isotopic mixed crystals as a function of temperature	45, 46 47, 48

Their importance is mainly due to the fact that the shifts of energy levels in the host crystal at X traps are typically of the same order of magnitude between 0 and 200 cm^{-1} as most of the transfer matrix elements which are listed or mentioned in Tables 1 and 2. - A very illustrative example of the influence of such defects on electronic properties of molecular crystals is given in Fig. 1, where an absorption line width of a highest purity Bridgman grown anthracene crystal is compared with that of an extremely thin sublimation flake. From all our experience it seems to be evident that the crystallographically best crystals which have been studied so far are such thin sublimation flakes which have been grown without a substrate, and that all crystals grown by other methods contain much higher concentrations of defects of different origin.

6. Directions of Future Work

In such an old and large field it is difficult and perhaps impossible to predict the lines of important new developments and discoveries. I want to mention, nevertheless, a few problems which are still open for good and interesting experimental work and, in my opinion, are of great importance in the field.

One is the question of exciton coherence and coherent energy transfer. Coherent exciton states have been unambigously demonstrated so far only in static experiments. Despite the fact that coherence of exciton transport has been discussed and studied in many papers evaluating and discussing experiments in this field, crucial and very direct measurements giving evidence for coherent energy transfer are, in my opinion, still missing. It would be nice to see standing exciton waves or other types of exciton interference in a crystal.

Another problem which has been discussed often and still leaves many simple questions to be answered is the detailed process both in the energy and time domain of exciton capture and localization in the process of energy transfer. Most of the experiments on exciton transfer measure only the sum of transfer and localization processes. More work must be done in order to disentangle such measurements.

Exciton-exciton interaction is also not really understood. The question whether two excitons attract each other or if the interaction is repulsive is not answered. Also the exciton-phonon scattering processes are not well understood in general.

Finally it seems highly desirable to apply the large amount of detailed knowledge on energy transfer processes to other more complex systems. A very appealing biological example which has been discussed for almost 50 years is the energy transfer from antenna molecules to the reaction center in photosynthesis. Already as early as 1936 GAFFRON [49] postulated the existence of a photosynthetic unit in the photosystem of algae and plants. This is the name for structures within the plants which contain several hundred chlorophyll molecules. These antenna molecules absorb the light and transfer excitation energy to the reaction center where photochemistry starts. Already 1938, FRANCK and

TELLER [7] discussed whether exciton transport within these units is responsible for the transfer to the reaction place. At this time it was assumed that a CO_2 molecule absorbed at the end of such a unit is the "reaction center" where light energy is transformed into chemical reaction. Now we know that the process is more complicated.

In the last few years, it has been shown by ESR and ENDOR that at least in photosynthetic bacteria the heart of the reaction center is the "special pair" of two chlorophyll molecules [12]. But the energy transfer mechanism from the antenna system to the reaction center is still much more an object of speculation rather than of real understanding. There are almost no experiments available which measure this transfer process. Only very recently, [13,14] it has been shown that the existence of energy transfer between the antenna and the reaction center system can be demonstrated by ODMR techniques. In this system, energy transfer is going on in both directions. In other words, the reaction center is apparently not a trap due to energy but due to fast reaction.

One can expect stimulation and progress in understanding of complex systems by applying the knowledge of excitation interaction and transfer which comes from studying molecular crystals as model systems to processes of solid-state and condensed phase photochemistry and to many problems in photobiology.

References

1. J. Becquerel, Le Radium 4, 328 (1907)
2. P. Pringsheim and A.Kronenberger, Z.Physik 40, 75 (1926)
3. see for instance I.W. Obreimow and A.F. Prichotjko, Physik. Z.Sowjetunion 1, 203 (1932)
4. A.S. Davydov, J.Exptl.Theor.Phys. (USSR) 18, 210 (1948)
5. A.S. Davydov, Theory of Molecular Excitons, engl.Translation, McGraw-Hill, New York (1962)
6. A. Winterstein, U. Schön and H. Vetter, Naturwiss. 22, 237, (1934)
7. J. Franck and E. Teller, J.Chem.Phys. 6, 861 (1938)
8. J. Franck and R. Livingston, Rev.Mod.Phys. 21, 505 (1949)
9. M. Pope and C.E. Swenberg, Electronic Processes in Organic Crystals, Clarendon Press (1982)
10. N. Karl, in Crystals, Vol. 4, Springer Verlag Berlin (1980)
11. A. Pröpstl and H.C. Wolf, Z.Naturforsch. 18a, 724 (1963)
12. J.R. Norris, R.A. Uphaus, H.L. Crespi, J.J. Katz, Proc.Natl.Acad.Sci.USA 68, 625 (1971)
13. H.J. den Blanken, G.R. van der Zwet and A.H. Hoff, Chem.Phys.Letters 85, 335 (1982)
14. J. Beck, J.U. v.Schütz and H.C. Wolf, Chem.Phys.Letters 94, 141, (1983)
15. D.M. Hanson, J.Chem.Phys. 52, 3409 (1970)
16. H. Port, D. Rund, H.C. Wolf, J.Chem.Phys. 60, 81 (1981)
17. M. Schwoerer and H.C. Wolf, Mol.Cryst. 3, 177 (1967)
18. G.C. Nieman and G.W. Robinson, J.Chem.Phys. 39, 1298 (1963)
19. D.M. Hanson, R. Kopelman and G.W. Robinson, J.Chem.Phys. 51, 212 (1969)

20. E.I. Rashba, Optika i. Spektroskopiya 2, 568 (1957)
21. V.L. Broude, E.I. Rashba and E.F. Sheka, Sov.Phys.Doklady 6, 718 (1962)
22. E.I. Rashba, Sov.Phys.Sol.State 5, 757 (1963)
23. S.D. Colson, D.M. Hanson, R. Kopelman and G.W. Robinson, J.Chem.Phys. 48, 2212 (1968)
24. R.M. Hochstrasser, T.Y. Li, H.N. Sung, J. Wessel and A.H. Zewail, Pure Appl. Chem. 37, 85 (1974)
25. A.H. Francis and C.B. Harris, Chem.Phys.Letters 9, 181,188 (1971)
26. R.M. Hochstrasser and J.D. Whiteman, J.Chem.Phys. 56, 5945 (1972)
27. H.C. Wolf, Adv.Atomic and Molec.Phys. 3, 119 (1967)
28. R.C. Powell and H.G. Soos, J.Lumin. 11, 1 (1975)
29. H.C. Wolf and K.W. Benz, Pure Appl. Chem. 27, 439 (1971)
30. G. Gallus and H.C. Wolf, phys.stat.sol. 16, 277 (1966)
31. P. Avakian and R.E. Merrifield, Phys.Rev.Letters 13, 541 (1964)
32. V. Ern, P. Avakian and R.E. Merrifield, Phys.Rev. 148, 862 (1966)
33. J.R. Salcedo, A.E. Siegman, D.D. Dlott and M.D. Fayer, Phys.Rev.Lett. 41, 131 (1978)
34. A. Braun, U. Mayer, H. Auweter, H.C. Wolf and D. Schmid, Z.Naturforsch. 37a, 1013 (1982)
35. D. Haarer and H.C. Wolf, Mol.Liq.Cryst. 10, 359 (1970)
36. R. Schmidberger and H.C. Wolf, Chem.Phys.Letters 25, 185 (1974)
37. G. Maier, U. Haeberlen and H.C. Wolf, Phys.Lett. 25A, 232 (1967)
38. H. Port, D. Rund, G.J. Small and V. Yakhot, Chem.Phys. 39, 175 (1979)
39. A.J. van Strien, J.F.C. van Kooten and J. Schmidt, Chem.Phys.Letters 76, 7 (1980)
40. D.M. Burland, U. Konzelmann and R.M. Macfarlane, J.Chem.Phys. 67, 1926 (1977)
41. P. Avakian and R.E. Merrifield, Mol.Cryst. 5, 37 (1968)
42. E.L. Frankevich and S.I. Kubarev, in Triplet State ODMR Spectroscopy ed. R.H. Clarke, Wiley, New York (1982)
43. J.U. v.Schütz, W. Steudle, H.C. Wolf and V. Yakhot, J.Chem.Phys. 46, 53 (1980)
44. V.A. Benderskii, V.Kh. Brikenstein, Ye. Ya Misochko, N.A. Vidmont and V.V. Korshunov, Mol.Liq.Cryst. 57, 47 (1980)
45. V.L. Broude and M.I. Onoprienko, Opt. and Spectrosc. 10, 133 (1961)
46. S.D. Colson and G.W. Robinson, J.Chem.Phys. 48, 2550 (1968)
47. K.E. Mauser, H. Port and H.C. Wolf, J.Chem.Phys. 1, 74 (1973)
48. A.H. Francis and R. Kopelman, in Topics Appl.Phys. Vol. 49, 238 (1981) Springer-Verlag
49. H. Gaffron and K. Wohl, Naturwiss. 24, 81,103 (1936)

Electronic Excitations in Molecular Solids

C.B. Duke
Xerox Corporation, 800 Phillips Road, Bldg. 114, Webster, NY 14580, USA

Abstract

This paper is a synopsis of the keynote theoretical lecture at the 1983 International Symposium on Electronic Excitations and Interaction Processes in Organic Molecular Aggregates. Models of charge and exciton motion in molecular solids are reviewed. The effects of both static and dynamic disorder on these excitations are noted. An important feature of certain molecular crystals is their "quasi-one-dimensional" character. Thus, the interplay between the dimensionality and the dynamics of electronic excitations is stressed. Specific examples of the consequences of disorder and collective phenomena are provided. The paper concludes with a list of several important unresolved issues.

1. Introduction

Molecular solids are characterized by two fundamental features: the retention of molecular identity in the solid state and the occurrence of relatively weak interactions between the individual molecular entities. The weak interactions lead to a high degree of disorder caused both by defects and by thermally induced motions of the molecules. In addition, the hopping integrals for an excess charge carrier to move from one molecule to its neighbors are small, typically about 0.01 eV, whereas the relaxation energy of these charges due to their polarization of both the molecules on which they reside and the surrounding dielectric medium is quite large, typically over 1 eV. The combination of large relaxation energies, small hopping integrals, and copious disorder suggests that charge carriers in organic solids form a Fermi glass of self-trapped polarons, i.e., they occupy localized states characterized by a high degree of local atomic and electronic polarization [1]. Similar expectations arise for excitons, although the values of the various parameters are quite different for charges, singlet excitons, and triplet excitons [2]. Therefore the fundamental structural and dynamical features of molecular solids suggest the probable localization of electronic excitations in these materials and hence the description of their motion in terms of statistical analyses of site-to-site hopping rather than of energy-band theory [1,3]. An important aspect of models of the dynamics of these excitations is whether the disorder is structural in orgin (e.g., in molecular glasses) or whether it arises from dynamic disorder induced by excitation-phonon coupling in a vibrating molecular medium.

A third fundamental feature of certain molecular solids, polymers and segregated-stack linear-chain charge transfer salts in particular [4], is the large asymmetry between electronic motion in one direction (e.g., along the macromolecular backbone) and that in the other two. This fact leads to their appellation as "quasi-one-dimensional" materials [4,5]. An important aspect of such quasi-1D systems is the occurrence of collective semiconductor (e.g., "charge-density-wave") low-temperature phases, together with their concomitant unusual excitation spectra [6]. Indeed, a major thrust of current research in this area is the suppression of these semiconducting collective states in order to obtain conducting or superconducting low-temperature phases [7].

Modern research on the theory of electronic excitations in organic molecular solids concerns, therefore, the consequences of disorder and collective phenomena in both the eigenvalue spectra and the dynamics of these excitations. This is a rich and complex area for research, as may be inferred from inspection of recent pertinent conference proceedings [4-7]. A systematic discussion of the ingredients, scope, and status of current models of the electronic properties of polymers is given elsewhere [8]. Herein, I attempt only to outline the structure of these models and to emphasize the subtlety of the interplay between disorder and dynamics by reminding you of how disorder produces polarons in typical molecular glasses and possibly crystals, but causes the breakdown of the low-temperature Peierls insulator state in doped polyacetylene. The details of exciton as well as charge carrier motion in both localized-state (e.g., glasses) and extended-state (e.g., conducting charge transfer salts) systems will be discussed by many other speakers at this symposium.

2. Models of Charge Carrier Motion in Organic Solids

To illustrate the nature of models of injected charges in organic solids, we consider the transfer of an injected ("extra") electron (or exciton, only the values of the parameters change) between molecular electronic states centered on each of two molecular sites labelled by i=1,2, respectively. A model Hamiltonian describing this process is given by

$$H = H_o + H_1, \tag{1a}$$

$$H_o = \varepsilon_1 a_1^+ a_1 + \varepsilon_2 a_2^+ a_2 + \sum_n g_{1n} \hbar \omega_n a_1^+ a_1 (b_n + b_n^+)$$

$$+ \sum_n g_{2n} \hbar \omega_n a_2^+ a_2 (b_n + b_n^+) + \sum_n \hbar \omega_n (b_n^+ b_n + 1/2), \tag{1b}$$

$$H_1 = t_o (a_2^+ a_1 + a_1^+ a_2). \tag{1c}$$

In Eqs. (1) a_i^+ designates the creation operator of an electron in an orbital of energy ε_i at site i, and b_n^+ that of a normal mode of vibration of energy $\hbar\omega_n$ coupled to the electronic state labelled i via the dimensionless coupling constant g_{in}. H_1 describes the electron transfer from one site to another via the electron overlap integral, t_o. The $\hbar\omega_n$ are the energies of the normal modes of vibration of the molecular medium in which we envision the two sites to be embedded. A simple extension of this Hamiltonian may be utilized to describe transport in molecular crystals. The electron sites are placed on a periodic lattice and a Hamiltonian is defined by summation over these sites, i.e.,

$$H = \sum_i \varepsilon_i a_i^+ a_i + \sum_{i \neq j} (t_{ij} a_i^+ a_j + \text{h.c.}) + \sum_{i,n} g_{in} \hbar \omega_n a_i^+ a_i (b_n + b_n^+)$$

$$+ \sum_n \hbar \omega_n (b_n^+ b_n + 1/2). \tag{2}$$

The eigenstates of this Hamiltonian become the well-known Bloch energy-band states in the limit that $g_{in} \to 0$ for all values of the normal mode index n. In this case for nearest-neighbor hopping the overlap integrals, t, would be related to the band width, W, via an expression of the form

$$W = 2zt \tag{3}$$

in which z is the coordination number of the (presumed identical) sites in the periodic lattice. Furthermore, the electron eigenstates are of the Bloch form

$$\psi_{nk}(r) = \exp(ik \cdot r)\, u_{nk}(r) \tag{4}$$

where the $u_{nk}(r)$ exhibit the periodicity of the crystalline lattice.

If the electron-vibration coupling constants, g_{in}, are not zero, then the electronic and vibrational states are coupled. Nevertheless, at low temperatures electronic motion still occurs via extended states built from periodic superpositions of local electronic wave functions accompanied by a local lattice distortion [9,10]. The entity consisting of the localized electron with its accompanying distortion of the lattice is called a polaron. At high temperatures the number of excited vibrational modes (phonons) changes when an electron moves from one site to another, leading to the idea of electron transport via hopping motion between localized states: the familiar small polaron hopping model [9,10]. Unfortunately these traditional types of motion do not often describe the behavior of injected charges in organic solids, so the model requires extension [1,8].

For polymers and molecular glasses the most important extension lies in the recognition that both the site energies $\{\varepsilon_i\}$ and the hopping integrals $\{t_{ij}\}$ in Eq. (2) form distributions [8]. Thus, the model Hamiltonian itself is specified statistically in terms of these distributions. This fact has profound consequences for the nature of the eigenstates of Eq. (2): a result which in recent years has come to be labelled a "Scientific Revolution" in the sense of Thomas Kuhn [11]. The qualitative features of interest to us can be defined in terms of a mean site energy ε, the rms deviation from this mean, $\Delta = [\langle(\varepsilon_i - \varepsilon)^2\rangle_{AV}]^{1/2}$, and a mean hopping integral t [2]. Variations in the site energies from the mean (described by Δ) are referred to as "diagonal disorder" whereas analogous variations of the hopping integrals are called "off-diagonal disorder." Similarily, if these variations are caused by local time-independent fluctuations in composition or structure we speak of "static disorder" while if they are generated by the time-dependent thermal vibrations we employ the term "dynamic disorder." Injected charges are localized, i.e., they form molecular cations or anions within the solid, if

$$\Delta > czt = cW/2 \tag{5}$$

in which c is a dimensionless number of the order of unity which depends both on the connectivity (i.e., dimensionality) of the molecular system and on the extent of off-diagonal disorder. Typically $c \sim 2.5$ for (isotropic) three-dimensional systems, 1.5 for two-dimensional systems, and zero for one-dimensional systems, although some confusion currently exists in the case of two dimensions [12]. Inequality (5) is believed to be satisfied in molecular glasses and in certain aromatic polymers [1,8]. In these cases electrons and holes injected into the material form local, molecular radical anions and cations, respectively, rather than extended mobile states like those characteristic of crystalline covalent semiconductors. Consequently, under these circumstances the electrons and holes are more accurately visualized as ions in solution which move, however, by carrier hopping rather than by ionic diffusion. On the other hand, static diagonal disorder should be zero in a (defect-free) bulk organic crystal, so that inequality (5) is not satisfied and extended band-like states are not prohibited by these considerations.

A second important extension of the model defined by Eq. (2) is the incorporation of dynamic off-diagonal disorder, i.e.,

$$H_2 = \sum_{i\neq j} \sum_{n} f_{ij,n} \hbar \omega_n (a_i^+ a_j + a_j^+ a_i)(b_n + b_n^+). \qquad (6)$$

This term is required both to describe the well-known temperature-independent electron and hole mobilities observed in many molecular crystals [1] as well as to obtain the collective semiconducting Peierls insulating state in segregated-stack linear-chain charge transfer salts [4,5] and polyacetylene [13,14]. It also has important applications for electron transfer in biological and electrochemical systems [15]. The sum of the Hamiltonians given by Eqs. (2) and (6) defines a general (schematic) model which, in one form or another, is utilized to describe the various systems considered at this symposium.

3. Polarons in Organic Glasses

Organic glasses constitute a class of well-documented examples of the localization of injected carriers by static diagonal disorder [1,8, 16-18]. Analyses of transport and thermally stimulated current measurements suggest that the widths of the distributions characterizing static, diagonal disorder are $\Delta \sim 0.1$ eV for charge carriers injected into suitably prepared materials [17,18]. Singlet excitons also are localized with $\Delta \sim 0.05$ eV [16,18]. Photoemission experiments give [8,19,20]

$$\Delta^2 = \Delta_o^2 + c^2 T \qquad (7)$$

in which $\Delta_o \sim 0.4$ eV and $c \sim 3x10^{-2} eVK^{-1/2}$. The increase in the value of Δ_o relative to the value measured by transport probably is caused by spatial dependence of the relaxation energy in the vicinity of a surface [1,8]. A detailed small-polaron model of photoemission [8,16], UV absorption [8,16], and electron-transfer [21] has been developed for these materials and shown to be in quantitative agreement with available photoemission and UV absorption data [8,16]. The model also is consistent with transport measurements in non-polar glasses [17,18] and early results on polar glasses [22], but recent measurements [23] on molecularly doped polar glasses have led to puzzles [18]. Overall, in these systems static diagonal disorder localizes electronic excitations whose subsequent dynamics are essentially those of small polarons.

4. Temperature-Dependent Dimensionality: Electron Mobilities in Naphthalene

Although there have been numerous measurements of carrier drift mobilities in organic molecular crystals [24], interest in the topic was greatly stimulated by the observation in 1978 of a rapid rise below 100K in the mobility along the c' direction of electrons in single crystals of naphthalene [25]. These data were interpreted at the time as evidence for the long-awaited [9,10] band-to-hopping transition in molecular crystals. Subsequently, this interpretation was challenged by Sumi who reported an analysis in which the observed mobilities were attributed to hopping transport between quasi-1D stacks of naphthalene molecules in the a direction [26]. A review of the topic [1] revealed that in addition to embodying copious adjustable parameters, this model is not consistent with the rotation with temperature in the ac plane of the principal axes of the mobility tensor. Moreover, no detailed energy-band model analyses of the low-temperature mobilities were available, so the presumption of the failure of that model might well be premature. Therefore in spite of the fact that considerable additional measurements of these drift mobilities have been reported since 1978 [24], the issue of the existence of a band-to-hopping transition for electrons in naphthalene is not yet resolved.

The tentative picture which is emerging from both recent measurements [27,28] and calculations [29] is that below $T_{c3} \simeq 100K$ naphthalene is a 3D band semiconductor with the electron mobilities in all directions being limited by scattering from defects and acoustical phonons. For temperatures in the range $T_{c3} \lesssim T \lesssim T_{c2}$, $T_{c2} \simeq 170K$, the electrons are localized in the ab plane and hopping occurs normal to this plane. Finally, for $T > T_{c2}$ the electrons are localized normal to the b axis as well, leaving their wave functions "extended" over multiple molecules along the a axis but localized along segregated stacks parallel to this axis. Thus, the Sumi model is recovered for $T > T_{c2}$ and the energy-band model for $T < T_{c3}$. This transition from a quasi-1D material at high temperatures to a 3D semiconductor at low temperatures is remarkably similar to those observed for a wide variety of charge-density-wave, spin-density-wave, and superconducting segregated-stack linear-chain charge transfer salts [30]. The occurrence of such a common characteristic in all of these electronically dissimilar materials illustrates well the delicate balance which exists between dynamic disorder and both energy-band and collective phenomena in organic molecular crystals.

5. Semiconductor-to-Metal Transitions in Polyacetylene

Thus far we have considered only examples in which disorder-induced localization was associated with semiconducting behavior. In the case of Peierls insulators, however, disorder can induce metallic behavior in an otherwise semiconducting material. We indicate the nature of this remarkable phenomenon by considering the specific example of the semiconductor-to-metal transition in iodine-doped polyacetylene [14,31].

Polyacetylene exhibits a doubly degenerate commensurate Peierls insulating ground state characterized by a bond-alternating geometry of the $(CH)_x$ backbone and a Peierls gap, E_P, in the electronic excitation spectrum. The introduction of charge transfer dopants initially induces domain walls (called "kinks" or (improperly) "solitons") between the two degenerate phases. For acceptor dopants like iodine each kink is associated with an (induced) empty mid-gap electronic state, $E_k = E_p/2$, above the top edge of the valence band. The spatial extent of these kinks is about $10\underline{a}$, where $\underline{a}$ is the average distance between two carbon species along the backbone. As long as these kinks do not overlap (i.e., dopant concentrations are $c < c^* \sim \underline{a}/(20\underline{a}) = 0.05$), the Peierls state is insensitive to disorder and acts like a semiconductor of gap E_p with a suitable number of empty mid-gap states with spin 0 and charge +e.

When $c \gtrsim c^*$ the ground state of an ideal Peierls insulator would be a uniform incommensurate charge density wave (CDW) with its associated distorted geometry of the polyacetylene backbone. Disorder of the dopants destroys this CDW state, however, by pinning the induced charges on the backbone at the random positions of the impurities. This pinning destroys the order parameter, thereby eliminating both the one-electron gap and the bond alternation. In this limit polyacetylene becomes a "dirty" metal.

The predictions of the detailed analysis of Mele and Rice [14] are in remarkable correspondence with a variety of measurements on iodine-doped polyacetylene [31]. First a sharp rise in the conductivity occurs at $c \sim 0.01$ associated with variable range hopping within (disorder-induced) electronic states in the gap just above the valence band edge. Finally, in the range $0.04 \leq c \leq 0.1$ the magnetic susceptibility rises because of the breakdown of the Peierls state and the concomitant elimination of the gap. In particular, for $0.01 \lesssim c \lesssim 0.05$ the remarkable phenomenon of conductivity without comparable susceptibility is predicted, in excellent accord with both recent data and the early data [13] that originally stimulated the "soliton" concept. Thus, doped polyacetylene affords an example in which disorder destroys a collective semiconducting ground state and produces an apparently "spinless" conductor as an intermediate stage characterized by only the partial breakdown of this state as reflected in a low but continuous distribution of electronic states in the Peierls gap.

6. Prospects and Issues

The above examples reveal that the polarizability of organic materials and their propensity to disorder produces expected phenomena like small polarons as well as exotic phenomena like temperature-dependent dimensionality and apparently spinless conductivity. Many puzzles and issues remain, however, including the establishment of limitations of the Bloch model via a quantitative analysis of the band-to-hopping transition (or lack thereof) in naphthalene; the interpretation of the lack of activation of carrier transport in certain polar organic glasses; the clarification of the nature of the spinless charge carriers in polymers like poly(p-phenylene) which are essentially different from polyacetylene in that they do not exhibit degenerate ground states; and the establishment of the interplay between superconductor, Peierls, and spin-density-wave instabilities in quasi-1D organic crystals. In particular, the description of disorder by statistically specified model Hamiltonians has led us into uncharted waters containing exotic new phenomena, only a few of which are considered herein. Weak intermolecular interactions usually lead to a disordered character of the organic solid state, and hence to its considerable interest as a test bed for the limitations of traditional models of periodic solids.

Acknowledgements

I am indebted to D. Costenoble for assistance in preparing this manuscript and to M.J. Rice and A.J. Epstein for stimulating discussions of the semiconductor-to-metal transition in polyacetylene.

References

1. C.B. Duke: "Organic Solids: Traditional Semiconductors or Fermi Glasses?" in Festkörperprobleme, J. Treusch, ed., Advances in Solid State Physics, Vol. 22 (Vieweg, Braunschweig, 1982), pp. 21-34
2. C.B. Duke: Surf. Sci. 70, 674 (1978)
3. G. Hohler, ed: Exciton Dynamics in Molecular Crystals and Aggregates, Springer Tracts in Modern Physics, Vol. 94 (Springer-Verlag, Berlin, 1982)
4. J.S. Miller and A.J. Epstein, eds: Synthesis and Properties of Low Dimensional Materials, Ann. New York Acad. Sci., Vol. 313 (New York Academy of Sciences, New York, 1978)
5. J. Ehlers, K. Hepp and H.A. Weidenmuller, eds: One-Dimensional Conductors, Lecture Notes in Physics, Vol. 34 (Springer-Verlag, Berlin, 1975)
6. J. Bernasconi and T. Schneider, eds: Physics in One Dimension, Springer Series in Solid State Sciences, Vol. 23 (Springer-Verlag, Berlin, 1981)
7. A.J. Epstein and E.M. Conwell, eds: Proceedings of the International Conference on Low-Dimensional Conductors, Mol. Cryst. Liq. Cryst., Vols. 77, 79, 81, 83, 85 and 86 (Gordon and Breach, London, 1982)
8. C.B. Duke: "The Electronic Structure of Semiconducting Polymers," in Extended Linear Chain Compounds, J.S. Miller, ed., Vol. 2 (Plenum, New York, 1982), pp. 59-125

9. T. Holstein: Ann. Phys. (N.Y.) 8, 343 (1959)
10. J. Yamashita and T. Kurosawa: Phys. Chem. Solids 5, 34 (1958): J. Phys. Soc. Jpn. 15, 802 (1960)
11. P.W. Anderson: "Lectures in Amorphous Systems", in Ill-Condensed Matter, R. Balian, R. Maynard and G. Toulouse, eds. (North Holland, Amsterdam, 1979), pp. 159-261
12. D. Weaire: "The Anderson Localization Problem", in Fundamental Physics of Amorphous Semiconductors, F. Yonezawa, ed., Springer Series in Solid State Sciences, Vol. 25 (Springer-Verlag, Berlin, 1981), pp. 155-163
13. W.P. Su, J.R. Schrieffer and A.J. Heeger: Phys. Rev. Lett. 42, 1698 (1979)
14. E.J. Mele and M.J. Rice: Phys. Rev. B 23, 5397 (1981)
15. C.B. Duke, "Concepts in Quantum Mechanical Tunneling in Systems of Biological and Chemical Interest", in Tunneling in Biological Systems, B. Chance, D.C. Devault, H. Frauenfelder, R.A. Marcus, J.R. Schrieffer and N. Sutin, eds. (Academic, New York, 1979), pp. 31-65
16. C.B. Duke, Mol. Cryst. Liq. Cryst. 50, 63 (1979)
17. H. Bässler: Phys. Stat. Sol. (b) 107, 9 (1981)
18. J. Lange and H. Bässler: Phys. Stat. Sol. (b) 114, 561 (1982)
19. W.R. Salaneck, C.B. Duke, W. Eberhardt, E.W. Plummer, and H.J. Freund: Phys. Rev. Lett. 45, 280 (1980)
20. N. Sato, K. Seki, H. Inokuchi, Y. Harada and T. Takahashi: Solid State Commun. 41, 759 (1982)
21. C.B. Duke and R.J. Meyer: Phys. Rev. B23, 2111 (1981)
22. W.D. Gill: "Polymeric Photoconductors", in Photoconductivity and Related Phenomena, J. Mort and D. Pai, eds. (Elsevier, Amsterdam, 1976), pp. 303-334
23. H. Bässler, G. Schonherr, M. Abkowitz and D. Pai: Phys. Rev. B26, 3105 (1982)
24. L.B. Schein and D.W. Brown: Mol. Cryst. Liq. Cryst. 87, 1 (1982)
25. L.B. Schein, C.B. Duke and A.R. McGhie: Phys. Rev. Lett. 40, 197 (1978)
26. H. Sumi: J. Chem. Phys. 75, 2987 (1981)
27. N. Karl and W. Warta: unpublished
28. L.B. Schein, W. Warta, A.R. McGhie and N. Karl: "Mobilities of Electrons and Holes in Naphthalene and Perdeuterated Naphthalene", in Program of the Tenth Molecular Crystal Symposium (National Research Council of Canada, Ottawa, 1982), pp. 245-248
29 J. Andersen and C.B. Duke: unpublished
30. D. Jerome: Mol. Cryst. Liq. Cryst. 79, 155 (1982)
31. A.J. Epstein, H. Rommelmann, R. Bigelow, H.W. Gibson, D.M. Hoffman and D.B. Tanner: J. Physique, in press

Part II

Interaction of Electronic Excitations with Electromagnetic Radiation

Excitation Spectroscopy of Triplet State Monomers, Aggregates and Excitons in Anthracene Crystals

H. Port

Physikalisches Institut Teil 3, Universität Stuttgart, Pfaffenwaldring 57
D-7000 Stuttgart 80, Fed. Rep. of Germany

1. Introduction

This paper surveys optical studies of electronic excitations and interaction processes in the lowest excited triplet state of an aromatic model compound (anthracene). It is concerned with spectroscopic properties of excited states of individual molecules and molecular aggregates as well as of excited states (excitons) in the matrix crystal. The scope of the paper is summarized in Fig. 1 schematically, providing both the basic concepts and the experimental approach. (The basic concepts have been known for a long time [1-4], but so far preferentially applied to the excited singlet state).

It is well established that in order to understand the excitonic spectra one has to start from the spectroscopic properties of the isolated molecule (monomer). Guest molecules in general are not soluble in a host matrix easily up to higher concentrations. Almost ideally suited, however, for a systematic study of the whole range of possible compositions are isotopic mixed host-guest systems, for instance of perdeuterated anthracene ($A\text{-}d_{10}$) and normal anthracene ($A\text{-}h_{10}$). Then in first approximation the guest excited state can be discussed in the framework of the "ideal mixed crystal" (IMC) model [2].
The basic assumptions of this model are as follows: The guest molecules are perfectly substituted in the host lattice. Guest and host only differ in their site excitation energies (deuteration shift U_0); the static and dynamic interactions (solvent shift D and exchange integral J) are invariant to isotopic substitution. Non-resonant interactions between guest and host molecules can be neglected, as long as U_0 is large as compared to the width B of the host exciton band ("deep trap limit").

The dimer as the smallest guest aggregate plays a particular role. In several respects the dimer can be regarded as "miniexciton" [5]. Guest pair levels are split with respect to the monomer level. In the IMC approximation the splitting is given by two times the resonance interaction between the guest pair molecules, which is identical to the resonance pair interaction in the pure crystal. By determining the dominant interactions J_{ij}, corresponding to different pair configurations (i,j) of translationally equivalent (AA) or inequivalent (AB) pairs, the excitonic band structure $\varepsilon(k)$ can be determined.

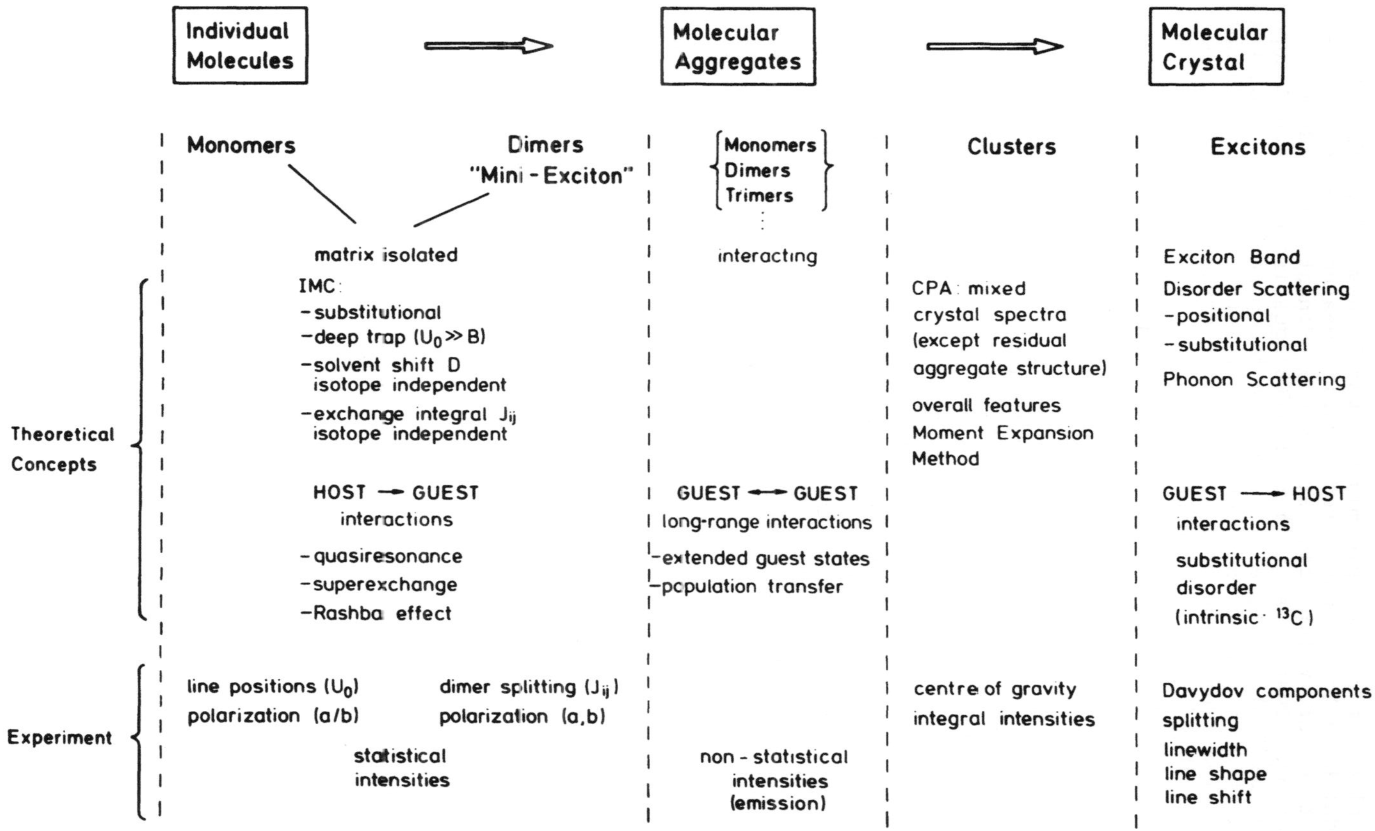

Fig. 1: Basic concepts and experimental approach, schematic

In actual cases of isotopic mixed crystals deviations from the IMC model have to be considered. Quasiresonance effects are expected to occur, which originate in nonresonant Host → Guest interactions. These lead to line shifts (quasiresonance and superexchange shifts) [2, 6] and changes in the polarization properties (Rashba effect) [3] of the guest lines relative to the IMC model. The relative importance of quasiresonance effects is determined by the ratio of host exciton band width B and deuteration shift U_o of the guest.

At intermediate guest concentrations the molecular aggregates can no longer be treated as well separated from one another. Due to Guest → Guest interactions extended states are expected to be formed, capable of energy transfer within the guest system. At higher concentrations increasingly larger guest clusters are formed. The cluster excited states gradually should go over into the exciton band of the pure guest crystal. For the theoretical description of highly concentrated mixed crystals different approximation schemes have been developed [4, 7].

The exciton band structure is required as basis for the understanding of the crystal properties. Two Davydov components, which correspond to optical transitions to the k = 0 states of the exciton band, are observed in crystals (like anthracene) with two molecules in the unit cell. Dynamical aspects of exciton motion are reflected in the spectral properties of the Davydov components, i. e. in their line shapes, widths and shifts. The line broadening can be homogeneous or inhomogeneous and can be due to different exciton interaction processes, e.g. scattering by phonons or scattering by positional and substitutional disorder.

In this paper the scheme of Fig. 1 is applied to anthracene as a model compound, since its triplet state T_1 has been most thoroughly studied in the last few years [8-12]. The relative

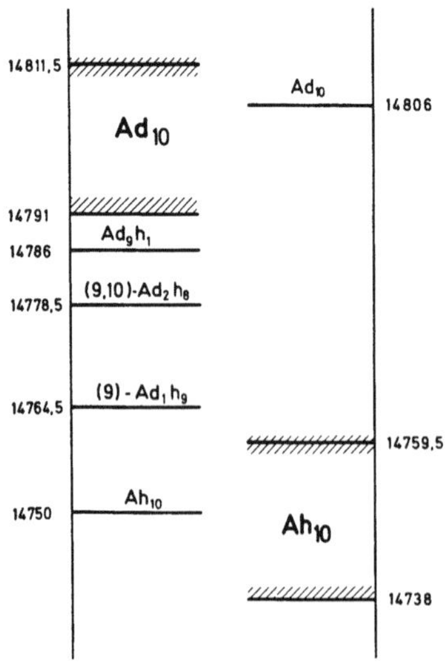

Fig. 2:
Triplet energy level scheme (0.0 region of T_1) of the anthracene mixed systems
Left: host A-d_{10},
Right: host A-h_{10}

energetic positions of the isotopic guest-host systems analyzed are given in Fig. 2. The experimental results are obtained by the method of excitation spectroscopy with narrow band cw dye lasers. With this method the relative absorption of the weak, spin-forbidden singlet-triplet transitions $T_1 \leftarrow S_0$ can be measured at very high spectral resolution and with high sensitivity, even in thin crystals grown by sublimation. The spectra to be presented are confined to the purely electronic 0.0 transitions of both host and guest and to helium temperatures.

Keeping within the scheme of Fig. 1 the experimental results are presented and discussed in three sections:

- Matrixisolated monomers and dimers in dilute isotopic mixed crystals.
- Higher aggregates and clusters in heavily doped crystals.
- Excitonic Davydov components.

2. Matrixisolated Monomers and Dimers in Dilute Isotopic Mixed Crystals.

Figure 3 provides an example of the excitation spectra of various isotopic guests (see Fig. 2) [10] together with the Davydov components H_a and H_b of the host A-d_{10}. The spectra are measured with light polarized parallel to the a and b crystal axis, respectively.

The main guest monomer lines are accompanied by weak satellites. At low guest concentrations this substructure is dominated by dimeric and further monomeric lines. This is shown in Fig. 4 for the guest A-h_{10} in enlarged energy scale. The dimer contribu-

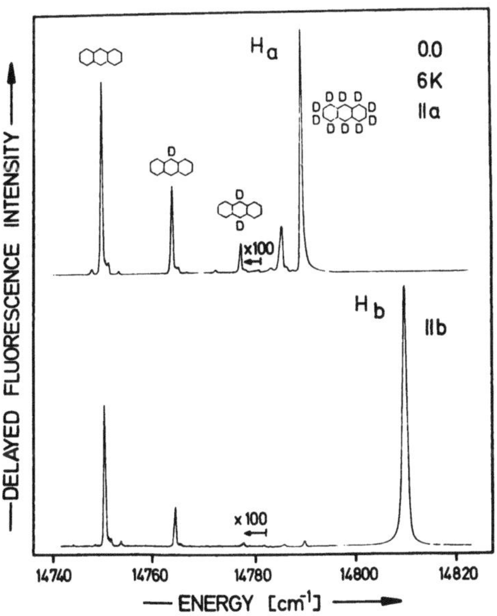

Fig. 3: Excitation spectra of A-d_{10} doped with A-h_{10}, (9)-A-d_1h_9 and (9,10)-A-d_2h_8 in light polarized parallel to the a and b crystal axis, resp

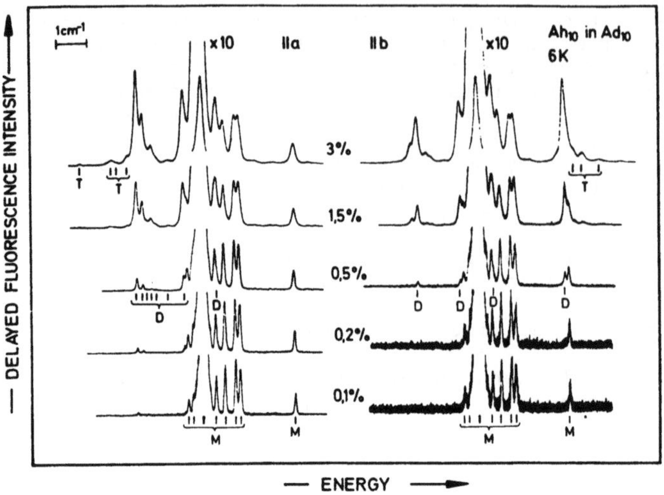

Fig. 4:
Concentration dependence of the A-h_{10} in A-d_{10} guest spectra in light polarized parallel to the a and b crystal axis, resp. All spectra are normalized to the main monomer (position and intensity)

tions can be separated by their concentration dependence. In the plot of Fig. 4 the intensities are normalized to the main monomer at 14750 cm^{-1}. The relative intensities of the dimers increase (in Fig. 4 linearly) with guest concentration corresponding to their relative statistical probabilities. These are different for AA and AB dimers, which are further distinguished by their polarization behaviour.

The monomeric satellites are due to natural abundant ^{13}C-substituted A-h_{10} isomers. They are identified by separate experiments on mixed crystals, in which the guest A-h_{10} is enriched selectively with the various ^{13}C isomers [11].

Full account of the analysis of the guest subspectra is given in the level scheme of Fig. 5a. In this scheme the energetic position of the main monomer is taken as the origin of the energy scale. Five dimer contributions are observed which are attributed to the pair configurations shown in Fig. 5b. In first approximation (IMC model) from the dimer splittings the pairwise resonance interactions (exchange integrals) J can be determined directly. The dominating pair interactions belong to the nearest-neighbour pairs within the ab plane of the crystal. The out-of-plane interactions are much smaller but not negligible. Using these J values, listed in Fig. 5, the triplet exciton band structure of anthracene is calculated (Fig. 6).

In order to prove the validity of the approach the spectra of various guests below (Fig. 3) as well as above (Fig. 7) the host exciton band are compared. This corresponds to a "variation of

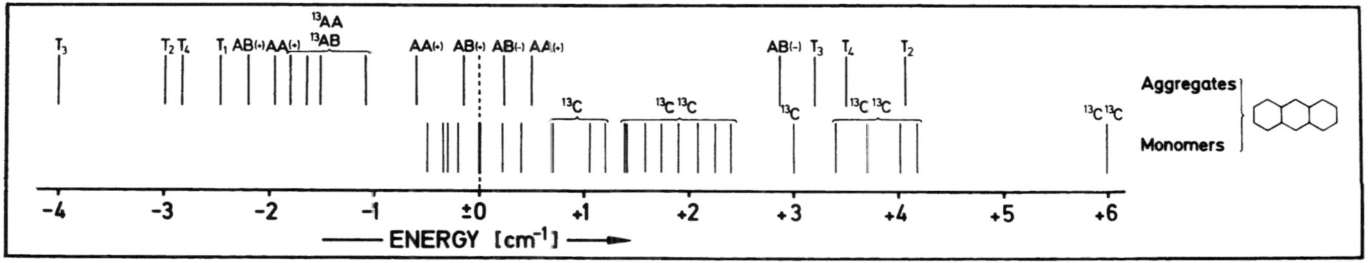

Fig. 5: a. (above): Guest aggregate and monomer level scheme.
b. (below): Guest dimer configurations and respective resonance interactions J

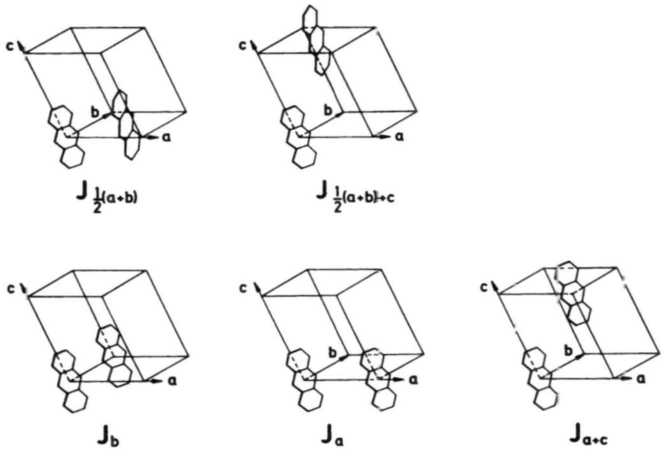

	dimer configuration	J $[cm^{-1}]$
AB	$\frac{1}{2}$ (a+b)	- 2.55
	$\frac{1}{2}$ (a+b)+c	- 0.2
AA	b	- 2.0
	a+c	- 0.6
	a	+ 0.5

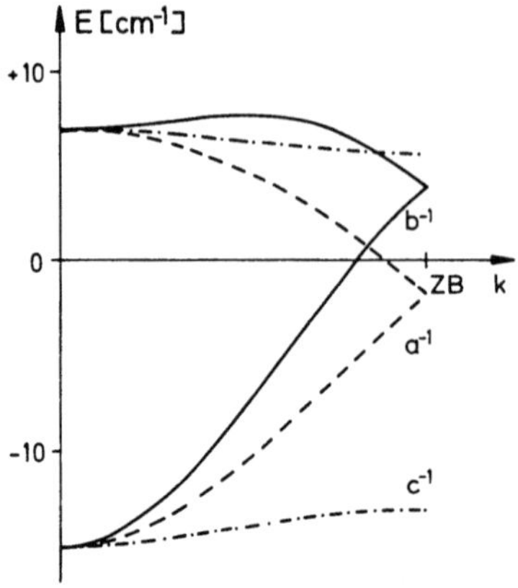

Fig. 6 : Triplet exciton band structure of anthracene (0.0 of T_1)

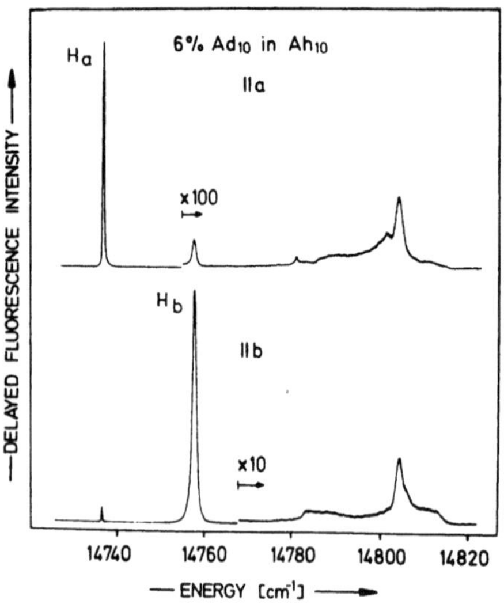

Fig. 7: Excitation spectra of A-h_{10} doped with A-d_{10} in light polarized parallel to the a and b crystal axis, resp.

the energy denominator" U_0 both in magnitude and sign. From the detailed quantitative analysis [10] it is concluded that for the magnitude of the J values determined from A-h_{10} in A-d_{10} the quasiresonance corrections are negligible. On the other hand from the comparative studies it is found that the resonance interactions are isotope independent but not the solvent shifts.

3. Higher Aggregates and Cluster States in Heavily Doped Crystals A-h_{10}:A-d_{10}.

At medium guest concentrations the excitation spectra can still be interpreted in terms of individual guest aggregates. In addition to the monomer and dimer lines, the lines of trimers and higher aggregates become more and more important. The pronounced polarization behaviour is maintained. At guest concentrations up to 9 % it is verified quantitatively that the discrete aggre-

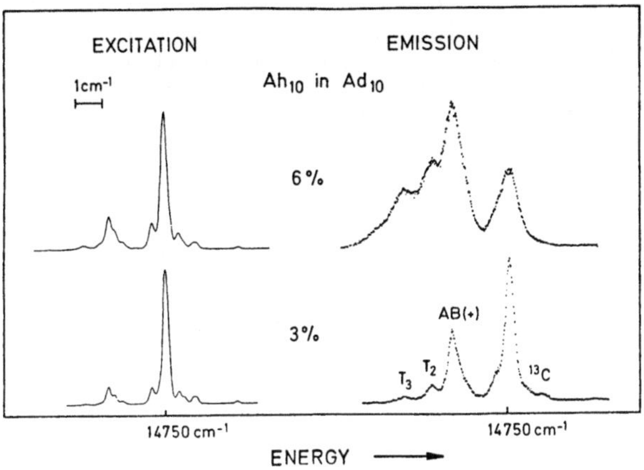

Fig. 8: Guest excitation and emission spectra of A-d_{10} doped with 3% and 6% A-h_{10}, resp. (E ∥ a; T = 2 K)

gates contribute to the excitation spectra according to their statistical probability and transition matrix elements. On the contrary, in the triplet emission spectra (guest phosphorescence) the relative intensities within the aggregate spectra are non-statistical above about 3 % (taking Boltzmann distribution into account). The difference to the intensity distribution within the excitation spectrum is very pronounced at 6 % (Fig. 8). This result indicates efficient communication and excitation energy transfer within the guest system.

At guest concentrations above 15 % the description of the guest spectra as a superposition of individual cluster contributions is no longer appropriate [12]. Certain substructures are observed throughout the concentration range of the mixed system A-h_{10}:A-d_{10} (Fig. 9). Details, however, cannot be specified any more. (In Fig. 9 the most intense lines in each subspectrum are normalized to equal height). The absorption of A-h_{10} and A-d_{10} remains separated by an energy gap at all concentrations. Each subband exhibits two components with opposite polarization behaviour in correspondence to the polarization behaviour of the Davydov components of the undoped crystals.

Relative to the energetic positions in the undoped crystals the lower and upper Davydov components of both A-h_{10} and A-d_{10} are shifted with increasing concentration to higher and lower energy, respectively. Consequently the splittings of the oppositely polarized components within the subbands are diminished successively (to about 15 cm^{-1} at 50 %).

The spectra of Fig. 9 qualitatively correspond to the expectations for a binary isotopic mixed crystal in the separate band case. Since the energy difference U_O between the mixing components of the system A-h_{10}:A-d_{10} exceeds the band width B of either pure crystal (U_O = 55 cm^{-1}, B = 22 cm^{-1}), two separated bands result, divided by an energy gap with zero density of states. There is no existing theory which could explain the spectral features including the cluster state contributions. Therefore for the quantitative description of the results in Fig. 9 the approximation scheme of the moment expansion method [7] is used.

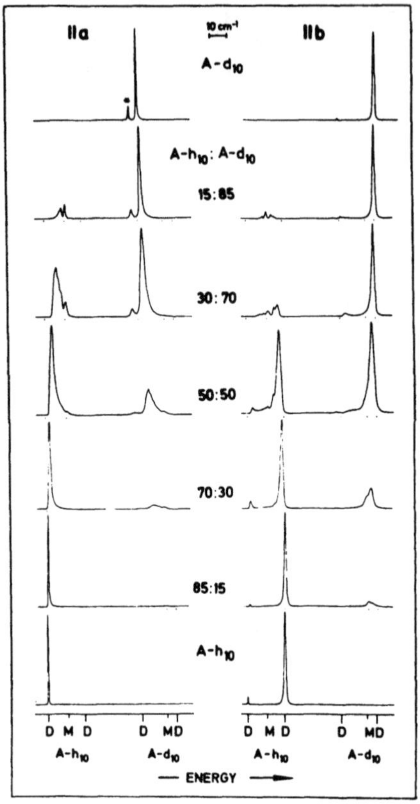

Fig. 9:
Concentration dependence of the a- and b-polarized excitation spectra of the A-h_{10}:A-d_{10} mixed crystals (6 K; spectral region of the 0.0 transition). The spectra of undoped A-h_{10} and A-d_{10} crystals are also given as references. The energetic positions of the Davydov components in the undoped crystal and of the guest monomer absorption of both A-h_{10} and A-d_{10} are marked "D" dn "M", respectively

All parameters required (J_{ij}, U_0 and the energetic positions of the Davydov components) are available from the measurements on the weakly doped (Sect. 2) and on the undoped (Sect. 4) anthracene crystals. In Figs. 10a and 10b the measured concentration

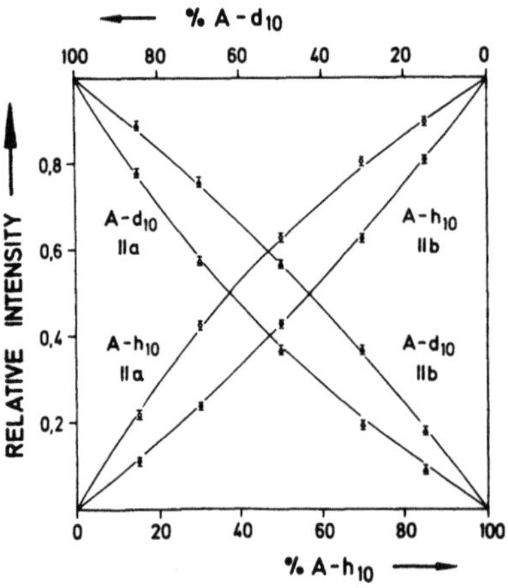

Fig. 10a:
Concentration dependence of the integrated intensities of the A-h_{10} and A-d_{10} subbands (normalized to the total intensity)

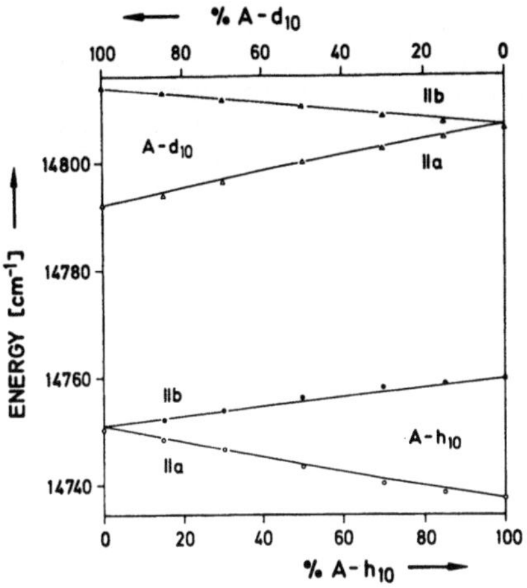

Fig. 10b:
Energetic position of the centres of gravity of the polarized subbands of A-h_{10} and A-d_{10}. Experimental points and calculated dependences

dependences of both the integral intensities and the centres of gravity of the polarized A-h_{10} and A-d_{10} subbands are plotted together with the calculated ones (full curves). The experimental results are described quantitatively by the theoretical calculations.

4. Excitonic Davydov components

For a typical A-h_{10} crystal of high quality, grown by sublimation, the Davydov components differ drastically in linewidth and lineshape at helium temperature (Fig. 11). The linewidth of the lower Davydov component at 14738 cm^{-1} is two orders of magnitude smaller than that of the upper component at 14759.5 cm^{-1}, 0.009 cm^{-1} and 0.9 cm^{-1}, respectively. The lineshape of the lower component is pronounced asymmetric with a long high energy tail, that of the upper one, however, is symmetric (close to a Lorentzian).

The principal difference between the lower and upper Davydov component [9] is due to the fact that in anthracene the optically accessible k = 0 states are located at the bottom and near the upper edge of the triplet exciton band (see Fig. 6). The

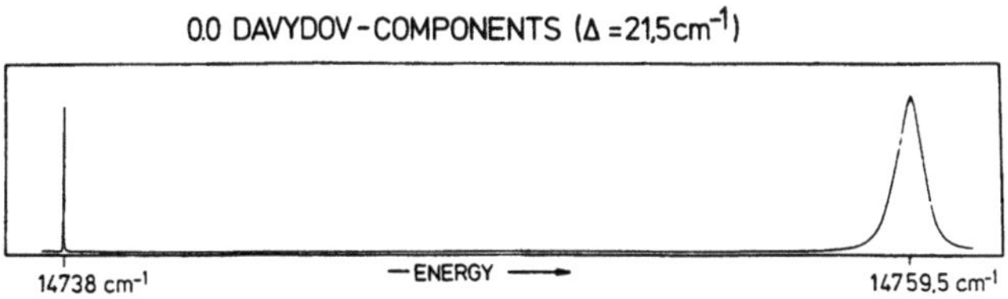

Fig. 11:
Excitation spectrum of the T_1 Davydov components of anthracene A-h_{10} at 1.8 K. Davydov splitting Δ = 21.5 cm^{-1}

large residual linewidth of the upper component observed at helium temperature is understood as a result of interbranch downward scattering via phonon emission (one-phonon direct process), which is still effective even in the limit $T \to 0$. This process leads to homogeneous line broadening and also can explain the Lorentzian lineshape. On the other hand broadening of the lower Davydov component by interbranch scattering can occur solely via phonon absorption and is hardly effective at low T, when kT is small as compared to the Davydov splitting. From this a very narrow lower component is expected with its linewidth tending to zero in the limit $T \to 0$. Since, however, the linewidth of the lower component rapidly sharpens upon decreasing the temperature down to 6 K, but is not further narrowed below, this width in its low temperature limit clearly is not determined by the exciton-phonon scattering process.
There are further experimental observations which lead to the conclusion that the source of the residual linewidth of the lower Davydov component and its asymmetry is the exciton scattering at structural defects.

Inhomogeneous line broadening of the lower Davydov component, related to the crystal quality, can very effectively be reduced by at least one order of magnitude using sublimation flakes (as in Fig. 11) instead of samples grown by the Bridgman method. An illustrative example of this effect is presented, in another context, in the first chapter, Fig. 1, of these proceedings.
Since the linewidth of the upper Davydov component is much larger in general, the reduction by the same amount does not lead to significant linewidth changes.

An example of another type of line broadening is given in Fig. 12, where upon doping with A-d_{10} substitutional disorder is introduced in the host A-h_{10} in a controlled manner. With increasing A-d_{10} concentration the lower A-h_{10} Davydov component is shifted and broadened monotonically, whereas the asymmetric lineshape is maintained. The lineshift can be described quantitatively by a theoretical model deduced by Lifshitz [13]. The Lifshitz theory predicts a linear dependence of the shift of the lower edge of the exciton band, which coincides for anthracene with the lower Davydov component, as function of the guest concentration.

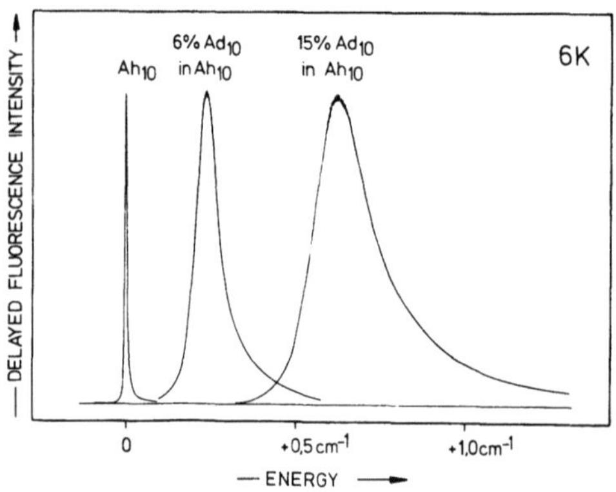

Fig. 12:
Lower Davydov components of the host A-h_{10} at different A-d_{10} guest concentrations

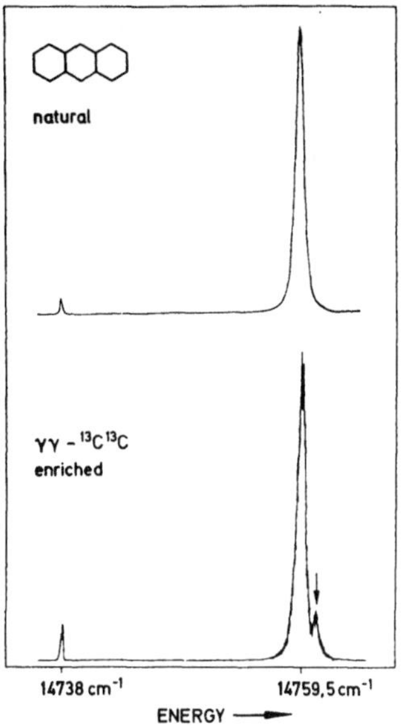

Fig. 13:
Davydov components of natural A-h_{10} (upper trace) and of A-h_{10} enriched with twofold $\gamma\gamma$-$^{13}C^{13}C$ substituted A-h_{10} (lower trace)-(E ‖ b; T = 4.2 K)

Under discussion, but without direct experimental proof so far, is the possibility of line broadening due to disorder introduced intrinsically in the undoped anthracene crystal, in particular by the naturally abundant ^{13}C-substituted isomers.

From the experiments mentioned in Sect. 2 [11] the isotope shifts U_O of these isomers are known. The U_O values of all the ^{13}C-monosubstituted isomers are below the critical value U_{cr} in anthracene [10], for which a guest term appears just outside the exciton band. Therefore the states of these monosubstituted isomers are amalgamated in the (^{12}C) exciton band.

The relative importance of the ^{13}C controlled disorder is expected to depend rather critically on the exciton band structure. For comparison purposes with the naphthalene crystal it is of interest whether the twofold at both central C atoms ^{13}C-substituted $\gamma\gamma$-$^{13}C^{13}C$ anthracene isomer because of $U_O > U_{cr}$ actually creates a bound disturbed level outside the exciton band. Fig. 13 confirms this expectation. The lower trace belongs to A-h_{10}, which is enriched with $\gamma\gamma$-$^{13}C^{13}C$ substituted A-h_{10}. As result of the doping a satellite line appears at the high energy side of the upper Davydov component. In the excitation spectrum of the undoped (natural) A-h_{10}, given for comparison in the upper trace of Fig. 13, this line is not observed because of the low natural concentration of $\gamma\gamma$-$^{13}C^{13}C$ molecules (about 0.01 %).

Acknowledgement

This paper is based largely on a close collaboration with D. Rund. Helpful discussions with Prof. H.C. Wolf and support by the Deutsche Forschungsgemeinschaft are gratefully acknowledged.

References

1. A.S. Davydov, Theory of Molecular Excitons
   a. Mc-Graw Hill, New York (1962)
   b. Plenum Press, New York (1971)
2. E.R. Bernstein, S.D. Colson, R. Kopelman and G.W. Robinson J. Chem. Phys. 48, 5596 (1968)
3. E.I. Rashba, Soviet Phys. Solid State 4, 2417 (1963)
4. J. Hoshen and J. Jortner, J. Chem. Phys. 56, 5550 (1972)
5. H.C. Wolf, J. Phys. C5a, 101 (1971)
6. H.-K. Hong and R. Kopelman, J. Chem. Phys. 55, 72 (1971)
7. a. B. Velicky, S. Kirkpatrick and H. Ehrenreich, Phys. Rev. 175, 747 (1968)
   b. J. Hoshen and J. Jortner, J. Chem. Phys. 56, 933 (1972)
8. H. Port, K. Mistelberger and D. Rund, Mol. Cryst. Liq. Cryst. 50, 11 (1979)
9. H. Port, D. Rund, G. Small and V. Yakhot, Chem. Phys. 39, 175 (1979)
10. H. Port, D. Rund and H.C. Wolf, Chem. Phys. 60, 81 (1981)
11. U. Doberer, H. Port, D. Rund and W. Tuffentsammer, Mol. Phys. 11, 1 (1983)
12. D. Rund and H. Port, Chem. Phys. in press
13. I.M. Lifshitz, Soviet Physics Uspekhi 7, 549 (1965)

Photoemission from Molecular Crystals, Bandstructure and Resonance Effects

E.E. Koch

Hamburger Synchrotronstrahlungslabor HASYLAB at DESY, Notkestrasse 85
D-2000 Hamburg 52, Fed. Rep. of Germany

1. Introduction

Two decades ago the only well-investigated electronic states of organic solids were those close to the top of the valence bands and at the bottom of the conduction band. Since then photoelectron spectroscopy in its various modes has been established as a very powerful tool for probing the electronic structure of atoms, molecules, solids and surfaces over large binding energy ranges. Different aspects of photoemission spectroscopy have been treated in a series of review articles [1,2]. In particular this spectroscopy has contributed significantly to our progress in understanding the electronic structure of molecular crystals and organic solids as recent review papers document (e.g. Ref. 3-6). Using conventional x-ray and VUV sources (e.g. Al K_α or He lines) a wealth of interesting information has been obtained concerning, for example, binding energies and widths of valence bands and core levels, ligand field splittings, chemical shifts and intermolecular relaxation phenomena in the solid state.

The availability of intense polarized synchrotron radiation from electron storage rings has strengthened, among other recent advances in experimental techniques, the applicability of the photoemission technique tremendously. Important extensions of the technique have only become possible by using this novel radiation source[7].

A comprehensive treatment of the major recent results from photoemission work on molecular crystals is out of the scope of the present paper. Instead we try to illustrate new and exciting research developments with selected examples including the use of atomic cross-section effects in bandstructure determinations (section 3), the mapping of conduction band density of states (section 4), resonant intershell interaction in photoemission (section 5) and the use of angular resolved photoemission from ordered films of organic compounds (section 6).

2. Band Formation in Aromatic Hydrocarbons

Although the question has been raised whether band theory is sufficient to describe the electronic structure of molecular crystals [8] , it has nevertheless been used quite sucessfully in many instances for a quantitative treatment of properties of such materials. Thus for example the valence and conduction band structure of rare gas solids have served as a testing ground for first principles band structure calculations for these prototype van der Waals crystals [9] .

For organic solids the evolution of solid state bands from molecular orbitals has been studied for a number of model substances including linear alkenes [10] and aromatic hydrocarbons [3]. The valence band spectra for solid benzene, naphthalene, anthracene and tetracene (Fig. 1) may serve as an example to illustrate the major general results. First, we note that although the structure of these molecules is still fairly simple, the number of valence electrons contributing to the valence band density of states is so large that ab initio calculations for molecular orbitals (MO's) and bands including all valence electrons become increasingly difficult. Thus even for gas phase photoemission discussion has been mainly restricted to the π-electron system and only a few attempts have been made to partially disentangle π and σ bands. To achieve a systematic de-

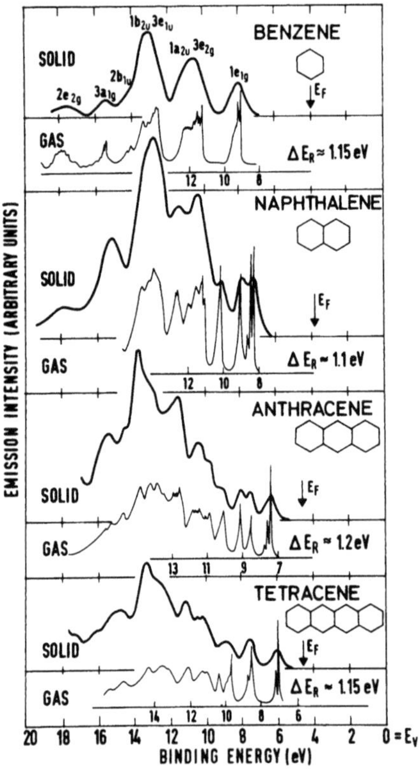

Fig. 1. Valence band spectra for solid aromatic hydrocarbons. The gas phase spectra, shifted by the relaxation energy ΔE_R, are shown for comparison. Note the close 1:1 correspondence of the main features in the gas and the solid phase (from Ref. 3)

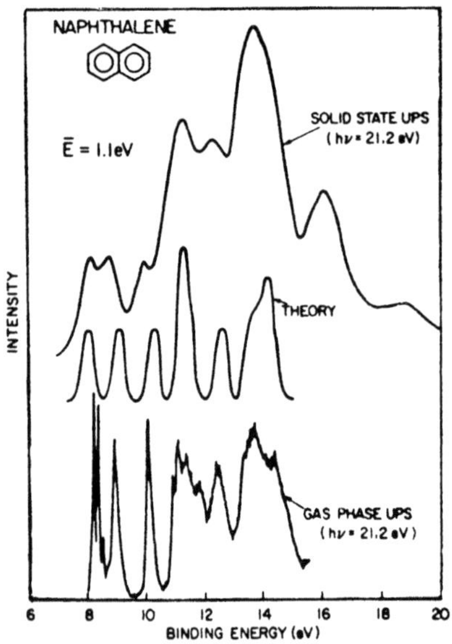

Fig. 2. Comparison of valence electron photoemission spectra of naphthalene in the gas phase and solid state with the density of valence electrons calculated using the CNDO/S3 MO model. The energy scale is the one associated with the gas phase spectrum (from Ref. 8)

scription of the electronic structure of these materials several semiempirical LCAO approaches have been developed including the HAM method [11] and a spectroscopic CNDO-level MO model (CNDO/S3) calculation to describe simultaneously the photoemission and optical absorption spectra of aromatic heterocycles containing hydrogen, carbon, nitrogen, oxygen and fluorine species [12] . An example is shown in Fig. 2.

Several general observations can be made concerning spectra of monomolecular organic solids such as displayed in Fig. 1 and 2: (a) while the energy distribution curves (EDC's) are still fairly simple for molecular crystals formed by small organic molecules, like for instance benzene, the spectra become more crowded with increasing size; (b) the solid state spectra exhibit a one to one correspondence to those of the gas phase. This is a consequence of the prevailing tight binding situation in molecular crystals where the individual molecules, held together by the weak van der Waals forces, are only slightly affected by the solid environment; (c) all bands in the solid state are shifted to lower binding energies relative to the gas phase by $\Delta E_R \approx$ 1-2 eV due to the intermolecular relaxation of the surrounding medium once a hole is created on one of the molecules during the photoemission process [13]; (d) the bands are considerably broadened in the solid state relative to the gas phase Several mechanisms have been discussed for this broadening, which is also observed for solid films formed by condensed simple di- and triatomic gases[14]; (e) the uppermost filled valence bands originate from pure π orbitals; at about 10 eV binding energy π and σ bands start to overlay; at even larger binding energies from about 15 to 19 eV we expect pure σ bands and finally at energies in excess of about 19 eV bands with essential 2s character are expected.

While these general trends have now been fairly well understood we remain with the severe problem of overlapping bands. For instance the two π bands in solid benzene ($1e_{1g}$ and $1a_{2u}$) develop in tetracene into two partly overlapping groups of nine π bands extending over a binding energy range from 5 to $\approx$ 15 eV where they also overlap with the uppermost σ bands. Thus more selective techniques are needed. In fact, only recently has it become possible to separate these bands with reasonable confidence in the overlapping range with the aid of angle resolved EDC measurements from oriented polycrystalline films [15].

3. Atomic Cross-Section Effects, Metal-Phthalocyanines

As shown within the one electron approximation by COOPER and FANO [16] the partial photoionisation cross sections of a subshell depend strongly on the overlap of the initial state and final state wavefunctions. The important point here is to note that by tuning the photon energy one can selectively enhance the visibility of partial density of states. For example, for d-initial states delayed cross-section maxima above threshold due to an increasing overlap of d-initial and f-final state wave functions are predicted and have been observed in a number of 3d metals and compounds.

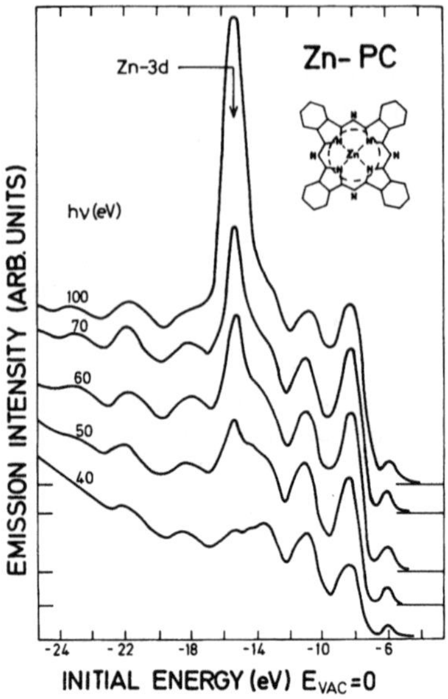

Fig. 3. A family of photoelectron energy distribution curves for Zn-Phthalocyanine in the valence band range for photon energies between 40 and 100 eV. The maximum of the Zn 3d density of states is located at a binding energy 15.2 eV below the vacuum level (from Ref. 17)

We use Zn-Phthalocyanine (Zn-Pc) as an example to illustrate this effect (Fig. 3) [17]. From the EDC's obtained with photon energies ranging from 30 to 120 eV the strong hν-dependent increase of the Zn 3d partial cross section is evident. The center of the 3d-derived states is located at 15.2 eV below the vacuum level and is 1.2 eV wide (fwhm). These results were compared with ab initio Hartree-Fock-LCAO calculations on ZnN_4 clusters and close agreement with experiment was found. This indicates that the ZnN_4 planar cluster is already a good first approximation. In particular the energy level scheme for ZnN_4 showed that the bands with Zn 3d character are well separated from groups of bands which have mainly N 2p or N 2s character.

Together with similar measurements on other 3d metal PC's [18, 19] a rather complete picture of the valence bands of the 3d transition metal PC's emerges (Fig. 4). For all metal PC's the EDC's show quite complex valence bands extending over a binding energy range from 5 to 35 eV (see, e.g., [20]).

In Fig. 4 the upper part of the metal-PC valence bands is shown schematically with H_2-PC for comparison. The location of the 3d bands with respect to the ligand bands changes quite remarkably over the 3d series. For H_2-PC the valence band density of states was determined in a combined photoemission- and soft x-ray emission experiment [20]. In these experiments the maximum of the nitrogen 2p contribution to the valence band density of states was located at 8.6 eV binding energy. Furthermore, bands with largely 2s character have been identified in the 15-30 eV binding energy range. It is of considerable interest to see the general trend in Fig. 4 which is in qualitative agreement with predictions from simple Hückel MO calculations [21]

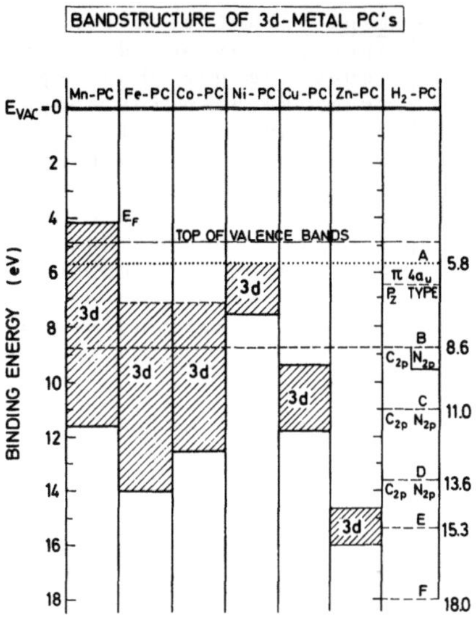

Fig. 4. Valence-bandstructure scheme for 3d-metal-phthalocyanines. On the right comparison is made with H_2-PC where the major contributions to the maxima in the EDC's are given. The hatched areas give the location and widths of the metal-derived 3d partial density of states

Furthermore the large range of overlap of metal-derived states with ligand-derived bands for Mn, Fe and Co having rougly half filled 3d shells is noteworthy. Mn-PC is of particular interest since there is a complete overlap of the highest occupied metal-derived band with the highest filled ligand-derived band A centered at 5.8 eV with the consequence that the top of the valence band has largely metal d character.

4. Conduction Band Density of States, Shape Resonances

Up to this point we have been concerned with the energetic structure of the *occupied* valence bands. In favorable cases one can obtain detailed information about the conduction band density of states by carefully measuring the intensity variation of primary emission as a function of excitation energies. It can be shown [2,22] that in this mode of photoemission, which resembles constant initial state spectroscopy, one can map the conduction band density of states, provided the initial bands show little or no dispersion and other factors influencing the photoemission intensity such as the photoelectron escape depth are slowly varying functions in the range of interest.

The partial photoemission cross sections for a number of molecular crystals including solid N_2, O_2, CO, CO_2, C_2H_2, CCl_4 and SF_6 have been determined recently close to threshold with the aim of studying final state effects [22-25]. For the latter two molecular crystals a very close similarity to gas phase cross-section data was found even close to threshold, while the cross sections for solid N_2, O_2, CO and CO_2 differ importantly from those of the gas phase in exhibiting pronounced maxima, which are located just above threshold, i.e., at roughly 3-6 eV above the vacuum level. These resonances are absent in the gas phase photoemission, or considerably weaker, or perturbed by autoioni-

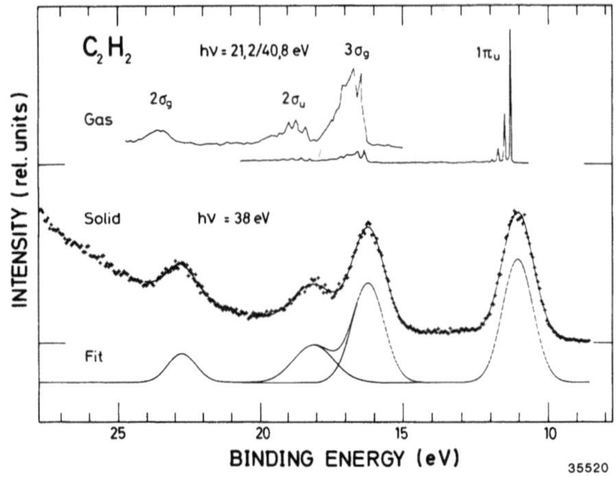

Fig. 5 Photoelectron spectra for solid acetylene. The gas phase spectrum is shown for comparison and the molecular assignment of the bands is given. In the lower part a fit of the experimental spectrum with 4 Gaussians is depicted after subtraction of a smooth background (from [25])

sation Rydberg states. They have been attributed in a solid-state band structure picture to effects of high density of final (conduction band) states. By comparison to electron scattering experiments, gas phase photoemission, and theoretical calculations,

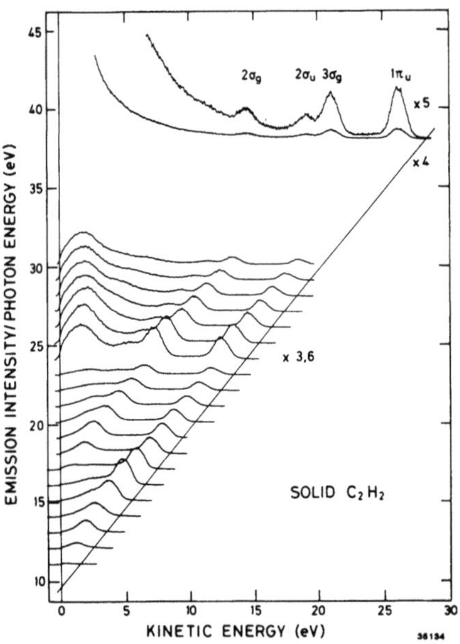

Fig. 6 Photoelectron energy distribution curves for polycrystalline solid C_2H_2 for excitation energies between 10 and 40 eV. In this plot the same initial states follow inclined lines (from [25])

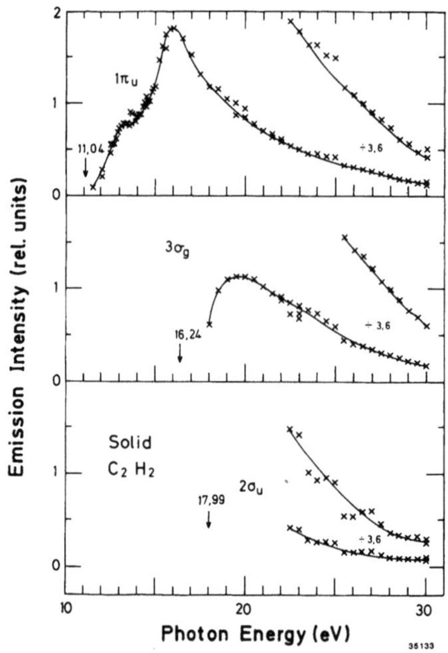

Fig. 7 Relative partial ionisation cross sections for the $1\pi_u$, $3\sigma_g$ and $2\sigma_u$ derived valence bands of solid C_2H_2. The arrows mark the onset for photoemission from these bands (from [25])

the molecular origin of these states contributing to the conduction band density of states has been traced back to shape resonances [24,26]. In the electron scattering experiments these shape resonances are associated with short lived (~ 10^{-14} s) negative ion states.

We discuss here the experimental results for solid C_2H_2 [25] as an example. In Fig. 5 the photoelectron spectra of gaseous and solid C_2H_2 are compared. Again we note a close 1:1 correspondence of both spectra and a relaxation shift of the vertical ionisation potentials of ΔE_R =.6 eV. In Fig. 6 a family of EDC's is shown at different photon energies. The area under each primary emission peak was determined by fitting the measured spectrum by Gaussians corresponding to the uppermost MO's (see Fig. 5). The strong $h\nu$ dependence of the partial photoionisation cross sections is already apparent from inspection of EDC's (Fig. 6) and displayed in more detail in Fig. 7. In particular, for the valence band derived from the $1\pi_u$-MO a pronounced cross-section maximum at 5.0 eV above threshold and a shoulder at 2.5 eV are observed. The molecular origin of these resonances in the continuum is in the case of C_2H_2 most probably the $1\pi_u \rightarrow \varepsilon\delta_g$ channel and a strong contribution of a $2\sigma_u \rightarrow 1\pi_g$ valence excitation which decays via autoionization. In this respect solid acetylene resembles quite closely the gas phase behavior.

5. Resonant Photoemission at the 3p Threshold of 3d Metal Phthalocyanines

Besides using photoelectron spectroscopy for the determination of partial cross sections over extended energy ranges and the possibility to map out the bandstructure,entirely new effects have been observed with tunable synchrotron radiation, as is the case of resonant photoemission at the onset of inner shell absorption thresholds. In 1977 a strong resonance enhancement in the photoelectron energy distribution curves from Ni was observed at the 3p threshold [27]. The intensity of a peak located at about 6 eV below the Fermi level increased dramatically (more than a factor of 10), when the photon energy was swept through the 3p threshold at about 66 eV. This result was explained as a Fano-type resonance being due to the interaction between a discrete excitation $3p^6 3d^9 4s \rightarrow 3p^5 3d^{10} 4s$ and the continuum channel $3p^6 3d^9 4s \rightarrow 3p^6 3d^8 4s\ \varepsilon\ell$ coupled via the Super-Coster-Kronig decay $3p^5 3d^{10} 4s \rightarrow 3p^6 3d^8 4s\ \varepsilon\ell$. In the meantime a large number of experiments have been made using synchrotron radiation to further explore these resonances. Most of the 3d metals (3p → 3d resonance), rare earth metals and compounds (4d → 4f resonances), and uranium (5d → 5f resonances) have been investigated.

Resonance photoemission and the observation of two-electron satellites in addition to the one-electron emission clearly demonstrates the importance of many electron effects in photoemission. Aside from a principle investigation of these effects in order to clarify the photoemission process, resonance effects have offered novel possibilities to study the electronic structure of more complicated materials.

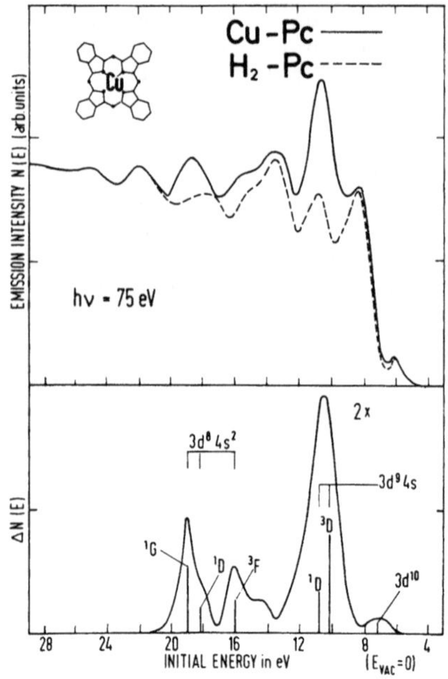

Fig. 8 Photoelectron energy distribution curves for Cu-Phthalocyanine and H_2-PC for a photon energy of $h\nu \cong 75$ eV at the 3p threshold show marked differences (upper panel) due to the Cu derived partial density of states. The difference spectrum at resonance (lower panel) can be assigned to a $3d^9 4s$ and a $3d^8 4s^2$ configuration (from Ref. 28)

We discuss this aspect using Cu-PC as an example. A photoelectron energy distribution curve for Cu in Cu-PC excited with photons at the Cu 3p threshold at $h\nu = 75$ eV appears in Fig. 8 [28]. A sharp 3d electron satellite with two peaks with initial energies of 5.5 and 8.5 eV below the center of the Cu 3d main peak is observed which exhibits a strong resonant enhancement in a photon energy range ≤ 5 eV around the 3p threshold. The one-electron photoemission process exciting the 3d electrons $3p^6 3d^{10} 4s + h\nu \rightarrow 3p^6 3d^9 4s + e^-$ results in the main lines ^3D, ^1D at $E_i = -10.5$ eV in the photoelectron distribution (Fig. 8). At resonance, the 3p → 4s excitation followed by a Super-Coster-Kronig decay creates a two-hole state with a $3d^8 4s^2$ final state: $3p^6 3d^{10} 4s + h\nu \rightarrow 3p^5 3d^{10} 4s^2 \xrightarrow{SCK} 3p^6 3d^8 4s^2 + e^-$. This process results in the resonant enhancement of the $3d^8 4s^2$ final state multiplet (Fig. 8). The multiplet structure and energetic separation of the satellite structure from the 3d emission peak (^3D, ^1D) is very close to the energy difference between $3d^9 4s$ and $3d^8 4s^2$ multiplets for atomic Cu^+ and was indeed corroborated by later resonant photoemission experiments for atomic Cu [29]. Thus the atomic assignment of the main configurations contributing to the photoelectron spectrum for "quasi-atomic" Cu in Cu-PC is justified (see Fig. 8). Similar results have been obtained for Ni-PC and Zn-PC [30,31].

Clearly from this type of experiment and a comparison of the PC results with those for the metal, other 3d compounds and free atoms, important information can be obtained concerning differences in the chemical environment, relaxation (screening) and configuration shifts in the organic solid. For example, the importance of empty s-like states becomes evident by comparing the relative intensities of the $3d^8$ two-hole configuration to the

$3d^9$ one-hole intensity. For Zn-PC at resonance the relative satellite intensity is about 30 % in Zn-PC and 6 % in Zn [31]. The corresponding relative satellite intensities are 36 % in Cu-Pc and 12 % in Cu, respectively. One can associate the weakness of the satellite in Zn metal with the smaller number of empty 4s states in Zn as compared to Cu. In both Cu-PC and Zn-PC the satellite intensity is enhanced compared to the metal since 4s electrons are transferred from the metal site to the ligand.

6. Angle-Resolved Photoemission from Oriented Organic Films on Metal Surfaces

As a final example for photoemission experiments on organic materials, we discuss here angle-resolved photoemission from oriented films of molecules. Strong variations of photoelectron emission intensity with the emission angle have been observed from adsorbates and ordered layers on surfaces [32-34]. The emission intensity also depends on the photon energy, angle of incidence, and the polarization of the incident light. The determination of such angular variations of photoelectrons from oriented films is of great interest for two interrelated reasons. The first is that one can obtain molecular orbital information about the symmetry assignments of the MO's involved. Secondly, the angular distribution of photoemission from an orbital of a molecule adsorbed on a surface contains information concerning the orientation of the molecule with respect to the surface [32-34]. This information in turn is of importance, e.g., in surface chemistry. Most investigations of this kind have so far been restricted to simple di- or triatomic adsorbates.

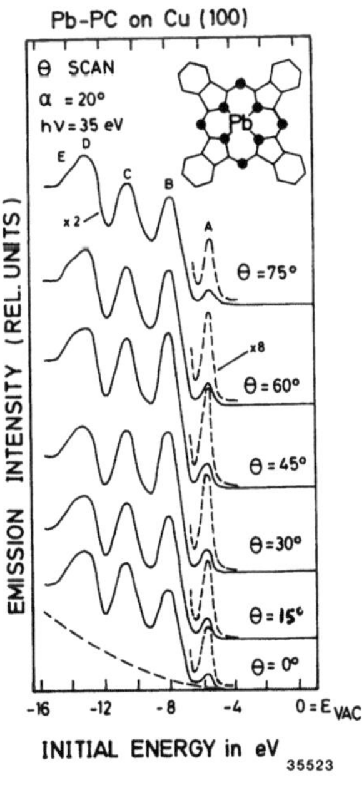

Fig. 9 Angle resolved EDC's from Pb-PC films on a Cu (100) surface with the emission angle Θ between Θ = 0° (normal emission) and Θ = 75°, hν = 35 eV and an angle of incidence α = 20°. For Θ = 0° the estimated background due to inelastically scattered electrons is indicated by the dashed line. The inset shows the Pb-PC molecule (dots represent N atoms) (from [35])

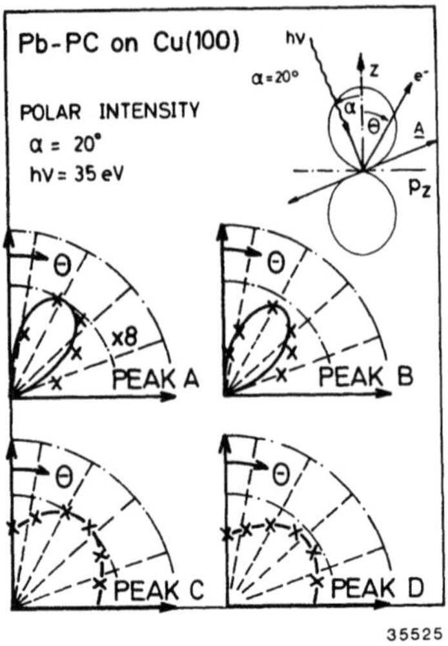

Fig. 10 Variation of the photoemission intensity of the valence band photoemission peaks A to D from a thin Pb-PC film with emission angle Θ for an angle of incidence α = 20° and hν = 35 eV. Crosses represent the experimental values (for peaks A and B an isotropic background has been subtracted). The solid curves for peaks A and B represent the expected intensity variation for a p_z-like orbital, taking α =20°. For peaks C and D the broken line connects the experimental points

Recently we succeeded in preparing oriented films of Pb-PC, a large metal-organic molecule, by in situ sublimation under well-defined conditions on a Cu (100) single crystal substrate [35]. The preparation was similar to the one in earlier LEED studies on metal-PC films[36]. In Fig. 9 the angle resolved EDC's are shown for a Pb-PC film at hν = 35 eV, measured with an angle of incidence α = 20°. For the EDC's the polar angle Θ, the direction of electron emission with respect to the normal of the sample, was varied between Θ = 0° and Θ = 75°. The spectra have been normalized to the number of incident photons and are displaced for clarity. For the uppermost valence band peaks A and B at an initial energy of -5.8 and -8.3 eV, respectively, an increase in intensity with increasing Θ up to Θ ≈ 45° and a subsequent decrease is observed.

From EDC's such as displayed in Fig. 9 the intensity of each emission band as a function of the polar angle Θ can be obtained and compared to simple model calculations in order to get information on the orientation and/or orbital character of the Pb-PC films (Fig. 10). In a simple MO approximation, where the final state wavefunctions in the photoemission process are represented as a single plane wave, one can predict the angular variation for p_z-type orbitals for the geometry of our experiment assuming π-type orbitals to be oriented perpendicular to the substrate. Under this assumption we note an excellent agreement of the experimental results with the shape of the calculated emission profile for the uppermost two valence bands (peaks A and B) once an isotropic background has been subtracted (Fig. 10) (An analysis of the angular patterns for the lower lying features in the EDC's is more cumbersome since it is difficult to account for a varying background of scattered electrons and since p_z- and σ-type orbitals overlap.) Thus the experiment is in excellent accord with LEED investigations [36] showing that the planar Pb-PC molecules are oriented parallel to the Cu (100) substrate. The successful preparation and characterization of such oriented films may offer new possibilities to follow surface experiments (e.g., catalytic reactions) on these films in detail by photoemission techniques.

References

1 K. Siegbahn: Nobel Lecture Dec. 1981, Uppsala Univ., Inst. Physis Report 1052 Dec. 81, and referencs therein
2 M. Cardona and L. Ley, (editors), Photoemission in Solids Vol. I, Vol. II, Topics in Applied Physics Vol. 26, 27, Springer Verlag Heidelberg, New York, 1978, 1979
3 W.D. Grobmann and E.E. Koch: in: Ref. 2, Vol. II, p. 261
4 A.I. Belkind: Photoelectronia Emissia is Organicheskich Twerdich Tel, Riga, SINATNE, 1979 (in russian)
5 W.R. Salaneck: in: Photon, Electron and Ion Probes of Polymer Structure and Properties, D.W. Dwight, T.J. Fabish and H.R. Thomas (editors) ACS Symposium Series Vol. 162, Am. Chem. Soc. Washington DC, 1981, p. 121
6 C.B. Duke: in: Extended Linear Chain Compounds, J.S Miller (editor), Plenum 1982, Vol. 2, p. 59
7 see e.g.: Handbook on Synchrotron Radiation Vol. I, E.E. Koch (editor), North Holland Publishing Company, Amsterdam, New York, 1983
8 e.g. C.B. Duke and L.B. Schein: Physics Today 33, 42 (1980); C.B. Duke in: Festkörperprobleme 22, 21 (1982)
9 U. Rössler: in Rare Gas Solids Vol. I, M.L. Klein and J A. Venables (editors), Academic Press, p. 505
10 J.J. Pireaux, S. Swensson, E. Basilier, P.Å. Malmqvist, U. Gelius, P. Caudano and K. Siegbahn: Phys. Rev. A14, 2133 (1976)
11 L. Åsbrink, C. Friedh and E. Linderholm: Chem. Phys. Letters 52, 63, 69, 72 (1977)
12 C.B. Duke, Int. J. Quant. Chem.: Quantum Chem. Symposium 13, 207 (1979)
13 For a recent extensive compilation and discussion see: N. Sato, K. Seki and H. Inokuchi, J. Chem. Soc Faraday Trans. 77, 1621 (1981)
14 F.J. Himpsel, N. Schwentner and E.E. Koch: Phys. Stat. Sol. b71, 615 (1975)
15 C.A. Feldmann and E E. Koch: to be published
16 U. Fano and J.W. Cooper: Rev. Mod. Phys. 40, 441 (1968)
17 E E. Koch, M. Iwan, K. Hermann and P.S. Bagus: Chem. Phys. 59, 249 (1981)
18 M. Iwan and E.E. Koch: Solid State Comm. 31, 261 (1079)
19 M. Iwan, E.E. Koch, T.C. Chiang, D.E. Eastman and F.J. Himpsel: Solid State Comm. 34, 57 (1980)
20 E. Tegeler, M. Iwan and E.E. Koch: J Electr. Spectr. Relat. Phenom. 22, 297 (1981)
21 see e.g.: R. Taube Z Chem. 6, 8, 1966
22 J.-H.Fock, H.J. Lau and E.E. Koch: preprint
23 H.-J.Lau, J.-H. Fock and E.E. Koch: Chem. Phys. Letters 89, 281 (1982)
24 J.-H.Fock and E.E. Koch: Proc. VII Internat. Conf. VUV Radiation Physics, Jerusalem, August 1983
25 J.-H.Fock and E.E. Koch: to be published

26 J.L. Dehmer, D. Dill: Symp. Electron-Molecule Collisions, Univ. Tokyo, Sept. 79, p. 95, ed. Shimamura and M. Matsuzawa; J.L. Dehmer, D. Dill and A.C. Parr, in Photophysics and Photochemistry in the VUV, ed. S. McGlynn, G. Findley and R. Huebner, Reidel Publ. Comp. Dordrecht, 1983
27 J.C. Guillot, Y. Ballu, J. Paigné, Lecante, K.P. Jain, P. Thiry, R. Pinchaux, Y. Petroff and L.M. Falicov: Phys. Rev. Letters 39, 1632 (1977)
28 M. Iwan, E.E. Koch, T.C. Chiang, D.E. Eastman and F.J. Himpsel: Solid State Comm. 34, 57 (1980)
29 R. Bruhn, E. Schmidt, H. Schröder and B. Sonntag: J. Phys. B15, L 441 (1982)
30 M. Iwan and E.E. Koch: Solid State Commun. 31, 261 (1979)
31 M. Iwan, E.E. Koch, T.C. Chiang and F.J. Himpsel: Physics Letters, 76 A, 177 (1080)
32 E.W. Plummer and W. Eberhardt: Adv. Chem. Phys. 49, 533 (1982)
33 J.W. Gadzuk: Phys. Rev. B10, 5030 (1974)
34 W.D. Grobman: Phys. Rev. B17, 4573 (1978)
35 T. Permien, R. Engelhardt, C.A. Feldmann and E.E. Koch: Chem. Phys. Letters, 1983, in press
36 J.C. Buchholz and G.A. Somorjai: J. Chem. Phys. 66, 573 (1977)
37 N.V. Smith: in Ref. 2, Vol. I, p. 237

New Type of Local Resonances in Thin Rough Films

V.M. Agranovich, V.E. Kravtsov, T.A. Leskova
Institute of Spectroscopy, Academy of Sciences USSR, Troitsk
Moscow obl. 142092, USSR

Nowadays surface electromagnetic waves (surface polaritons SP) are finding increasing application in the study of optical properties of surfaces and thin films. The reason for this is that SP dispersion law is determined not only by bulk dielectric constants of contacting media, but also bears some information on the properties of the transition layer.

The effect of the transition layer on SP dispersion law proves to be most prominent at SP frequencies $\omega \approx \omega_0$ close to those of dipole vibrations in the transition layer (or a thin film on substrate). Namely, a gap is formed in the SP frequency spectrum at $\omega \approx \omega_0$, and the width of this gap Δ is proportional to $\sqrt{d/\lambda}$ (d is the thickness of the transition layer, $\lambda = 2\pi c/\omega$). Such an effect and, in particular, the square root dependence $\Delta(d)$ was theoretically predicted in [1] and experimentally observed in different frequency ranges [2,3] .

It was recently shown by one of the authors [4] that at frequencies just below the gap an additional surface wave can also arise. This effect results from the long-range Coulomb interaction of polarization at different points x and x′ on the surface. As a result, a strong nonanalytic k dependence appears in the dispersion law

$$\varepsilon_1(\omega)\left[1-\varepsilon_1(\omega)\right]^{-1} + |\vec{k}|d\left[\varepsilon_2-\varepsilon_1(\omega)\right](\varepsilon_2+1)^{-1}. \qquad (1)$$

Here $\varepsilon_1(\omega)$ is the effective dielectric constant of a transition layer (or a thin film on the surface), ε_2 and 1 are the bulk dielectric constants of contacting media, $\vec{k}$ is SP wave vector.

In the vicinity of dipole vibration frequencies ω_0 the dielectric constant of the transition layer $\varepsilon_1(\omega)$ either

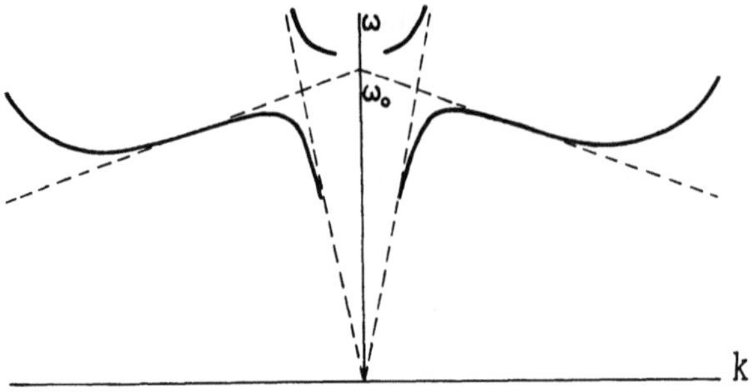

Fig.1 Dispersion curve for surface waves in thin films

goes to zero ($\varepsilon_1 \sim (\omega - \omega_0)$) or becomes arbitrarily large ($\varepsilon_1 \sim (\omega_0 - \omega)^{-1}$). One can obtain from (1) that in both cases the dispersion law is linear in $|\vec{k}|$ at $|\vec{k}| \gg \omega/c$, and the slope is negative, if $\varepsilon_2 < -1$. (See Fig.1).

In addition to this peculiar branch of the dispersion curve at $\varepsilon_2 < -1$ there exists, of course, an ordinary SP with dispersion

$$k^2 = \frac{\omega^2}{c^2} \frac{\varepsilon_2}{1 + \varepsilon_2} . \qquad (2)$$

The interaction between these two SP branches results in a gap in the frequency spectrum near the intersection point of dispersion curves of (1) and (2). Below the gap two SPs may exist at the same frequency ω . One of them is an ordinary SP with $k \gtrsim \omega/c$ and the other is a short-wavelength additional SP.

The appearance of an additional surface wave may be considered in a certain sense as an effect of spatial dispersion due to long-range dipole-dipole interaction.

Note that all the above said is valid, if we neglect the short-range polarization correlations which lead to the k dependence of dielectric permeability $\varepsilon_1(\omega, \vec{k})$. The simplest way to allow for nonlocal effects in linear response is to introduce the dependence of the resonance frequency ω_0 on the wave vector k

$$\omega_0(k) = \omega_0 + \alpha k^2 .$$

As a result, the dispersion law of (1) takes the form:

$$\omega = \omega_0 - \tilde{\Delta} |\vec{k}| d + \alpha k^2 \qquad (3)$$

where $\tilde{\Delta} = \varepsilon_2 (1+\varepsilon_2)^{-1} \left(\frac{d\varepsilon_1}{d\omega}\Big|_{k=0,\,\omega=\omega_0} \right)^{-1}$ $\qquad (\varepsilon_1(\omega,0) \to 0)$

$$\tilde{\Delta} = (1+\varepsilon_2)^{-1} \left(d\varepsilon_1^{-1}/d\omega \,\Big|_{k=0,\,\omega=\omega_0} \right)^{-1} \qquad (\varepsilon_1(\omega,0) \to \infty).$$

Equation (3) shows that at $\alpha > 0$ the initial decrease changes into the increase when the wave vector k increases (Fig.1). It is clear that one more surface wave appears in such a case and therefore three SPs of the same frequency may exist.

The appearance of additional waves due to spatial dispersion is well known in bulk crystallo-optics (see [5]). In the frequency region where additional waves exist the problem of light reflection from an interface becomes rather difficult because the Maxwell's boundary conditions prove to be insufficient to solve this problem unambiguously. To avoid this ambiguity additional boundary conditions (ABC) are required, which depend on the detailed structure of an interface.

The problem of SP transformation at the impedance step contains an inherent complication, which is connected with the possibility of SP transformation into bulk radiation. Therefore, the diffraction at the impedance step is to be considered.

Such a problem has been solved analytically in [6-8] . A thin film (the thickness d $\ll \lambda$) with dielectric constant $\varepsilon_1(\omega, \vec{k})$ has been supposed to occupy a half plane (x > 0) on the metal surface. The energy transformation coefficients for SP incident normal to the boundary of the film x = 0 have been calculated. At frequencies close to the resonance with vibrations in the film all SP energy transformation coefficients have been shown to increase. In particular, the excitation coefficients of the additional SPs by long-wavelength incident SP show narrow resonance peaks at $\omega \approx \omega_0$. Similar frequency dependence has been predicted in [6-8] for the reflection coefficient and the intensity of the bulk cylindrical wave, appearing due to SP diffraction. The latter effect means that at the incidence of white (with the broad frequency spectrum) surface radiation on the impedance step the spectrum of bulk radiation must contain frequencies in the narrow interval near $\omega = \omega_0$. The above effect is in a certain sense analogous to the formation of residual rays when white light is reflected from ionic crystals. In the vicinity of the maximum the part of the incident SP energy transformed into bulk radiation is as much as 90% for $\varepsilon_2=-20$ and increases with $|\varepsilon_2|$. The angular dependence of bulk radiation intensity has two péaks. The more intense peak at small angles corresponds to diffraction in nearly forward direction and the second one at angles close to 180° corresponds to back diffraction.

The study of bulk radiation frequency spectrum seems to be one of the possible methods to determine the eigenfrequencies in a transition layer or a thin film on a surface. Another method may be based on the investigation of the frequency dependence of the reflection coefficient of bulk radiation incident upon the surface in the region of the impedance step. As it has been shown in [7] at observation angles (the angle between the normal to the surface and the perpendicular from the observation point onto the impedance step) close to the angle of mirror reflection the frequency dependence of the reflection coefficient R has a dip at the frequency $\omega \approx \omega_0$. The width of this dip is very small at almost normal and grazing incidence ($\delta\omega/\omega_0 \sim 10^{-3}$ at $\varphi = 5°$) and $R_{min} \approx 0$.

As it has been shown in [6,7] the energy transformation coefficients are universal functions of the frequency and do not depend on the shape of the film edge provided that the width of the transition region at the edge is much smaller than the SP wavelength. However, such a situation occurs only in the case when the k dependence of the dielectric function in the film ε_1 is negligible. When the spatial dispersion of the dielectric function $\varepsilon_1(\omega, \vec{k})$ is taken into account, the local edge modes may exist [8] which result in additional resonance peculiarities in the frequency dependence of the energy transformation coefficients. Since the frequencies of the edge modes strongly depend on the film profile function at the edge, the energy transformation coefficients must also depend strongly on the shape of the film edge.

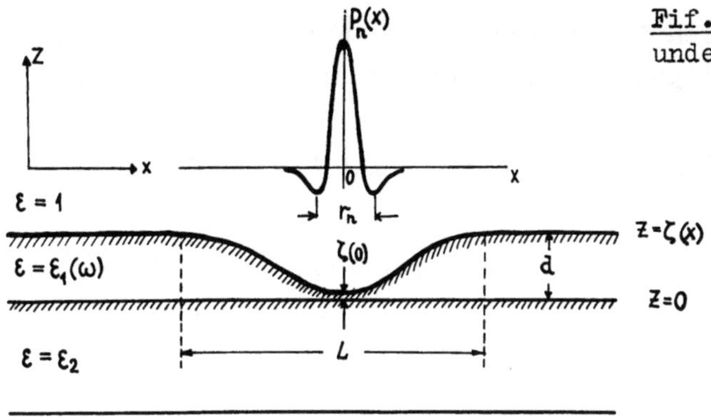

Fif.2 Surface profile under consideration

It is clear, however, that the local modes may exist not only at the edge of a thin film on the surface, but also if there is a pit or a bump on the film.

We now consider a rough film and we assume that its thickness is a function of one coordinate only, that we choose to be x. The surface profile function is supposed to be $\zeta(x) = d - \tilde{\zeta}(x)$, where $\tilde{\zeta}(x)$ is an even function of x that vanishes rapidly as $|x|/L \gg 1$ (L is a characteristic size of the roughness in the film plane). When $\tilde{\zeta}(x)>0$, there is a symmetrical pit and, when $\tilde{\zeta}(x)<0$ - a symmetrical bump on the film (Fig.2).

At frequencies near the resonance frequency ω_0, given by $\varepsilon_1(\omega_0)=0$, the equation for E_z component of the vacuum electric field (see Fig.2) takes the form

$$\tilde{\varepsilon}_1(\omega)P(\vec{k}) = -|\vec{k}|\int \frac{d\vec{q}}{(2\pi)^2} P(\vec{q})\,\zeta_{\vec{k}-\vec{q}} \tag{4}$$

where
$$\tilde{\varepsilon}_1(\omega) = (1+\varepsilon_2)\,\varepsilon_1(\omega)/\varepsilon_2 \tag{5}$$

$$P(\vec{k}) = \int E_z(\vec{r},0)\, e^{i\vec{k}\vec{r}}\, d^2\vec{r}. \tag{6}$$

Equation (4) has been obtained in [9] for an arbitrary film profile function
$$\zeta(\vec{r}) = \int \zeta_{\vec{k}}\, e^{-i\vec{k}\vec{r}} \frac{d^2\vec{k}}{(2\pi)^2}$$
provided that $|\zeta(\vec{r})| \ll \lambda$, $|\vec{\nabla}\zeta| \ll 1$. Retardation effects in (4) have been neglected. The physical nature of the right-hand side of (4) is rather simple. Since the polarization in

the film per unit area is proportional to $P(\vec{r})\zeta(\vec{r})$, the contribution of dipole-dipole interactions to the energy is given by

$$H_{int.} = \int P(\vec{r})\zeta(\vec{r})\,|\vec{r}-\vec{r}'|^{-3}\,P(\vec{r}')\zeta(\vec{r}')\,d^2\vec{r}\,d^2\vec{r}'. \quad (7)$$

It means that the additional term proportional to $\delta H_{int.}/\delta(P\zeta)$ appears in the equation of motion for $P(\vec{r})$ which apparently coincides with the right-hand side of (4). The factor $(1+\varepsilon_2)/\varepsilon_2$ in (5) arises because of image forces effects. As it is well known, polarization in the substrate may be taken into account by means of introduction of "image dipoles" $d_{im.} = d(\varepsilon_2-1)/(\varepsilon_2+1)$. Therefore, the effective dipole is $[2\varepsilon_2/(\varepsilon_2+1)]\,d$ and not d.

We shall seek the solutions of (8) which correspond to the local states of the size $r_n \ll L$. For the sake of simplicity we shall consider here only the one-dimensional case, when $\zeta(\vec{k}) = \zeta(k_x)\,2\pi\delta(k_y)$, $P(\vec{k}) = P(k_x)\,2\pi\delta(k_y)$.

Since the function $\tilde{\zeta}(k)$ goes rapidly to zero at $|k| > L^{-1}$, the main contribution to the right-hand side of (4) is determined by integration over the narrow region $|q-k| \lesssim L^{-1} \ll r_n^{-1}$, where the Fourier component $P(k)$ is smoothly varying. Thus, using for $P(k)$ its Taylor series expansion at $q=k$ we find

$$\int\frac{dq}{(2\pi)}\,\zeta_{k-q}\,P(q) = \zeta(0)\,P(k) - \frac{1}{2}\,\tilde{\zeta}''(0)\,\frac{d^2P}{dk^2} + \cdots . \quad (8)$$

If we keep only the first two terms in the expansion (8) and substitute them in (4), we obtain the resulting equation for $P(k)$ that coincides with the Schrödinger equation for an s state (l=0) of a hydrogen atom

$$-\frac{1}{2m}\,\frac{d^2P_n}{dQ^2} - \frac{g}{|Q|}\,P_n(Q) = E_n\,P_n(Q) \quad (9)$$

where $Q = kr_n$, $E_n = -\zeta(0)\,|\tilde{\zeta}''(0)|/\tilde{\varepsilon}_1^{\,2}(\omega_n)$,

$r_n = |\,\tilde{\varepsilon}_1(\omega_n)/\tilde{\zeta}''(0)\,|$, $m = -\,\mathrm{sign}(\tilde{\zeta}''(0))$, $g = -\,\mathrm{sign}(\varepsilon_1)$.

As it is seen from (9), the considered local resonances (note that we seek localized polarization vibrations with $r_n \ll L$) may exist only in the frequency region $\omega < \omega_0$ ($g > 0$) at $\tilde{\zeta}''(0) < 0$ ($m > 0$), i.e. when there is a pit in the film. The local resonance spectrum in this case is determined from a well-known condition $E_n = -1/2n^2$.

In the vicinity of $\omega = \omega_0$ the dielectric constant $\tilde{\varepsilon}_1 = \frac{\omega - \omega_0}{\tilde{\Delta}}$ so that the local resonance spectrum is an equidistant set of frequecies

$$\omega_n = \omega_0 - n\Omega \tag{10}$$

$$\Omega = \tilde{\Delta}\sqrt{2\zeta(0)\,|\tilde{\zeta}''(0)|}\,.$$

The spatial size of the localized state is

$$r_n = \sqrt{\frac{2\zeta(0)}{|\tilde{\zeta}''(0)|}}\; n \approx \sqrt{\frac{2\zeta(0)}{d}}\; L\, n\,. \tag{11}$$

It is seen from (11) that strongly localized ($r_n \ll L$) states exist if the film thickness on the pit bottom $\zeta(0)$ is much less than that far away from the pit and $n \lesssim (d/\zeta(0))^{1/2}$.

It should be stressed that the frequencies ω_n fall in the region where surface waves with a continuous spectrum exist (3) (see also Fig.2). This circumstance leads to radiative broadening of local resonances. The calculation of the resonance linewidth Γ_n has been made in [9] and the result is

$$\Gamma_n/\Omega \simeq 2\pi\, \frac{\zeta(0)}{d}\, n^2. \tag{12}$$

The equation (12) shows that at $n \lesssim (d/\zeta(0))^{1/2}$ the linewidth Γ_n is less than the distance Ω between two neighbouring resonance frequencies and the resonance peaks may be experimentally observed.

References

1. V.M.Agranovich, A.G.Malshukov: Opt.Commun. 11, 169 (1974)
2. V.A.Yakovlev, V.G.Nazin, G.N.Zhizhin: Opt.Commun. 15, 293 (1975)
3. T.Lopez-Rios, F.Abeles, G.Vuye: J. de Phys. 39, 645 (1978)
4. V.M.Agranovich: Zh.Exp.Teor.Fiz. 77, 1124 (1979)
5. V.M.Agranovich, V.L.Ginzburg: Crystaloptics with Spatial Dispersion and the Theory of Excitons (Springer-Verlag Berlin 1982)

6. V.M.Agranovich, V.E.Kravtsov, T.A.Leskova: Zh.Exp.Teor.Fiz. 81, 11 (1981)
7. V.M.Agranovich, V.E.Kravtsov, T.A.Leskova: Solid State Commun. 40, 687 (1981)
8. V.M.Agranovich, V.E.Kravtsov, T.A.Leskova: Zh.Exp.Teor.Fiz. 84, 103 (1983)
9. V.M.Agranovich, V.E.Kravtsov, T.A.Leskova: to be published in Solid State Commun.

Part III

Electronic Excitations and Spin Dynamics

Electron Spin Echo Spectroscopy of One-Dimensional Excitons

J. Schmidt

Huygens Laboratorium, University of Leiden, Postbox 9504
NL-2300 RA Leiden, The Netherlands

1. Introduction

In the last few years it has been possible to obtain detailed information about scattering of triplet excitons in the molecular crystal of 1,2,4,5-tetrachlorobenzene (TCB) by the application of electron spin echo (ESE) spectroscopy in combination with pulsed, narrow-band, tunable dye lasers. The attraction of the TCB system is that the intermolecular interaction in good approximation is one-dimensional which makes it a prototype for investigating the properties of tightly bound excitons [1,2].

The experimental and theoretical results to be discussed here apply to the low-temperature regime where the "coherent" description of the triplet excitons applies. Here the collective excitations of the linear chain are characterized by a wave vector k and a well-defined band structure exists. We shall show that by pulsed laser excitation it is possible to prepare the triplet exciton system in a non-equilibrium situation where only exciton states with a very small k value are populated. The subsequent evolution towards thermodynamical equilibrium then is observed by the pulsed ESE method. The experiments allow us to draw conclusions about the scattering mechanisms and about the spatial delocalization of the excitons in the crystal. In particular it will appear that in the TCB crystal studied by us the scattering is an impurity "assisted" process. This means that the exciton-phonon scattering can be explained by considering only the conservation of energy. This is in contrast with current theoretical models [3,4] in which one always assumes that both energy and quasi-momentum must be conserved in the exciton-phonon scattering process.

2. The System TCB

The attraction of the system TCB is that the interaction energy between two molecules, which for triplet excitations depends on the overlap of the molecular wavefunction and consequently is non-zero only for neighbouring molecules, has an appreciable value only in one dimension. The crystal structure of TCB is monoclinic at room temperature with two molecules A and B per unit cell [5,6]. As inferred from the crystal structure the intermolecular exchange interaction predominantly is along the short $\vec{a}$ axis. The generally accepted value is $\beta = 0.34\ cm^{-1}$ [7]. The intermolecular interaction in the other directions is estimated to be at least an order of magnitude smaller. Consequently there are two types of exciton chains (A and B) between which only a very weak interchain interaction exists. Obviously the A and B chains are indistinguishable in zero-magnetic field.

The energy dispersion for a linear chain exciton system like TCB with a nearest-neighbour interaction of β is given by

$$E(k) = E_o + 2\beta \cos ka \tag{1}$$

where k is the wave vector, E_o is the energy of the centre of the band and a the intermolecular distance. Since β is positive in TCB the k = 0 states are at the top of the band. Each of the exciton k states, however, is split into three spin components and since we are dealing with translationally equivalent molecules we would expect that this splitting is the same as for isolated TCB molecules (see Fig. 1).

About a decade ago Francis and Harris showed that the zero-field splitting of the triplet exciton is slightly k dependent as a result of spin-orbit coupling (SOC) between triplet and singlet excitons with the same value of k [7]. In Fig. 1 we have indicated by broken lines the positions of the T_z and T_y triplet bands with respect to T_x as a result of this SOC. Via a straightforward perturbation calculation Francis and Harris derived the following expression for the resonance frequency of the T_z-T_x and T_y-T_x zero-field transition

$$\nu(k) = \nu_o + 2\mu \cos ka \tag{2}$$

with

$$\mu = \zeta^2 (\beta_S - \beta_T)/(E^o_{ST})^2 . \tag{3}$$

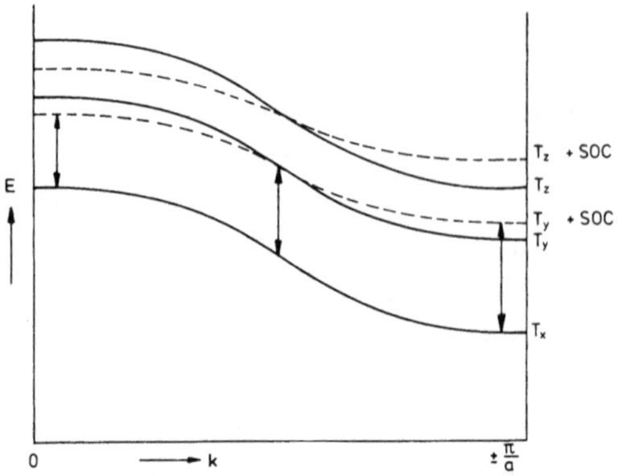

Fig. 1. A schematic depiction of the k dependence of the zero-field splitting in the linear chain triplet exciton system of TCB as a result of selective SOC. The positions of the triplet sublevels in the absence of SOC are indicated by the full lines. The dashed lines indicate their position including SOC. Note that the zero-field splittings and the exciton dispersion are not drawn to scale for illustrative purposes

Here ζ is the SOC constant and β_S and β_T are the singlet and triplet molecular interaction energies and E^o_{ST} is the separation between triplet and singlet states in the absence of intermolecular interactions.

The steady-state lineshapes of the T_y-T_x and T_z-T_x zero-field transitions can be calculated using expression (2) for the k dependence of the zero-field frequencies and assuming a Boltzmann population distribution over the k states. In Fig. 2 we show results of ODMR exeriments on the T_y-T_x and T_z-T_x transitions taken from the original paper of Francis and Harris together with the theoretically predicted lineshapes. This theoretical curve has been derived using 4β = 1.36 cm^{-1} for the triplet exciton bandwidth, μ^{yx} = -2.5 MHz for the T_y-T_x and μ^{zx} = -3.9 MHz for the T_z-T_x transition respectively. The spectra show two peaks of different intensity which correspond with the maxima in the density of states function at the top and the bottom of the exciton band. The difference in height of the two maxima is related to the Boltzmann factor across the band [7].

The beautiful correspondence between the optically detected and theoretical lineshapes of the exciton zero-field transitions confirms that in the liquid helium temperature range the "coherent" description applies for

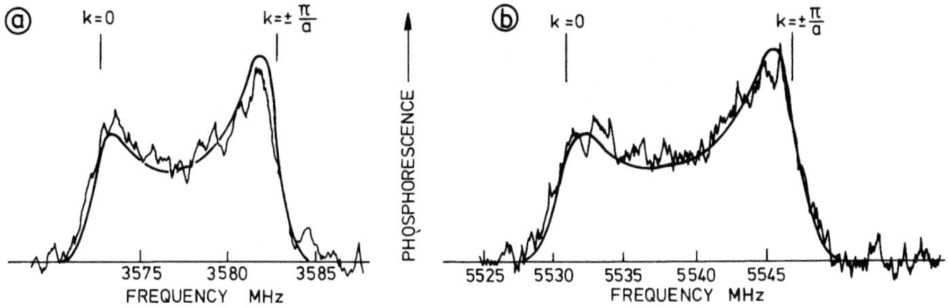

Fig. 2. a) The optically detected T_y-T_x zero-field transition of the triplet excitons in TCB. The solid line is the theoretically predicted curve at T=3.2 K.

b) The optically detected T_z-T_x zero-field transition of the triplet exciton in TCB. The solid line again is the theoretically predicted curve at T=3.2 K. After reference [7]

the triplet excitons in TCB. The important consequence of the model of Francis and Harris is that every k state is uniquely defined by its zero-field resonance frequency. We will show in the next section how this property can be used with advantage to study the dynamic properties of the exciton wave packets.

3. The Dynamic Behaviour of Triplet Excitons in TCB

a. *Experimental results in zero-magnetic field*

In 1980 Van Strien et al. [8] performed a time-resolved electron spin echo (ESE) experiment on the triplet excitons in TCB following optical excitation with a pulsed tunable dye laser directly into the 0-0 band of the $T_0 \leftarrow S_0$ absorption. The sequence of the optical and microwave pulses is indicated in Fig. 3. The echo signals are generated by two microwave pulses resonant with the T_y-T_x zero-field transition with an interval τ = 1.5 μs and a delay t_d between the exciting laser pulse and the first microwave pulse. In the first experiment t_d is given its minimal value of 0.5 μs and the echo signal, which is proportional to the population difference between T_y and T_x, is used to monitor the lineshape at this short delay. The result, displayed in Fig. 4, shows that a very strong signal is present around 3572.5 MHz, the resonance frequency of the k ≈ 0 excitons, indicating that the optical selection rule leads to an overpopulation of the

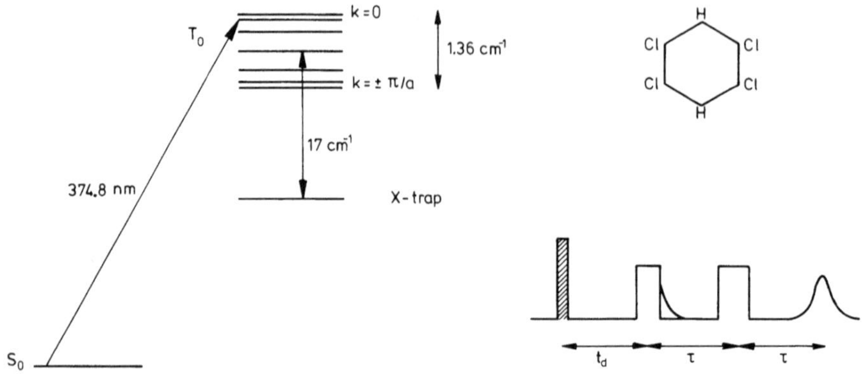

Fig. 3. A simplified energy level scheme of TCB with the position of the S_0 ground state, the lowest triplet exciton band and the major structural X trap. In addition we have indicated the sequence of the laser flash at 374.8 nm, which excites the system directly into the k ≃ 0 region of the triplet exciton band, and the two microwave pulses resonant with the T_y-T_x transition

k ≃ 0 excitons. On the other hand we also observe a weaker signal around 3582.5 MHz, the resonance frequency of the k ≃ ±π/a excitons. We shall see hereafter that the corresponding population of the k ≃ ±π/a excitons is caused by scattering processes in the exciton band. It is important to note

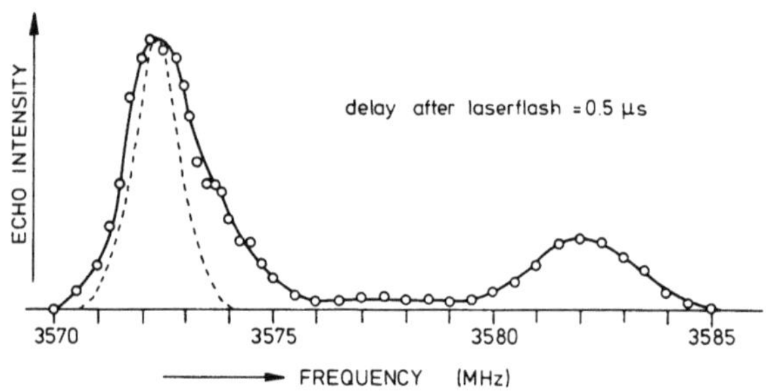

Fig. 4. The ESE detected lineshape of the T_y-T_x triplet exciton zero-field transition in TCB following laser flash excitation in the $T_0 \leftarrow S_0$ absorption origin at 374.8 nm. The open circles indicate the echo intensity. The laser bandwidth is 0.2 cm^{-1}, the pulse duration is 15 ns and the pulse energy 3 mJ. The microwave pulses have a peak power of 0.25 W and the pulse length is 0.5 μs. The dashed line indicates the spread in microwave frequency corresponding to the bandwidth of 0.2 cm^{-1} of the laser

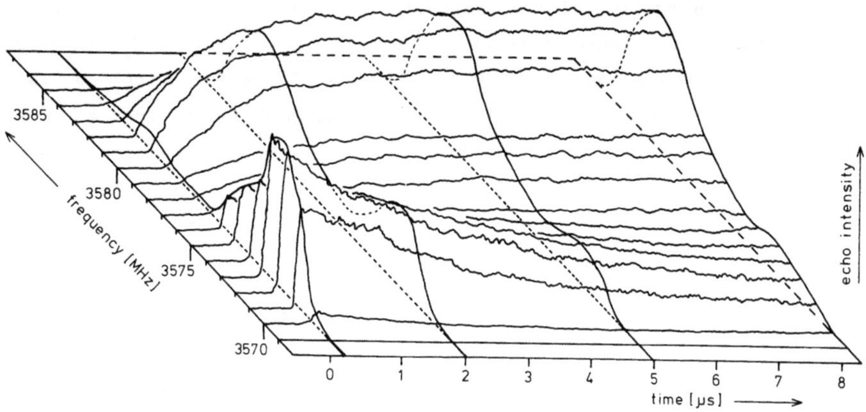

Fig. 5. A three-dimensional representation of the time evolution of the ESE detected lineshape of the T_y-T_x triplet exciton zero-field transition in TCB following laser flash excitation into the absorption origin at 374.8 nm. The laser bandwidth is 0.2 cm^{-1}, the pulse duration is 15 ns and the pulse energy is 3 mJ. The microwave pulses have a peak power of 1 W and a duration of 250 ns. The time is the time interval t_d between the laser flash and the first microwave pulse. The interval between the microwave pulses is 1.5 μs

that the ESE signals correspond to a strong overpopulation of the T_y spin level as compared with T_z and T_x. This effect arises because only the $T_y \leftarrow S_0$ carries an appreciable oscillator strength [9]. This spin alignment is of great practical importance because it leads to very large ESE signals after laser excitation.

In Fig. 5 we show in a three-dimensional representation the evolution of the non-Boltzmann population distribution created with the laser flash. The curves represent the intensity of the echo signal at a given resonance frequency as a function of the delay t_d between the laser flash and the first microwave pulse with fixed interval τ = 1.5 μs between the microwave pulses. It is seen that the overpopulation of the $k \simeq 0$ excitons has reached its thermal equilibrium value in about 10 μs and that the $\pm\pi/a$ excitons grow in roughly at the same rate. The curve also shows why in Fig. 4 we observe a signal at 3582.5 MHz corresponding to the $\pm\pi/a$ exitons. Even at this short delay scattering has already established an appreciable population of the $k \simeq \pm\pi/a$ excitons.

To study this problem of the scattering of the excitons in more detail we next measured the temperature dependence of the decay rate $T_s^{-1}(k \simeq 0)$ of the initial excess population of the $k \simeq 0$ excitons following the laser flash. This rate represents the total scattering probability from $k \simeq 0$ to all the other exciton k states. In Fig. 6 we show the results from 1.15 to 2.5 K. At higher temperatures the phase memory time T_M becomes so short that the echo signals can no longer be observed. It is seen that the scattering rate increases with increasing temperature.

b. The theoretical model for exciton scattering

The experimentally observed $k \rightarrow k'$ scattering in the triplet exciton band of TCB does not indicate any restriction in the scattering probabilities. This gives rise to doubts as to whether direct exciton phonon coupling is responsible for the observed scattering process, since according to the laws of conservation of energy and quasi-momentum scattering should occur solely in limited regions of k space. To account for the experimental observations we have developed a theory which is based on impurity-assisted scattering by one-phonon processes and therefore is valid only at low temperatures where k_BT is small compared with the exciton bandwidth. The resulting expressions for the scattering probability are [10]

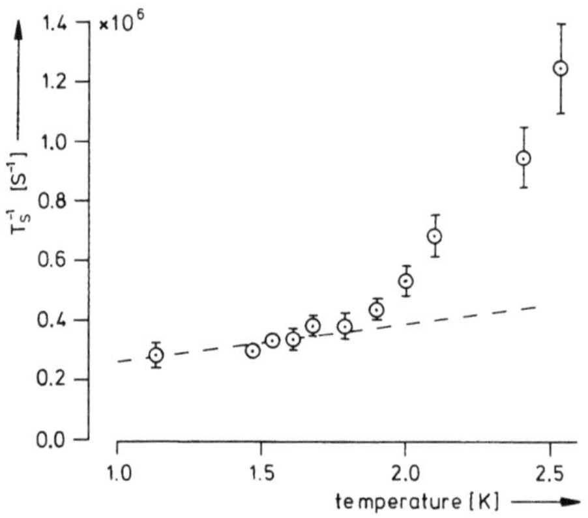

Fig. 6. The experimentally determined temperature dependence of the scattering rate T_S^{-1} ($k \simeq 0$), as obtained from the decay of the excess population of the $k \simeq 0$ excitons at various temperatures. The dashed line indicates the calculated temperature dependence of T_S^{-1} ($k \simeq 0$) using the impurity-assisted exciton-phonon scattering model

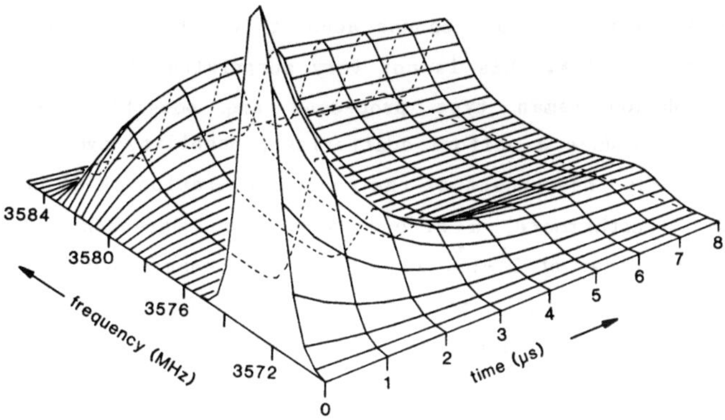

Fig. 7. A three-dimensional representation of the computer simulation of the experiment depicted in Fig. 6, using the impurity-assisted exciton-phonon scattering model

$$W_{k'k} = \frac{W}{N}(\Delta E)^3[\bar{n}(\Delta E) + 1] = \frac{W}{N}(\Delta E)^3 \frac{\exp(\Delta E/k_BT)}{\exp(\Delta E/k_BT) - 1} \quad (4)$$

for $E_k > E_{k'}$ and

$$W_{k'k} = \frac{W}{N}(\Delta E)^3 \bar{n}(\Delta E) = \frac{W}{N}(\Delta E)^3[\exp(\Delta E/k_BT) - 1]^{-1} \quad (5)$$

for $E_k < E_{k'}$. Here W and N are constants and $\Delta E = E_k - E_{k'}$.

As a first test of the model we performed computer calculations of the evolution of the population distribution over the exciton k states assuming that the scattering probability is determined by (4) and (5). In Fig. 7. the result is shown and it is clear that a very good reproduction is obtained of the experimentally observed evolution of the T_y-T_x lineshape [10].

As a second test we calculated the temperature dependence of T_s^{-1} ($k \simeq 0$). The result is indicated in Fig. 6 by the dashed line, which very nicely reproduces the experimentally observed temperature dependence of T_s^{-1} ($k \simeq 0$) up to 2 K (where k_BT is almost equal to the bandwidth 4β of the triplet exciton band). This is an important support for the impurity-assisted scattering model which seems to apply up to a temperature where k_BT is equal to the bandwidth.

The scattering model discussed above cannot account for the temperature dependence of T_S^{-1} above $T \approx 2$ K. This is not very surprising since above this temperature two-phonon Raman - type processes are possible. Such processes are expected to induce a uniform scattering probability over the band, since they are roughly independent of the energy separation ΔE of the two k states involved. Furthermore, a strong temperature dependence is predicted, usually of the order of T^5-T^7. In the limited temperature range between 2 and 2.5 K it is difficult to ascertain the exact temperature dependence. However, we do have an important additional piece of information, from the work of Botter et al. [11]. They have shown that at 5 K a motional averaging of the whole T_y-T_x zero-field lineshape sets in. As explained by them at this temperature the average scattering rate $\overline{W}$ must be of the order of the total linewidth, i.e. $\overline{W} \approx 10^8$ s^{-1}. Comparing this value with the estimated contribution of two-phonon processes to the scattering rate T_S^{-1} at 2.5 K, we derive a temperature dependence between these two temperatures of $T^{(6.4 \pm 0.6)}$. We are therefore led to believe that Raman-type processes dominate the scattering in this temperature range.

c. Intra- or interchain scattering?

Our observations may be explained by intra- as well as interchain scattering. In our experiments however which are performed in zero-magnetic field it is impossible to distinguish between the two processes. However, several years ago it was suggested that interchain hopping is responsible for the spin-lattice relaxation (SLR) in triplet exciton systems [12]. Therefore a study of the SLR rate could, in principle, provide a means of ascertaining the interchain hopping rate, and its relative importance in the scattering process.

In a detailed study of the orientational dependence of the SLR rate Van Strien could prove that the SLR is caused by interchain hopping [13]. From the calculated hopping rate it appears that the scattering in the zero-field experiments may be caused to a large extent by this process. So in contrast to the one-dimensionality of the exchange interaction it seems that the scattering process is not restricted to one dimension.

d. The width Δk of the triplet exciton wave packets

An important problem, not discussed so far, is the width Δk of the triplet exciton wavepackets. Since the exciton k states correspond to different zero-field splittings we expect that this width Δk gives rise to a well-defined spread in resonance frequency in the zero-field transitions. So from the homogeneous linewidth $(\pi T_M)^{-1}$ of the triplet spins, obtained from a measurement of their phase memory time T_M, we may expect to derive information about the width of the exciton wave packets.

The measurements of T_M have been performed by Van Kooten et al. [14] and the result shows that a very characteristic variation of T_M over the T_y-T_x zero-field lineshape obtains. This variation can be explained nicely by assuming that it is caused by a k-independent width Δk of the exciton wave packet with in addition a contribution from the scattering through the exciton band. The resulting value for Δk represents about 0.5% of the first Brillouin zone, indicating that the triplet exciton extends over about 200 molecules.

4. Conclusion

The experiments discussed in this chapter, to our knowledge, represent the first case where it has been possible to study the dynamic properties of triplet excitons in molecular crystals. The most important result is that the $k \rightarrow k'$ scattering of the excitons at very low temperatures is an impurity-assisted process. In the case of TCB we have been able to prove that chemical impurities do not play an important role. In recent experiments on new crystals of TCB in which the concentration of the dominant chemical impurity ($C_6H_2Cl_3Br$) is only 20 ppm, i.e. 50 times lower than in the crystals studied so far, we have observed the same scattering rate as in the "old" crystals. Hence we believe that the scattering is caused by the structural defects (X traps) or by isotopic impurities (^{13}C or ^{2}H containing molecules).

It appears that one-phonon processes dominate the scattering at temperatures where k_BT is smaller than the width of the exciton band and that at higher temperatures two-phonon Raman-type processes take over. It would be interesting to check whether this is a property which applies generally to exciton systems. Further, the results of our experiments on SLR indicate that the $k \rightarrow k'$ scattering is not restricted to one chain. The question arises whether this applied only to TCB or if it is a general property of one- (or two-) dimensional exciton systems in molecular crystals.

We hope that these questions will get an answer in the coming years.

Acknowledgements

The author acknowledges stimulating discussions with Professors D. Polder and R. Silbey. This work was performed as part of the research program of the "Stichting voor Fundamenteel Onderzoek der Materie (F.O.M.)" with financial support from the "Nederlandse Organisatie voor Zuiver Wetenschappelijk Onderzoek (Z.W.O.)".

References

[1] J. Frenkel: Phys. Rev. 37, 17; 1276 (1931)

[2] A.S. Davydov: "Theory of Molecular Excitons" (McGraw Hill, New York 1962)

[3] A.S. Davydov: Phys. Stat. Sol. 20, 143 (1967)

[4] A.S. Davydov: "Theory of Molecular Excitons" (Plenum Press, New York London 1971)

[5] C. Dean. M. Pollack, B.M. Craven and G.A. Jeffrey: Acta. Cryst. 11, 710 (1958)

[6] F.H. Herbstein: Acta Cryst. 18, 997 (1965)

[7] A.H. Francis and C.B. Harris: Chem. Phys. Lett. 9, 181; 188 (1971)

[8] A.J. van Strien, J.F.C. van Kooten and J. Schmidt: Chem. Phys. Lett. 76, 7 (1980)

[9] A.H. Francis and C.B. Harris: J. Chem. Phys. 57, 1050 (1972)

[10] A.J. van Strien, J. Schmidt and R. Silbey: Mol. Phys. 49, 151 (1982)

[11] B.J. Botter, A.I.M. Dicker and J. Schmidt: Mol. Phys. 36, 129 (1978)

[12] D. Haarer and H.C. Wolf: Phys. Stat. Sol. 33, K117 (1969); Mol. Cryst. Liq. Cryst. 10, 359 (1970)

[13] A.J. van Strien and J. Schmidt: Chem. Phys. Lett. 86, 203 (1982)

[14] J.F.C. van Kooten, A.J. van Strien and J. Schmidt: Chem. Phys. Lett. 90, 337 (1982)

Energy Transfer in Molecular Crystals and Its Influence on Spin Resonance

Robert Silbey

Department of Chemistry and Center for Materials Science and Engineering
Massachusetts Institue of Technology, Cambridge, MA 02139, USA

1. Introduction

In the past ten years, excitation transfer processes between molecules in condensed phases have been studied in a variety of systems and by a number of methods. However, the naphthalene dimer systems first studied experimentally by Schwoerer and Wolf [1] and theoretically by Haken and Reineker [2] remain the most beautiful model system for these processes. The elegant electron spin echo (ESE) experiments of van der Waals, Schmidt and coworkers [3] on this system were the final word on the experimental side, at least at low temperatures; however, there are still a few theoretical loose ends to clear up, as discussed recently by Schmid and Reineker [4]. In this paper, I discuss these model systems, focussing on the ESE experiments.

The dimers we are thinking about come in two types (inequivalent pairs and equivalent pairs) because there are two translationally inequivalent (but crystographically equivalent) molecules in the unit cell. Since the first excited triplet state excitation energy in $C_{10}D_8$ is ~100 cm^{-1} larger than that of $C_{10}H_8$, the latter molecules in their first triplet state are isolated from their deuterated neighbors at low temperatures. Near neighbor pairs of $C_{10}H_8$ will therefore be isolated dimers, formed from equivalent or inequivalent molecules. Since the matrix element for excitation transfer between molecules A and B (J) is large ($J_{EQN} \approx 0.5\ cm^{-1}$, $J_{NEQ} \approx 1.25\ cm^{-1}$) compared to the zero field splitting parameters for naphthalene ($D \approx 0.1\ cm^{-1}$, $E \simeq -0.01\ cm^{-1}$), the eigenstates for the pair are approximately given by

$$|\pm\rangle = 2^{-1/2}\{|A\rangle \pm |B\rangle\} \quad , \quad E_\pm \simeq \pm J \qquad (1)$$

where $|A\rangle$ is the state which describes a triplet excitation on molecule A.

Since the principal axes and elements of the fine structure tensor of the naphthalene triplet state are well known from the pioneering work of Hutchison [5], the fine structure tensor of the equivalent and inequivalent pairs can be found theoretically [6] and checked experimentally [1]. The final picture of the dimer triplet states is then that for each dimer: there are two triplet states (given by (1)) each of which has a distinct fine structure tensor. In the inequivalent dimer, $|+\rangle$ and $|-\rangle$ have fine structure tensors very different from the monomer and slightly different from each other. In the equivalent dimer, we might expect the triplet states to have fine structure identical to the monomer, but happily there is a small difference, and ESE experiments suggest that the $|+\rangle$ and $|-\rangle$ states also are slightly different.

If all relaxation processes could be turned off, we would expect to see electron resonance transitions for each dimer state we have discussed. However, phonon processes will scatter excitations from $|+\rangle$ to $|-\rangle$ and vice versa, broadening and moving the lines. From the temperature dependence we can learn about these scattering processes.

2. Model

In order to describe the dimer system, we must solve the equations of motion of the reduced density matrix, which is the trace over lattice states of the total density matrix. Since we expect the electronic and spin states of the dimer to be weakly coupled to the phonon states, we use the Redfield equations [7] for this problem. Since the dimer has six states, $|\pm,\alpha\rangle$ $\alpha = 1,2,3$, the reduced density matrix is a 6×6 and the relaxation matrix is a 36×36. The equations of motion are then

$$\dot{\rho}_{ij} = -i\omega_{ij}\rho_{ij} - \sum_{k,l} R_{ij;kl}\,\rho_{kl} \tag{2}$$

where all the usual assumptions have been made, and we are in the representation which diagonalizes H_0, the Hamiltonian in the absence of the phonons and other relaxation phenomena. In order to simplify the problem, without removing the important points, we will concentrate only on the dimer states which are involved in a particular spin resonance transition. We therefore have an equation similar to (2), but for which i, j, k, l are only allowed to have the values of the states involved in the ESR. This reduces the problem to eight variables, $\rho_{i\alpha,i\alpha'}$, i = + or -and $\alpha,\alpha' = 1,2$ for example. In the absence of relaxation, we write ($\sigma_{\alpha\alpha'} = \rho_{+\alpha,+\alpha'}$, $\Sigma_{\alpha\alpha'} = \rho_{-\alpha,-\alpha'}$)

$$\dot{\sigma}_{\alpha\alpha'} = -i\omega_{\alpha\alpha'}\,\sigma_{\alpha\alpha'} \tag{3a}$$

$$\dot{\Sigma}_{\alpha\alpha'} = -i\Omega_{\alpha\alpha'}\,\Sigma_{\alpha\alpha'}\,. \tag{3b}$$

Note that because the lower (+) dimer state and the upper (-) dimer state have slightly different fine structure tensors, the spin eigenstates (α and α') in the -manifold are rotated from those in the + manifold (the angle is $\theta \approx 2\pi/180$). If we rewrite (3) so that the two dimer states are in the same spin basis, and take the usual linear combinations we find

$$\dot{\sigma}_{12}+\dot{\sigma}_{21} = -i\omega_{12}^{0}(\sigma_{12}-\sigma_{21})$$

$$\dot{\sigma}_{12}-\dot{\sigma}_{21} = -i\omega_{12}^{0}(\sigma_{12}+\sigma_{21})$$

$$\dot{\sigma}_{11}-\dot{\sigma}_{22} = 0 \tag{4a}$$

and

$$\dot{\Sigma}_{12}+\dot{\Sigma}_{21} = -i\omega_{12}^{e}\cos 2\theta(\Sigma_{12}-\Sigma_{21})$$

$$\dot{\Sigma}_{12}-\dot{\Sigma}_{21} = -i\omega_{12}^{e}\cos 2\theta(\Sigma_{12}+\Sigma_{21}) + i\omega_{12}^{e}\sin 2\theta(\Sigma_{11}-\Sigma_{22})$$

$$\dot{\Sigma}_{11}-\dot{\Sigma}_{22} = -i\omega_{12}^{e}\sin 2\theta(\Sigma_{12}-\Sigma_{21}) \; . \tag{4b}$$

Now, we must add relaxation processes and the only one we consider is the phonon-induced scattering from $|+\rangle$ to $|-\rangle$. We will assume that this occurs without change in spin state, so that the final equations are (in the basis of (4))

$$\dot{\sigma}_{ij} = \text{same as 4a} - W\sigma_{ij} + \tau^{-1}\Sigma_{ij} \tag{5a}$$

$$\dot{\Sigma}_{ij} = \text{same as 4b} + W\sigma_{ij} - \tau^{-1}\Sigma_{ij} \tag{5b}$$

so that W represents the phonon-induced transition rate from $|+\rangle$ to $|-\rangle$ and τ^{-1} is the phonon-induced transition rate from $|-\rangle$ to $|+\rangle$. From the Boltzmann thermal equilibrium of the phonons,

$$W\tau = \exp(-\Delta/k_B T) \equiv \exp(-\beta\Delta) \tag{6}$$

where Δ is the difference in energy between the $|+\rangle$ and $|-\rangle$ states ($=2|J|$). If we now convert σ_{ij} and Σ_{ij} into the same notation used by Verbeek and Schmidt [8] we find the modified Bloch equations used by those authors:

$$\dot{r}_1 = -\omega_0 r_2 - W r_1 + \tau^{-1} R_1$$

$$\dot{r}_2 = \omega_0 r_1 - W r_2 + \tau^{-1} R_2$$

$$\dot{r}_3 = -W r_3 + \tau^{-1} R_3 \tag{7a}$$

$$\dot{R}_1 = -(\omega_e\cos 2\theta)R_2 - \tau^{-1}R_1 + W r_1$$

$$\dot{R}_2 = +(\omega_e\cos 2\theta)R_1 - (\omega_e\sin 2\theta)R_3 - \tau^{-1}R_2 + W r_2$$

$$\dot{R}_3 = (\omega_e\sin 2\theta)R_2 - \tau^{-1}R_3 + W r_3 \; . \tag{7b}$$

Here, r_i are the FVH [9] variables of the lower state and R_i those of the upper state (but in the coordinate axes of the lower state). These equations are a convenient starting point for the calculation of the spin resonance parameters. The solutions to (7) are linear combinations of 6 exponentials, 3 corresponding to "slowly" decaying variables and 3 to quickly decaying variables. We can associate the three slow exponentials with the spin resonance frequencies, line widths $(1/T_2)$ and the spin lattice relaxation rate $(1/T_1)$ of the observed lower triplet state. To make this clear, we look first at the case $\theta = 0$. (a) $\theta = 0$: In this case the 6 eigenvalues of the RHS of (7a) and (7b) are

$$\lambda_\pm = \{-\frac{i}{2}(\omega_0+\omega_e)-(W+\tau^{-1})\} \pm \{W\tau^{-1}+(\frac{1}{4})(-i\omega_0+i\omega_e-W+\tau^{-1})^2\}^{1/2} \tag{8}$$

$$\lambda_\pm^*,\ 0,\ W+\tau^{-1} \; .$$

These can be interpreted as the T_1^{-1} of the upper triplet $(W+\tau^{-1})$, resonance frequency of the lower triplet (Im λ_+), resonance frequency of the upper triplet (Im λ_-), T_2^{-1} of lower triplet (Re λ_+) and T_2^{-1} of upper triplet. In the limit

$$\frac{|\omega_0-\omega_e|W}{(\omega_0-\omega_e)^2+\tau^{-2}} \ll 1$$

then

$$\lambda_+ = -i\omega_0 + \frac{i(\omega_0-\omega_e)\tau^{-2}e^{-\beta\Delta}(1+e^{-\beta\Delta})}{\tau^{-2}(1+e^{-\beta\Delta})^2+(\omega_0-\omega_e)^2} - \frac{(\omega_0-\omega_e)^2\tau^{-1}e^{-\beta\Delta}}{\tau^{-2}(1+e^{-\beta\Delta})^2+(\omega_0-\omega_e)^2} \quad (9)$$

which in the limit $\tau^{-2}\gg(\omega_0-\omega_e)^2$ (fast exchange), valid for naphthalene, is

$$\lambda_+ = -i\omega_0+i(\omega_0-\omega_e)\frac{e^{-\beta\Delta}}{(1+e^{-\beta\Delta})} - (\omega_0-\omega_e)^2\frac{\tau e^{-\beta\Delta}}{(1+e^{-\beta\Delta})^2} . \quad (10)$$

At high enough temperature so that $\beta\Delta\ll1$ this agrees with the usual fast exchange limit [10]; as it stands it agrees with the exchange model of van 't Hof and Schmidt [11] and Harris and co-workers [12] which has been applied to many relaxation problems.

At low T, $e^{-\beta\Delta}\ll1$, and (10) predicts that both the frequency and the line width or inverse phase memory time ($T_M^{-1} = T_2^{-1} = \mathrm{Re}\,\lambda_+$) will show an activated behavior with activation energy $\Delta = 2J$. However, as T rises, the full temperature dependence of λ_+ will be evident. For example, even if τ is independent of temperature, we expect T_M^{-1} to saturate at the value $(\omega_0-\omega_e)^2\tau/4$ as kT becomes much larger than Δ. In addition, it is clear from dynamical studies that τ changes with temperature becoming shorter as T increases. Thus we expect that T_M^{-1} will have a maximum as T increases, and thus the line width will first broaden and then narrow with increasing temperature.

In order to calculate the temperature dependence of τ, we must model the process of phonon-induced scattering from the upper to the lower state. If we assume this to be a one-phonon emission process, we find, from second-order perturbation theory,

$$\tau^{-1} = \lambda^2[n(\Delta)+1]\rho(\Delta) \quad (11)$$

where λ is the electronic matrix element, $\rho(\omega)$ is the phonon density of states, and $n(\omega)$ is the Boson distribution function, $n(\omega) = (e^{\beta\omega}-1)^{-1}$, so that

$$\tau^{-1} = \lambda^2\rho(\Delta)(1-e^{-\beta\Delta})^{-1}. \qquad (12)$$

With this form of τ^{-1}, the maximum in T_2^{-1} should occur for $\exp(-\beta\Delta) = 1/3$, or $T \cong 0.9\ \Delta/k$ which is 1.5K for the equivalent dimer and 3.2K for the inequivalent dimer. The experimentally observed [3] maxima occur at ≈3K for the equivalent and 1.6K for the inequivalent in sharp disagreement with the theory.

If we assume that two-phonon (Raman) processes play a role in τ^{-1}, then we must add to (9) a term proportional to $(kT/\Delta)^6$ (actually a more complicated form is possible [13], but this will suffice to illustrate our point). There will now be two mechanisms for scattering the excitation from $|+\rangle$ to $|-\rangle$ and we do not know the relative magnitudes of the matrix elements. If the two-phonon mechanism is the more effective in both equivalent and inequivalent, we would again predict that the maximum in T_M^{-1} for the inequivalent dimer should occur at higher temperature than for the equivalent dimer. Since this contradicts experiment, we are left with the unhappy conclusion that the phonon processes responsible for scattering in the equivalent dimer and in the inequivalent must have very different temperature dependences.

From (10), we also see that the spin resonance frequency shifts from ω_0 to $(\omega_0+\omega_e)/2$ as temperature increases. In this case no maximum is predicted or seen in the experiments [3] on the inequivalent dimer.

(b) $\theta \neq 0$: In this case, the excitation is scattered from $|+\rangle$ to $|-\rangle$ with no change in magnetization, stays in the upper state for a short time, evolving according to a rotated spin Hamiltonian. When the excitation drops back into the lower state, the spin populations have changed due to the rotation. In order to examine this, we must find the eigenvalue of the RHS of (7a) and (7b) corresponding to this.

A perturbation calculation of $1/T_1$ has been done by Verbeek and Schmidt [8] and by Levinsky and Brenner [14]. Calculations on a similar system (spin lattice relaxation due to a local librational mode which rotates the spin axes) have been done by Vollmann [15] and Dietz et al. [16]. Although the results of these are similar, they do not agree in all details. We present a perturbation analysis, valid for low temperatures, which gives a result in total agreement with Levinsky and Brenner [14].

Starting from (7a) and (7b), which we write as

$$\begin{pmatrix} \dot{\underset{\sim}{r}} \\ \dot{\underset{\sim}{R}} \end{pmatrix} = \underline{L}\begin{pmatrix} \underset{\sim}{r} \\ \underset{\sim}{R} \end{pmatrix}, \qquad (13)$$

we note that we are interested in finding the eigenvalues of $\underline{L}$. Since these are invariant to unitary transformation, we first transform to a set of variables which would diagonalize $\underline{L}$ in the absence of W and τ^{-1} (these are respectively the principal axes of each triplet for appropriate sub-space of $\underline{L}$). Then $\underline{L}$ may be written as

$$
\begin{matrix}
-W-i\omega_0 & 0 & 0 & \tau^{-1}(\frac{1+c}{2}) & \tau^{-1}(\frac{-1+c}{2}) & \tau^{-1}s/\sqrt{2} \\
0 & -W-i\omega_0 & 0 & \tau^{-1}(\frac{-1+c}{2}) & \tau^{-1}(\frac{1+c}{2}) & \tau^{-1}s/\sqrt{2} \\
0 & 0 & -W & -\tau^{-1}s/\sqrt{2} & -\tau^{-1}s/\sqrt{2} & \tau^{-1}c \\
W(\frac{1+c}{2}) & W(\frac{-1+c}{2}) & -Ws/\sqrt{2} & -\tau^{-1}-i\omega_e & 0 & 0 \\
W(\frac{-1+c}{2}) & W(\frac{1+c}{2}) & -Ws/\sqrt{2} & 0 & -\tau^{-1}+i\omega_e & 0 \\
Ws/\sqrt{2} & Ws/\sqrt{2} & Wc & 0 & 0 & -\tau^{-1}
\end{matrix}
\tag{14}
$$

where $c \equiv \cos 2\theta$, $s = \sin 2\theta$. Note that if $\theta = 0$ this matrix breaks up into three 2×2 matrices leading to the results of the last section. In the limit of zero temperature, the three eigenvalues corresponding to the lower triplet are $\pm i\omega_0$ and 0. The latter is the spin lattice relaxation rate. For nonzero but low temperatures such that $W \ll \tau^{-1}$ ($e^{-\beta\Delta} \ll 1$), we can find these eigenvalues approximately by perturbation theory,

$$
\lambda_1 \cong -\frac{Ws^2(\omega_e^2\tau^2)}{1+\omega_e^2\tau^2} \equiv -\frac{1}{T_1}
\tag{15}
$$

$$
\begin{aligned}
\lambda_2 = \lambda_3^* \cong\ & -i\{\omega_0 + \frac{(\omega_e-\omega_0)e^{-\beta\Delta}}{1+(\omega_e-\omega_0)^2\tau^2}(\frac{1+c}{2})^2 + \frac{(\omega_e+\omega_0)e^{-\beta\Delta}}{1+(\omega_e+\omega_0)^2\tau^2}(\frac{-1+c}{2})^2 + \frac{\omega_0 e^{-\beta\Delta}}{1+\omega_0^2\tau^2}(\frac{s^2}{2})\} \\
& -W\{1 - \frac{(\frac{1+c}{2})^2}{1+(\omega_e-\omega_0)^2\tau^2} - \frac{(\frac{-1+c}{2})^2}{1+(\omega_e+\omega_0)^2\tau^2} - \frac{(s^2/2)}{1+\omega_0^2\tau^2}\} \\
& \equiv -i\omega - \frac{1}{T_2}
\end{aligned}
\tag{16}
$$

in agreement with ref. [14] and disagreeing in some details with ref. [8].

For higher temperatures, one must resort to numerical computation in order to find the eigenvalues; however, the major qualitative conclusions are already contained in equations (15) and (16).

The physical mechanism for $1/T_1$ and $1/T_2$ which we are discussing is that jumps between a lower triplet state and a higher triplet state can cause an accumulation of phase errors and population relaxation if the principal axes are rotated. In the limit that the lifetime of the upper state (τ) is very small compared to inverse Larmor frequencies (ω_0, ω_e), both $1/T_1$ and $1/T_2$ become very small, as is seen in equations (15) and (16). This makes physical sense, since if the upper state lifetime is very small, each jump from $|+\rangle$ to $|-\rangle$ and back to $|+\rangle$ will cause a very small phase error. In the limit of large $\tau\omega_e$, $1/T_1 = + Ws^2$ and $1/T_2 = W$ so that with each jump, phase memory is lost.

One-Dimensional Triplet Excitons: We now turn to another problem involving energy transfer in aggregates, that of the decay of electron spin echos in tetrachlorobenzene (TCB) single crystals. These experiments have been done by Schmidt and co-workers [17] over the last few years. The experimental details and much of the theory has been discussed by Schmidt in these proceedings [18]. We will discuss some recent theoretical developments in the scattering of excitons by phonons in this system, after a brief review of the notation, etc.

The electronic Hamiltonian for a linear array of N identical molecules with an energy transfer matrix element between nearest neighbors can be written

$$H_e = \sum_n \varepsilon|n\rangle\langle n| + \sum_n{}' J\{|n\rangle\langle n+1|+|n+1\rangle\langle n|\} = \sum_\rho \varepsilon(\rho)|\rho\rangle\langle\rho| \qquad (17)$$

where $|n\rangle$ represents a state in which molecule n is excited to its triplet state and all other molecules are in their ground state, and $|\rho\rangle$ is an exciton state (which diagonalizes H):

$$|\rho\rangle = \sum_\rho c_{n\rho}|n\rangle \quad . \qquad (18)$$

When we include phonons and their interactions with excitations, we must add a zeroth-order phonon Hamiltonian, H_{ph}, and the interaction term V to H_e. We will first assume that V arises from an expansion of ε and J to terms linear in phonon variables. Secondly, we know that the phonons are less anisotropic than the excitons; therefore we will assume that the phonons can be described as anisotropic but three-dimensional excitations with wavevector q defined in the usual manner. This leads to a Hamiltonian (in which the phonon band index is suppressed)

$$H = \sum_\rho \varepsilon(\rho)|\rho\rangle\langle\rho| + \sum_q \omega(q)b_q^+b_q + \sum_{\rho,\rho' q} G_{\rho\rho'}(q)\{|\rho\rangle\langle\rho'|(b_q+b_{-q}^+)\} \ . \qquad (19)$$

The scattering rate of electronic excitations from state $|\rho\rangle$, $W_{\rho\rho'}$, can be computed to second order in the $G_{\rho\rho'}$ by using standard theory.

In the case of TCB, the triplet exciton band is very narrow (~1.3cm^{-1}), so narrow that one-phonon scattering events from one exciton state to another could not occur if the total wavevector (electronic + phonon) of the system is conserved. Since the low-temperature experiments clearly show some scattering, with a weak temperature dependence characteristic of one-phonon processes, there must be a breakdown of this conservation, coming about from lack of translational symmetry. We [19] have examined three possible mechanisms for this within the framework of the Hamiltonian of (19). The first, which we call disruptive scattering, comes about if there are impurities or distorted molecules at both ends of an N_{ex} molecule linear array, forcing the excitation to be localized on the array. This corresponds to a linear array embedded in a larger medium; since there is no translational symmetry, one-phonon scattering will be allowed just as for the dimer. The second mechanism which we call interstitial occurs if a distorted or impurity molecule interacts with the linear array adjoining it in a manner different from all the other molecules. This can change for example the site diagonal excitation-phonon at the site nearest the impurity. The last mechanism we consider, called impurity-induced scattering, occurs when the impurity on a chain changes the $c_{n\rho}$'s in (18) due to the change in ε or J because of the presence of the impurity.

We have calculated the scattering rate in all three cases and find that at low impurity concentrations, the interstitial mechanism is largest. This mechanism favors large jumps in energy, and is the one examined by van Strien et al. [17] in their study of TCB. As impurity concentration increases, the other mechanisms become more important and these are more democratic in the size of jumps, so that both large and small energy jumps are possible. In no case do small energy jumps become favored.

When the temperature dependence of the decay rate of the initial population of $\tilde{k}=0$ triplet excitons in TCB was measured [17], it was found that between 1 and 2K, one-phonon scattering fitted the data very well. Above 2K, which is approximately equal to the exciton bandwidth, there is a significant deviation from this fit. The simplest explanation is that two-phonon Raman events become important. In a recent study, Jackson et al. [20] studied these processes specifically for the case of TCB. In this study, both linear phonon terms taken to second order and quadratic phonon coupling were considered. Using the anisotropic three-dimensional phonon model as before, these authors treated phonon scattering for both impurity-induced and pure cases. It was found that site diagonal quadratic phonon terms in the pure crystal interaction Hamiltonian, i.e.,

$$V = N^{-1} \sum_{k,k',\lambda,\lambda'} \eta(\lambda,\lambda') a_{k'}^{+} a_k (b_\lambda + b_{-\lambda}^{+})(b_{\lambda'} + b_{-\lambda'}^{+}) \delta(k'-k-\lambda-\lambda') \tag{20}$$

give rise to the largest terms. In this, $\eta(\lambda,\lambda')$ is the coupling constant to phonon modes λ and λ' assumed in the deformation potential approximation to be equal to $\eta\omega_\lambda^{1/2}\omega_{\lambda'}^{1/2}$. The probability of a jump from k_z to k_z' (z is the linear array direction) is found to be

$$W_{k'_z \leftarrow k_z} = \text{const} \times \eta^2 \times e^{-\beta D} [(k_B T)^6 + a_1 (k_B T)^5 + \dots + a_4 (k_B T)^2] \tag{21}$$

where $\beta = (k_B T)^{-1}$ and

$$D = \frac{1}{2} (c_\parallel |\Delta k_z| - |\Delta\varepsilon|), \tag{22}$$

$\Delta\varepsilon$ being the energy jump, Δk_z the change in exciton k_z, and $c_\parallel$ the speed of sound along the array. This form strongly favors small energy jumps and higher temperatures. The impurity-assisted rate is similar to this term, except it has a T^7 term and a smaller constant term (proportional to impurity concentration).

In the TCB experiments [17], it was found that the scattering rate increased with temperature as $T^{6.4}$ above T=2K; our theoretical results are in qualitative agreement with this. We therefore have shown that the temperature dependence of the scattering rate can be satisfactorily understood at the lowest temperatures, at least.

A related problem, reviewed by Fayer [21], is the width of the optical line as a function of temperature. For temperatures above ~10 K the line-width of the k=0 exciton can be fitted by a Raman process; however, at very low temperatures the line width appears to become independent of temperature and corresponding to a decay time of ~10 picoseconds, approximately 10^3 shorter than found in the ESE experiments. It has been suggested [21] that picosecond scattering out of k=0 is responsible for the optical line-width, but that only very small steps in energy are possible, leading to a much longer time for the ESE relaxation. Both the theoretical and experimental work mentioned above fail to support this view, and it is possible that the optical line width is dominated by inhomogeneous broadening at low T.

References

[1] M. Schwoerer and H. C. Wolf, Mol. Cryst. 3, 177 (1967).

[2] P. Reineker and H. Haken, Z. Physik 250, 300 (1972); P. Reineker in Exciton Dynamics in Molecular Crystals and Aggregates (Springer Verlag, N.Y. 1982).

[3] B. Botter, C. Nonhof, J. Schmidt, J. van der Waals, Chem. Phys. Lett. 43, 210 (1976); B. Botter, A. van Strien, and J. Schmidt, Chem. Phys. Lett. 49, 39 (1977).

[4] U. Schmid and P. Reineker (to be published).

[5] C. A. Hutchison and B. Mangum, J. Chem. Phys. 34, 908 (1961).

[6] J. King (thesis, U. of Chicago, 1973); C. Hutchison and J. King, J. Chem. Phys. 58, 392 (1973).

[7] A. Redfield, Adv. Mag. Res. 1, 1 (1965).

[8] P. Verbeek and J. Schmidt, Chem. Phys. Lett. 63, 384 (1979); see also H. McConnel, J. Chem. Phys. 28, 430 (1958).

[9] R. Feynman, F. Vernon, R. Hellwarth, J. Appl. Phys. 28, 49 (1957).

[10] A. MacLachlan and A. Carrington, Introduction to Magnetic Resonance, (Harper and Row, 1967).

[11] C. van t 'Hof and J. Schmidt, Chem. Phys. Lett. 36, 460 (1975); 42, 73 (1976).

[12] C. Harris, J. Chem. Phys. 67, 5607 (1977).

[13] B. Jackson and R. Silbey, J. Chem. Phys. 77, 2763 (1982).

[14] H. Levinsky and H. Brenner, Chem. Phys. 40, 111 (1979).

[15] W. Vollmann, Chem. Phys. Lett. 57, 157 (1981).

[16] F. Dietz, U. Konzelmann, H. Port and M. Schwoerer, Chem. Phys. Lett. 58, 565 (1978).

[17] A. J. van Strien, J. Schmidt, R. Silbey, Mol. Phys. 49, 151 (1982); A. J. van Strien and J. Schmidt, Chem. Phys. Lett. 86, 203 (1982); J. F. C. van Kooten, A. J. van Strien and J. Schmidt, Chem. Phys. Lett. 90, 337 (1982).

[18] J. Schmidt, these proceedings.

[19] H. Benk and R. Silbey, J. Chem. Phys. (submitted).

[20] B. Jackson and R. Silbey, J. Chem. Phys. 77, 2763 (1982).

[21] M. Fayer, in Spectroscopy and Excitation Dynamics of Condensed Molecular Systems ed. by V. Agranovitch and R. Hochstrasser (North Holland, NY, 1983).

Ensemble Averaged Spin Pair Dynamics of Doublet and Triplet Molecules

Klaus Schulten

Physik-Department, Technische Universität München
D-8046 Garching, Fed. Rep. of Germany

Summary

The dynamics of the spin transitions governing the magnetic field dependence of radical pair processes, triplet quenching by radicals and triplet - triplet annihilation in condensed media is considered. The spin dynamics are induced by the hyperfine interaction and by the zero field splitting and include the Zeemann interaction with an external magnetic field. The theory presented takes the average over all nuclear spin orientations as well as over all histories of either nuclear spin realignments by exchange or rotational diffusion.

1. Introduction

Ordinarily one does not expect that molecular processes in condensed matter at higher temperatures can be influenced by magnetic interaction energies which are much smaller than thermal energies. If, however, a molecular reaction at a decisive intermediate stage is not subject to fast thermal relaxation, i.e. proceeds by a coherent quantum process, then magnetic interactions can indeed decide the outcome of a reaction. In the following article we consider three bimolecular processes which are influenced by external magnetic fields in the range of a few Gauss to a few Tesla: radical pair recombination, triplet quenching by radicals, and triplet - triplet annihilation. These processes involve molecules with an electron spin of doublet and triplet character.

2. Radical Pair Recombination [1]

The first example involves two doublet molecules (radicals) ${}^2D_1 + {}^2D_2$. Such a pair can be generated thermally or photochemically and then will be born in a pure two-electron spin state, either singlet or triplet. Since the recombination products also assume a definite singlet (S) or triplet (T) spin state, the actual recombination process is selective with respect to the spin alignments. In a typical application [2] one considers a radical pair born in a singlet alignment which produces fast (nanosecond) triplet products by means of magnetic interactions. The reaction entails a spin transition by means of magnetic interactions and is followed by the actual reaction step:

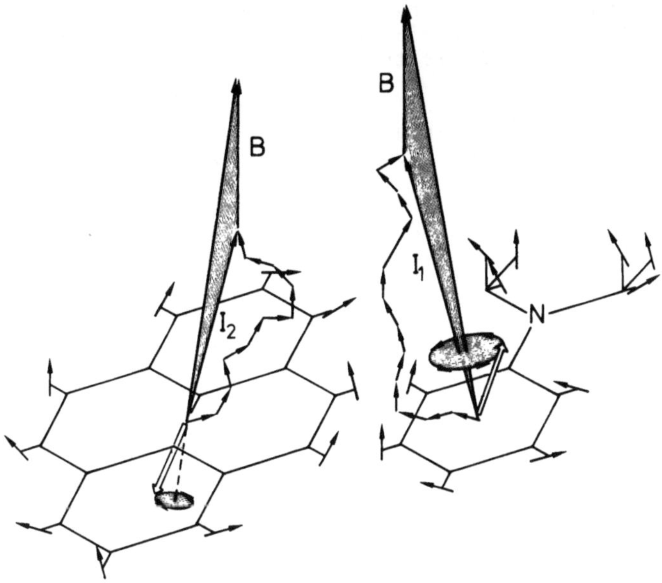

Figure 1: Schematic representation of the hyperfine-induced spin motion in the radical pair $^2D_1 + {}^2D_2$ ($^2D_1$ = pyrene, $^2D_2$ = dimethylaniline). The two electron spins are shown to precess around the axes given by $\sum_k a_{ik}\underline{I}_k^{(1)} + \underline{B}$ (pyrene) and $\sum_k a_{ik}\underline{I}_k^{(2)} + \underline{B}$ (dimethylaniline). The underlying approximation (2.3) treats these axes as constant classical vectors and applies as long as large numbers of nuclear spins contribute to the Hamiltonian (2.2) [3]

$$^1({}^2D_1 + {}^2D_1) \xrightarrow{p_T(t)} {}^3({}^2D_1 + {}^2D_2) \longrightarrow {}^3T^* . \quad (2.1)$$

singlet alignment triplet alignment triplet product

The reaction is governed by the probability $p_T(t)$ that the initial singlet pair assumes triplet character after time t and before the end of the bimolecular encounter. The dominant magnetic interactions of organic doublet molecules are the Zeeman and the isotropic hyperfine interaction. These interactions are collected in the Hamiltonian

$$H_{DD} = H_D^{(1)} + H_D^{(2)}, \quad H_D^{(i)} = \sum_k a_{ik}\, \underline{I}_k^{(i)} \cdot \underline{S}_i + g\mu\underline{B}\cdot\underline{S}_i . \quad (2.2)$$

The sum extends over the nuclear spins $\underline{I}_k^{(i)}$ of molecule i with hyperfine coupling constants a_{ik} and electron spin $\underline{S}_i$, g is the relevant g value, μ the Bohr magneton of the electron and $\underline{B}$ the external magnetic field.

The processes monitored in any observation involve ensembles of molecules with all possible initial nuclear spin orientations. The triplet probability in (2.1) is then [3]

$$p_T(t) = \langle \mathrm{tr}\, Q_T(t)\, Q_S(0) \rangle \tag{2.3}$$

where < > denotes the ensemble average over all nuclear spin orientations. tr denotes the trace over the electron spin states of the pair and

$$Q_S = \frac{1}{4} - \underline{S}_1 \cdot \underline{S}_2 \, , \qquad Q_T = \frac{3}{4} + \underline{S}_1 \cdot \underline{S}_2 \tag{2.4}$$

are the operators which project onto the singlet and triplet subspaces. The hyperfine-induced electron spin motion is illustrated in Fig. 1.

Inserting (2.4) into (2.3) one obtains the triplet probability

$$p_T(t) = \frac{3}{4} - {}_1\underline{\underline{U}} : {}_2\underline{\underline{U}} \, . \tag{2.5}$$

The spin correlation tensors ${}_i\underline{\underline{U}}$ describe the motion of the electron spins of the individual doublets and have the elements

$$ {}_iU_{\alpha\beta} = \langle \mathrm{tr}\, S_\alpha(t)\, S_\beta(0) \rangle \, , \qquad \alpha,\beta = +, -, z \, . \tag{2.6}$$

The tensor product : had been defined in [3]. There it had been shown that the electron spin motion averaged over all nuclear spin orientations yields to a simple analytical description. In fact, the spin transition in (2.1) is governed by the two time constants

$$\tau_i^{-2} = \frac{1}{6} \sum_k a_{ik}^2\, I_k^{(i)}(I_k^{(i)}+1) \, , \qquad i = 1,2 \tag{2.7}$$

which measure the overall strength of the hyperfine coupling in each doublet. In the case of low magnetic field the triplet probability is

$$p_T(t) = \frac{3}{4} - \frac{3}{4}\, {}^2c^{(o)}(t/\tau_1)\, {}^2c^{(o)}(t/\tau_2) \qquad \underline{\text{low field}} \tag{2.8}$$

and in the case of high fields

$$p_T(t) = \frac{1}{2} - \frac{1}{2}\, {}^2c^{(o)}(t/\tau_1)\, {}^2c^{(o)}(t/\tau_2) \qquad \underline{\text{high field}} \, . \tag{2.9}$$

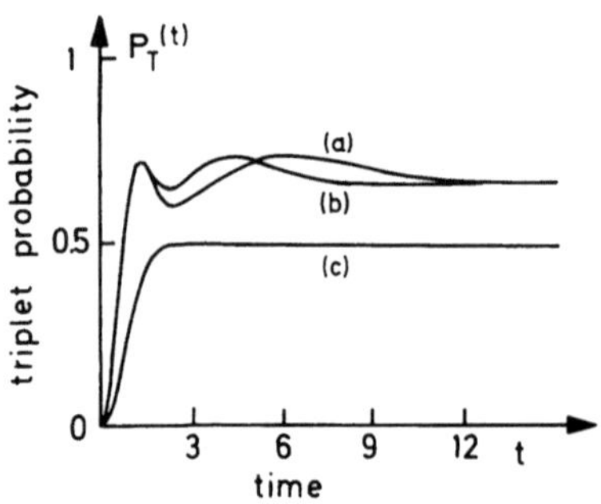

Figure 2: Triplet probability of doublet - doublet encounter pairs starting in a singlet state; the hyperfine coupling corresponds to $\tau_1 = 1$ and $\tau_2 = 5$ in arbitrary units; (a) at zero field, (b) at an intermediate field, (c) at high field; the asymptotic values 2/3 for (a) and 1/2 for (c) result from the asymptotic values of (2.10,11)

In these expressions the correlation functions

$$^2C^{(o)}(x) = \frac{1}{3}[1+2\exp(-x^2)-4x^2\exp(-x^2)] \qquad \underline{\text{low field}} \qquad (2.10)$$

$$^2C^{(o)}(x) = \exp(-x^2) \qquad \underline{\text{high field}} \qquad (2.11)$$

account for the spin motion in the individual doublets. The time development of the triplet probabilities at low fields (2.8), at high fields (2.10) and at an intermediate field is shown in Fig.2.

The spin motion is altered when a doublet undergoes diamagnetic - paramagnetic exchange [3]

$$^2D_1 + {}^1M_1' \rightarrow {}^1M_1 + {}^2D_1' \qquad (2.12)$$

whereupon the unpaired electron spin randomly visits several molecular sites and experiences altered nuclear spin orientations. This situation is illustrated in Fig. 3. In this situation the ensemble average for an evaluation of the spin correlation functions (2.10,11) must involve the nuclear spin orientations as well as all possible histories of exchange events. In case of fast exchange with rate constant $1/\tau_e$ the correlation functions at low and high fields are

$$^2c(t/\tau_i) = \exp(-4t\tau_e/\tau_i^2) \qquad (2.10')$$

$$^2C(t/\tau_i) = \exp(-2t\tau_e/\tau_i^2) \ . \qquad (2.11')$$

The resulting time development of the triplet probability is presented in Fig. 4.

The influence of diamagnetic - paramagnetic exchange processes demonstrated here only for the limiting cases of low and high magnetic fields is actually more pronounced at intermediate fields. This can be demonstrated if one considers the triplet

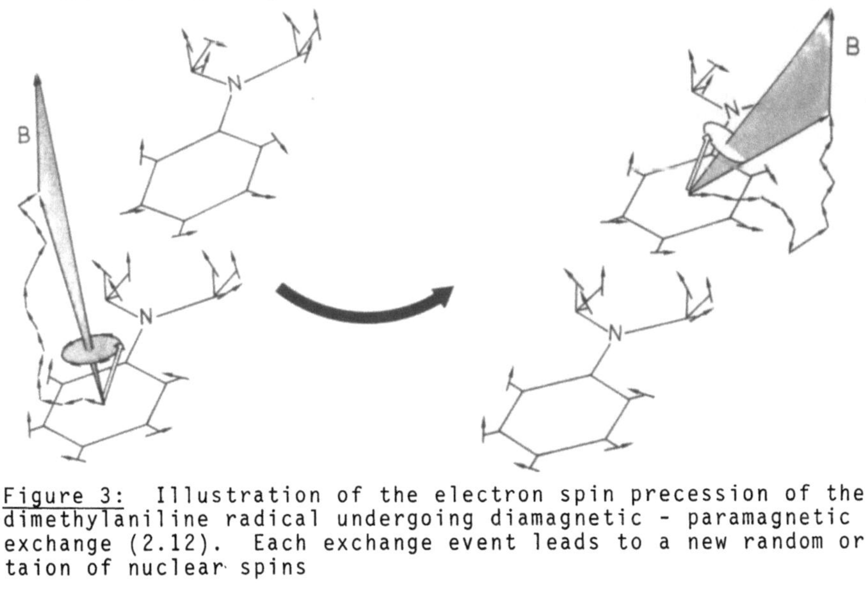

Figure 3: Illustration of the electron spin precession of the dimethylaniline radical undergoing diamagnetic - paramagnetic exchange (2.12). Each exchange event leads to a new random orientaion of nuclear spins

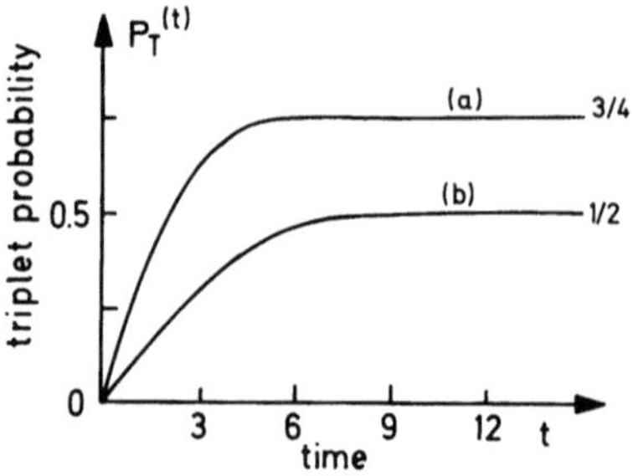

Figure 4: Triplet probability as in Fig. 2, but for a pair with one partner undergoing fast exchange (2.12), τ_e = .1; the evaluation employed the correlation functions (2.10), (2.10'), (2.11) and (2.11'); (a) low field, (b) high field

probability $p_T(t,B)$ for arbitrary magnetic fields B and determines the reaction yield of triplet products (2.1) following [3]. Figure 5 compares the magnetic field dependence of the reaction yields for various exchange rates. The results show that an exchange process (2.12) alters the socalled $B_{\frac{1}{2}}$ value, i.e. the field which produces half of the asymptotic decrease of the triplet yield. This prediction opened the possibility to monitor exchange processes (2.12) by an analysis of the magnetic field dependent reaction kinetics [4]. A typical observation of the concentration dependence of the $B_{\frac{1}{2}}$ value is shown in Fig. 5b.

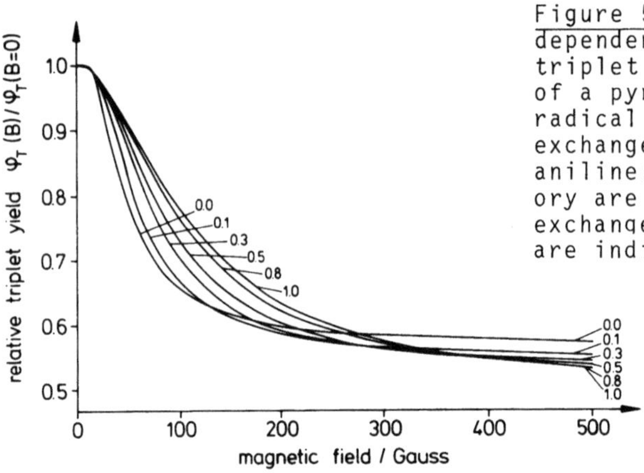

Figure 5a: Magnetic field dependence of the relative triplet recombination yield of a pyrene - dimethylaniline radical pair in case of an exchange involving dimethylaniline; details of the theory are given in [3]; the exchange rates in units 1/ns are indicated

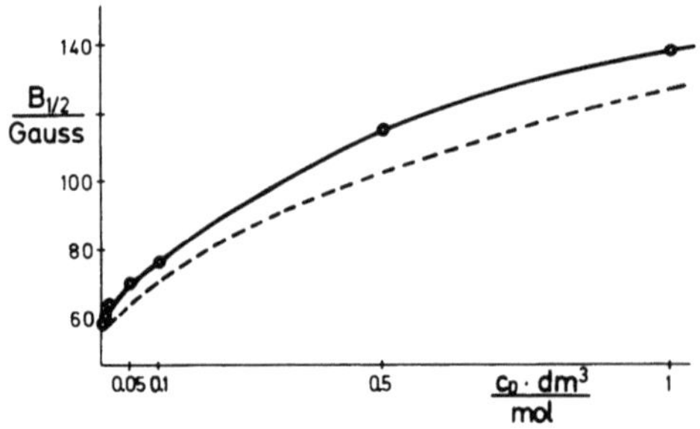

Figure 5b: Dependence of the $B_{\frac{1}{2}}$ value of the yield of triplet products of a pyrene - dimethylaniline radical pair on the concentration c_D of dimethylaniline; from [4]

3. Triplet Quenching by Radicals [5]

The second example we consider concerns triplet quenching by radicals [6] and involves the encounter of a triplet molecule and a doublet $^3T_1 + {}^2D_2$. The encounter pair is either in a quadruplet or in a doublet overall spin state. The quenching process to the ground state $^1S_1 + {}^2D_2$ can proceed only from the doublet alignment of the triplet - radical pair. The fraction of pairs initially in a doublet state populates the ground state immediately. The remaining fraction of quadruplet pairs has to acquire doublet character by means of magnetic interactions before it can engage in the actual quenching step

$$\underset{\text{quadruplet}}{{}^4({}^3T_1 + {}^2D_2)} \longrightarrow \underset{\text{doublet}}{{}^2({}^3T_1 + {}^2D_2)} \xrightarrow{\text{quenching}} \underset{\text{product}}{{}^1S_1 + {}^2D_2} \quad . \tag{3.1}$$

The dominant magnetic interactions are the zero field splitting of the triplet, the isotropic hyperfine coupling of the doublet, and the Zeeman interaction acting on both molecules. The interactions are collected in the Hamiltonian

$$H_{TD} = H_T^{(1)} + H_D^{(2)}; \qquad H_T^{(i)} = D_i S_{iz}^2 - E_i(S_{ix}^2 - S_{iy}^2) + g\mu \underline{S}_i \cdot \underline{B} \tag{3.2}$$

where $H_D^{(i)}$ has been defined in (2.2). D_i and E_i are the zero field splitting constants and $\underline{S}_i$ is the electron spin operator in the molecular frame. In the laboratory frame rotated by the Eulerian angle $\Omega = (\phi, \theta, \psi)$ relative to the molecular frame the triplet Hamiltonian is rather

$$H_{TL}^{(i)} = \frac{2}{3} D_i + \sum_{m,m'} h_m D_{m'm}^{(2)}(\Omega)\, T_{2m'} + g\mu \underline{S}_i \cdot \underline{B} \tag{3.3}$$

where $h_o = \sqrt{2/3}\, D_i$, $h_{\pm 1} = 0$, $h_{\pm 2} = -E_i$, $D_{m'm}^{(2)}$ are the elements of the rotation matrix and T_{2m} second rank tensor operators. (3.3) shows that the triplet spin motion depends on the orientation of the molecules. The probability $p_D(t)$ that a triplet - doublet pair initially in a quadruplet state assumes doublet character at time t, i.e. the probability relevant for the process (3.1), can be written in analogy to (2.3)

$$p_D(t) = \frac{1}{4} < \mathrm{tr}\, Q_D(t)\, Q_Q(0) > \quad . \tag{3.4}$$

This probability has to be averaged over all molecular orientations Ω of the triplet molecule and all nuclear spin orientations of the doublet. < > denotes this average. Q_Q and Q_D are the projection operators onto the manifold of doublet and quadruplet pair states, respectively:

$$Q_D = \frac{1}{3} - \frac{2}{3} \underline{S}_1 \cdot \underline{S}_2 \; ; \qquad Q_Q = \frac{2}{3} + \frac{2}{3} \underline{S}_1 \cdot \underline{S}_2 \quad . \tag{3.5}$$

The doublet probability can then be expressed again through the tensor operators (2.6)

$$p_D(t) = \frac{1}{4} - \frac{1}{9}\, {}_1\underline{\underline{U}} : {}_2\underline{\underline{U}} \; . \tag{3.6}$$

In this case ${}_1\underline{\underline{U}}$ represents the spin correlation tensor (2.6) for the triplet molecule. The dynamics of the triplet molecule can again be cast into scalar correlation functions. For the low field situation this is $[\omega_\pm = (D \pm E)/\hbar]$

$$ {}^3c_1^{(o)}(t) = \frac{1}{3}\cos\omega_+ t + \frac{1}{3}\cos\omega_- t + \frac{1}{3}\cos(\omega_+ - \omega_-)t \qquad (3.7) $$

and for high fields

$$ {}^3c^{(o)}(t) = \langle \exp[it(-D + 3D\cos^2\theta - 3E\sin^2\theta\cos 2\psi)/2\hbar] \rangle . \qquad (3.8) $$

In the special case E = 0 an analytical expression can be derived from (3.8)

$$ {}^3c^{(o)}(t) = \sqrt{\pi\hbar/3Dt}\,\exp(-iDt/2\hbar)\,[\,C(\sqrt{3Dt/\pi\hbar}) + iS(\sqrt{3Dt/\pi\hbar})\,] \qquad (3.9) $$

where C(z) and S(z) denote the Fresnel integrals. If one replaces D by the effective D value $D^* = \sqrt{D\cdot D + 3\,E\cdot E}$, (3.9) should provide a good approximation also in the case $E \neq 0$.

The doublet probability (3.6) in terms of the correlation functions introduced is at low fields

$$ p_D(t) = \frac{1}{3} - \frac{1}{3}\,{}^3c_1^{(o)}(t)\;{}^2c^{(o)}(t/\tau_2) \qquad (3.10) $$

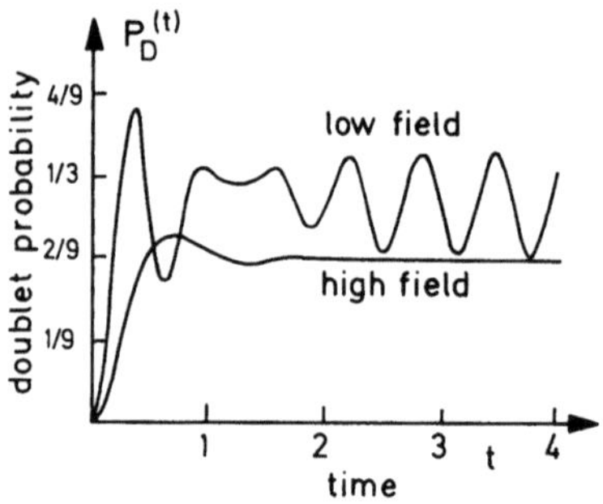

Figure 6a: Doublet probability of a triplet - doublet encounter pair starting in a quadruplet electron spin state for a low and a high field situation with $\tau_1 = \hbar/\sqrt{D\cdot D + 3E\cdot E} = .1$ and $\tau_2 = 1$

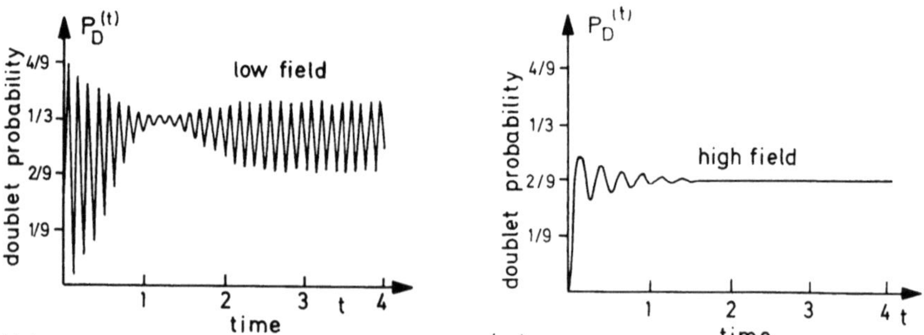

(b) (c)

Figure 6 b,c: Same as Fig. 6a, but with $\tau_1 = .02$

and at high fields

$$p_D(t) = \frac{2}{9} - \frac{2}{9}\,\mathrm{Re}[{}^3c^{(o)}(t)]\;{}^2c^{(o)}(t/\tau_2). \qquad (3.11)$$

The time development of (3.10) and (3.11) is illustrated in Fig. 6. Figure 6a shows a situation when the zero field splitting and the hyperfine coupling are of comparable strength. The low field probability asymptotically oscillates around a value 1/3, which lies above the asymptotic high field probability 2/9. This latter value is approached after a few damped oscillations. Figures 6b,c show a situation when the zero field splitting is much stronger than the hyperfine coupling, a situation which should prevail most often. The oscillations exhibit a shorter period in this case, otherwise the behaviour of the doublet probability is not much altered compared to the situation of Fig.6a.

In case the doublet undergoes paramagnetic - diamagnetic exchange (2.12) or the triplet is being reoriented randomly, one needs to replace the correlation functions (2.10,11) or (3.7,8) by their dynamic counterparts, in complete analogy to the treatment in Section 2. In case of fast rotational relaxation one replaces (3.7) and (3.8) by

$$^3c_1(t) = \exp[-2t\tau_r(\omega_+^2 + \omega_-^2 - \omega_+\omega_-)/3] \qquad (3.7')$$

$$^3C(t) = \exp[-4t\tau_r(\tfrac{3D^*}{2h})^2/45]. \qquad (3.8')$$

In these expressions τ_r denotes the relaxation time of the triplet reorientation. The resulting doublet probabilities are presented in Fig. 7. This figure demonstrates that the rotational relaxation process damps the oscillatory behaviour of the doublet probability in the low field case and, thereby, leads to an asymptotic value 1/3.

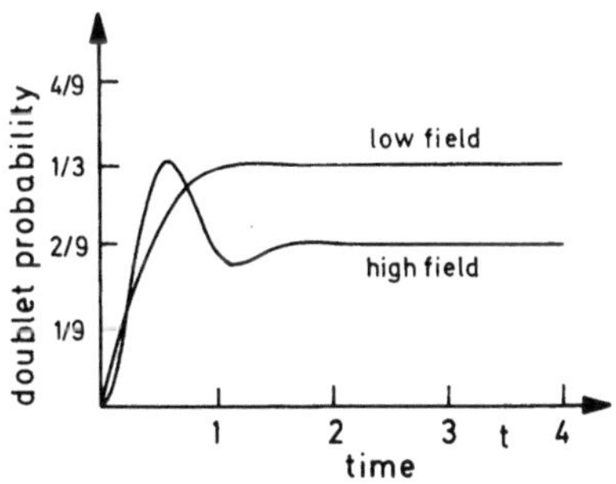

Figure 7: Doublet probability as in Fig. 6a, but for an ensemble of triplet - doublet pairs in which the triplet undergoes rotational relaxation with τ_r = .025

4. Triplet - Triplet Annihilation [5]

Our third example concerns the triplet - triplet annihilation [7] after a random encounter of two triplets. The triplet pair is initially in either a singlet, a triplet or a quintet overall spin state with relative populations 1:3:5. The annihilation step proper yields the product $^1S_1 + {}^1S_2^*$ with overall singlet character. Hence, the reaction route selects the pairs in the singlet spin alignment. The initial fraction 1/9 of random encounter pairs in the singlet state reacts very fast to the annihilation product. The remaining fraction 8/9 of triplet and quintet pairs can annihilate only after change of the overall spin state:

$$^{3,5}(^3T_1 + {}^3T_2) \longrightarrow {}^1(^3T_1 + {}^3T_2) \xrightarrow{\text{annihilation}} {}^1S_1 + {}^1S_2^{*}. \quad (4.1)$$

triplet, quintet alignment — singlet alignment — annihilation products

The spin motion results from the interaction

$$H_{TT} = H_T^{(1)} + H_T^{(2)} . \quad (4.2)$$

The process (4.1) is governed by the probability $p_S(t)$ that the pairs initially in either a triplet or in a quintet alignment assume singlet character at time t

$$p_S(t) = \frac{1}{8} < \mathrm{tr}\, Q_S(t)\; Q_{T+Q}(0) >. \quad (4.3)$$

In this case the average < > extends over all orientations of the triplet molecules including a possible history of random reorientations. The projection operators onto the singlet and triplet + quintet substates are

$$Q_S = -\frac{1}{3} + \frac{1}{3}(\underline{S}_1\cdot\underline{S}_2)^2; \qquad Q_{T+Q} = \frac{4}{3} - \frac{1}{3}(\underline{S}_1\cdot\underline{S}_2)^2. \quad (4.4)$$

Inserting these expressions into (4.3) yields

$$p_S(t) = \frac{1}{3} - \frac{1}{72}\; {}_1\underline{\underline{V}} :: {}_2\underline{\underline{V}} \quad . \quad (4.5)$$

The tensor product :: has been defined in Ref.5. The elements of the spin correlation tensors are in this case

$${}_1V_{\alpha\beta\gamma\delta} = < \mathrm{tr}\, S_\alpha(t)\; S_\beta(t)\; S_\gamma(0)\; S_\delta(0) >. \quad (4.6)$$

The singlet probability can again be expressed in terms of correlation functions. In the case of low fields one obtains

$$p_S(t) = \frac{1}{9} - \frac{1}{24}\, {}^3c_1^{(o)}(t)\; {}^3\hat{c}_1^{(o)}(t) - \frac{5}{72}\, {}^3c_2^{(o)}(t)\; {}^3\hat{c}_2^{(o)}(t). \quad (4.7)$$

The correlation functions are given by (3.7) and by

$${}^3c_2^{(o)}(t) = \frac{2}{5} + \frac{1}{5}\cos\omega_+ t + \frac{1}{5}\cos\omega_- t + \frac{1}{5}\cos(\omega_+ - \omega_-)t \; . \quad (4.8)$$

^ denotes the correlation functions of the second triplet. The high field singlet probability can be expressed in terms of the correlation function (3.8)

$$p_S(t) = \frac{1}{18} - \frac{1}{18}\,\mathrm{Re}[{}^3c^{(o)}(t)\; {}^3c^{(o)}(t)]. \quad (4.9)$$

The time development of the singlet probability is presented in Fig. 8 for the cases E = 0 and E ≠ 0. The low field probability exhibits an oscillatory behaviour which continues into the asymptotic region. The maximum values reached in the low field case are about 1/10. This value lies again above the constant high field asymptotic value 1/18.

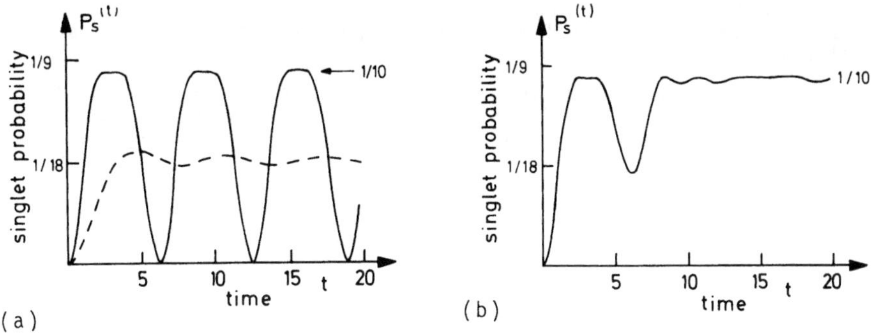

Figure 8: Singlet probability of a triplet - triplet encounter pair starting in either a triplet or a quintet overall spin state at low field (———) and at high field (- - -); (a) $\tau_1 = \tau_2 = \hbar / D$ and $E = 0$; (b) same as (a) but with $\tau_1 = \tau_2 = h / D^*$ and $E = D / 10$

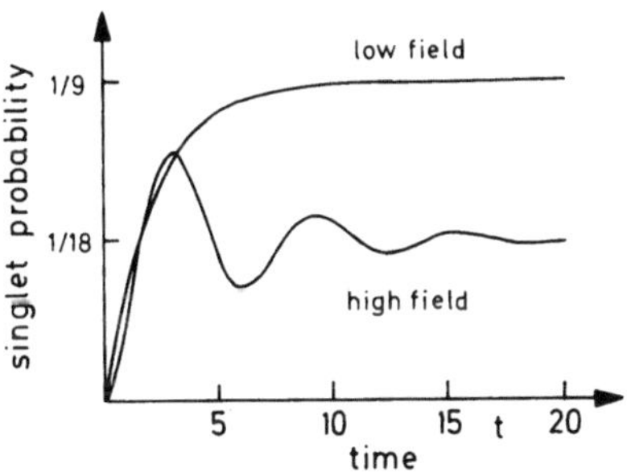

Figure 9: Singlet probability of a triplet - triplet encounter pair as in Fig. 8b for an ensemble in which both triplets undergo rotational relaxation with $\tau_r = .5$

If one wishes to describe triplets which undergo rotational relaxation one needs to replace the correlation functions in (4.7) and (4.9) by their dynamic counterparts, i.e. in the case of fast relaxation by (3.7'), (3.8'), as well as by

$$^3c_2(t) = \exp[-2t\tau_r(\omega_+^2 + \omega_-^2 - \omega_+\omega_-)/5] \ . \qquad (4.8')$$

The resulting singlet probability is shown in Fig. 9. The relaxation process induces the behaviour that constant asymptotic probability values 1/9 and 1/18 are quickly approached.

5. Summary

The magnetic field dependence of bimolecular processes in liquids originates from the spin motion of doublet and triplet molecules. We have outlined a theory which describes the spin dynamics for the proper ensemble of doublets with all possible nuclear spin polarizations and of triplets with all possible molecular orientations. The ensemble averaged spin motion has a simple long time behaviour and, thereby, provides simple estimates for the magnitude of magnetic field effects. These effects will be summarized now.

In case of radical pair processes to triplet products (2.1) the relevant triplet probability at low and high external magnetic fields approaches asymptotically the values 2/3 and 1/2, respectively (c.f. Fig. 2). The maximum field effect should be then a reduction of the yield of triplet products by about 17 percent. The field effect will be altered when the radicals undergo diamagnetic - paramagnetic exchange. The triplet probability assumes then the asymptotic values 3/4 and 1/2 at low and high fields (c.f. Fig. 4). The resulting relative field effect should be a maximum reduction by 25 percent. However, the most sensitive probe for the exchange process (2.12) is not furnished through a comparision of triplet yields at low and high fields but rather through an observation of the yield at all intermediate fields (c.f. Fig. 5a,b).

In case of doublet - triplet pair processes, i.e. triplet quenching by radicals (3.1), the relevant doublet probability at low and high fields approaches asymptotically the average values 1/3 and 2/9, respectively. The low field probability shows an indefinite oscillation (c.f. Fig. 6). In the special case $E = 0$ of Fig. 6a the long time average probability value at low field is 8/27 rather than 1/3. Figure 7 demonstrates that rotational relaxation alters considerably the time dependence of the doublet probability. In analogy to the radical pair reaction one can expect that the magnetic field dependence of the triplet quenching is influenced by rotational relaxation.

The singlet probability which governs the magnetic field dependence of triplet - triplet annihilation approaches at low field the maximum values 1/10 and at high field the constant value 1/18. The time development at low fields is oscillatory over long times (c f. Fig. 8). Rotational relaxation has a pronounced influence on the time dependence of the singlet probability in that constant asymptotic values 1/9 and 1/18 develop (c f. Fig. 9). One has to expect, therefore, that rotational diffusion of triplets will contribute to the magnetic field dependence of triplet - triplet annihilation.

Acknowledgements

The author wishes to thank A. Weller for a fruitful collaboration on the issue of this lecture. This work has been supported by the Deutsche Forschungsgemeinschaft (SFB-143C1).

References

1. For a most recent review on magnetic field effects on doublet processes in chemistry and biology see K. Schulten, Advances in Solid State Physics, XXII, 61, J. Treusch (ed.), Vieweg, Braunschweig (1982)
2. K. Schulten, H. Staerk, A. Weller, H.-J. Werner and B. Nickel, Z. Phys. Chem. NF101, 371 (1976); M.E. Michel-Beyerle, et al. Chem. Phys. 17, 139 (1976)

3. K. Schulten and P.G. Wolynes, J.Chem.Phys. 68, 3292 (1978); E.-W. Knapp and K. Schulten, J.Chem.Phys. 71, 1878 (1979)

4. A. Weller, F. Nolting, and H. Staerk, Chem. Phys. Lett. 96, 24 (1983)

5. The major part of this Section will be found in K. Schulten, J. Chem. Phys. (in press)

6. L.R. Faulkner and A.J. Bard, J.Am.Chem.Soc. 91, 6497 (1969)

7. J. Spichtig, H. Bulska, and H. Labhard, Chem.Phys. 15, 279 (1976); K. Lendi, P. Gerber, and H. Labhard, ibid.18,449(1976)

Part IV

Interaction of Electronic Excitations with Lattice Vibrations

Localization and Delocalization of an Exciton in the Phonon Field

Yutaka Toyozawa

Institute for Solid State Physics, University of Tokyo
Roppongi, Minato-ku, Tokyo 106, Japan

Abstract

Dynamics of an exciton in the phonon field of various dimensionalities, and its manifestations in the absorption and emission spectra, are described in the nearly adiabatic situation, with particular attention to localization and delocalization. The steepness coefficient of the Urbach tail is correlated with the exciton-phonon coupling constant, and this correlation is compared with experiments on self-trapping. One expects an essential difference between the weak and strong coupling cases of the phonon-structured lineshape of zero temperature absorption spectrum though there is no such difference at high temperatures. The tunneling resonance of nearly degenerate free (F) and self-trapped (S) states is proposed to explain the anomaly in the F-S coexistence luminescence in $AgBr_{1-x}Cl_x$ alloys.

1. Introduction

The optical excitation of the electronic system in an aggregate of like molecules (or atoms) causes two effects whose competition plays a crucial role in the subsequent behavior of the system. One is the transfer of excitation energy from site to site due to the intermolecular resonance. The other is the inbalance of the forces upon the relevant atoms caused by the electronic charge redistribution, seeking for new equilibria of intra- and inter-molecular configuration coordinates. The latter stabilizes the localized electronic excitation, destroying its resonance to the neighboring sites. The competition between the two dispositions, one towards delocalization and the other towards localization, starts already with the creation of electronic excitation as is reflected in the corresponding absorption spectra, and then leads, as the system relaxes, to a variety of stable and metastable states which will manifest themselves in the emission spectra of electronic de-excitation therefrom, or which may be new photo-chemical products.

The purpose of this paper is to describe the interplay of the two competing dispositions of electronic excitation before and after the relaxation as are reflected in the absorption and emission spectra. While this problem has already been reviewed in various contexts [1,2,3], a new aspect of the present article is the incorporation of the recent developments in theoretical and experimental studies, including the Urbach rule on the low-energy tail of the absorption spectra which enables us to correlate better the theory before and after the relaxation [4].

2. Systems Under Consideration and Their Simplification

Let the ground electronic state of the aggregate be nondegenerate and energetically well separated from the excited states. With atomic position coordinates Q_ℓ and their conjugate momenta P_ℓ, we denote the adiabatic potential energy and the atomic kinetic energy by U(Q) and K(P), respectively. On the tight-binding basis, we consider the electronic excitations, with energy $E_{n,i}$, within each site n which can be an atom, a molecule, a dimer or a unit cell of a crystal, as the case may be. These excitations can be transferred to a neighboring site m through transfer energies $t_{ni,mj}$ $(n \neq m)$. The electronic excitation energy can then be written in the Hamiltonian form:

$$H_e(Q) = \sum_{n,i} |ni > E_{n,i}(Q) < ni| + \sum_{n,i} \sum_{m,j} |ni > t_{ni,mj}(Q) < mj| \tag{2.1}$$

where the Q dependences of both site-diagonal (E) and site-off-diagonal (t) energies are considered.

Measuring Q_ℓ from their equilibrium positions in the ground electronic state, $H_e(Q)$ is expanded in powers of Q and written as $H_e(0)+\delta H_e(Q) \equiv H_e+H_{ep}$. The Hamiltonians for the ground and the excited state manifolds can then be written, respectively, as

$$H^{(g)} = H_p \equiv K + U \quad , \tag{2.2}$$

$$H^{(e)} = H_e + H_{ep} + H_p \quad . \tag{2.3}$$

H_{ep}, which will be called the exciton-phonon interaction (inclusive of the vibronic interaction within a molecule), consists of site-diagonal and site-off-diagonal parts as mentioned above.

We will now study the dynamics of this exciton-phonon system and its reflection in the optical spectra in detail by confining ourselves to periodic systems (crystals) and under further simplification and idealization into the "standard" system. Thorough study of such a system not only serves to clarify the basic physics common to similar systems, but also provides a useful starting point for its extension to more general systems. The simplification is as follows: i) a single mode of electronic excitation per site (indices i, j omitted), ii) a single and iii) harmonic mode of lattice vibration per site (Einstein model), iv) linear and v) site-diagonal electron-phonon interaction of vi) contact (short-range) type (ℓ=n). One can then write

$$H_e = \sum_n |n > E_a < n| + \sum_{n \neq m} \sum |n > t_{n,m} < m| \tag{2.4}$$

$$H_{ep} = -\sum_n |n > cQ_n < n| \tag{2.5}$$

$$H_p = \sum_n \frac{1}{2}(\omega^2 P_n^2 + Q_n^2) = K + U \tag{2.6}$$

choosing suitable sign and normalization of Q_n.

The above simplification is in fact not as restrictive as it looks. While (2.5) with (2.6) models after intramolecular vibrations, one can apply it also to the exciton-acoustic phonon interaction of the deformation potential type by taking for Q_n the local dilation of the lattice [5]; the resulting non-diagonality of K in P_n's causes no difficulty within the adiabatic approximation. Site-off-diagonal exciton-phonon interaction can be effectively incorporated into the site-diagonal one as far as the exciton scattering by long wavelength phonons is concerned [1]. One can take a sandwich type dimer (as found in pyrene and α-perylene) for the unit of site, the bonding excited state of it for the single excitation mode and the decrement of inter-monomer (within a dimer) distance for Q_n, in order to apply (2.4)~(2.6) to the excimer formation (a remaining extension to a pair of dimers per unit cell is trivial).

3. Localization and Delocalization Limits

With vanishing transfer $t_{n,m}$, the localized excitation $|n\rangle$ turns out to be an eigenstate of $H^{(e)}$ with excitation energy $E_n(Q)=E_a-cQ_n$, and hence with the adiabatic potential

$$W^{(en)}(Q) = (E_a - E_{LR}) + (Q_n - c)^2/2 + \sum_{n'}^{(\neq n)} Q_{n'}{}^2/2 \tag{3.1}$$

$$E_{LR} \equiv c^2/2 \tag{3.2}$$

as shown schematically in Fig. 1(a) together with that of the ground state, $W^{(g)}(Q) \equiv U(Q)$, in one-dimensional configuration coordinate (C.C.) space Q_n (other $Q_{n'}$'s are irrelevant). Within the Franck-Condon approximation and with the classical Boltzmann distribution for Q_n: $\exp(-Q_n{}^2/2\kappa_B T)$, the line shape of the absorption spectra for this localized excitation is given by a Gaussian

$$A(E) = (2\pi D^2)^{-1/2}\exp[-(E - E_a)^2/2D^2] \tag{3.3}$$

with dispersion

$$D^2 = \langle V_n{}^2 \rangle = 2E_{LR}\cdot\kappa_B T \tag{3.4}$$

where $V_n(t) \equiv -cQ_n(t)$ is the fluctuating part of the localized excitation energy.

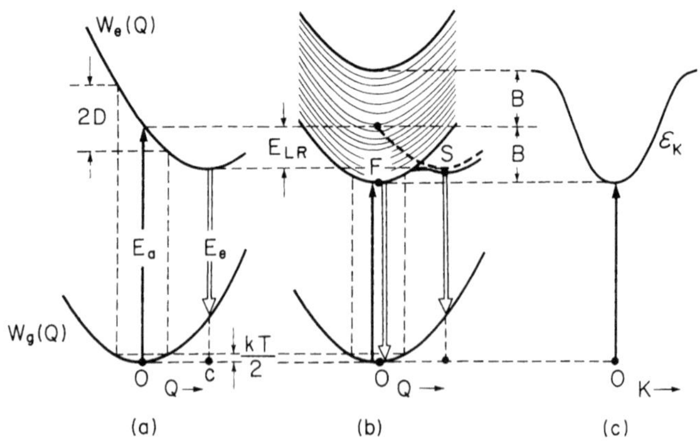

Fig.1 Configuration coordinate models for localized excitation (a) and exciton (b), and excitonic band in rigid lattice (c)

The emission of a photon usually takes place from the relaxed excited state which is lower than the Franck-Condon excited state E_a by E_{LR} — the lattice relaxation energy. The emission band has the same lineshape as (3.3), except for a Stokes shift by $2E_{LR}$.

In the limit of vanishing c with finite t_{nm}'s, the eigenstate of the electronic excitation is a free exciton: $|K\rangle = N^{-1/2}\Sigma_n \exp(iK\cdot R_n)|n\rangle$ with energy $E_K = E_a + \sum_m \exp[iK\cdot(R_m-R_n)]t_{n,m}$. This exciton band is assumed to extend from E_a-B to E_a+B' (B, B' > 0). B is the energy gain by delocalization, or the kinetic energy required for localization. For a simple lattice with only the nearest-neighbor transfer t being considered, we have $B = B' = \nu|t|$, where ν is the coordination number, and negative t corresponds to the direct gap case in which the state k = 0 is at the band bottom. The absorption spectrum to create an exciton consists of a line at $E_{K=0}$ because of the K-selection rule (see Fig. 1(c)).

One of our problems is to study how these two different modes of stabilization — the localization with lattice relaxation and the delocalization through intersite resonance — compete or compromise with each other under the coexistence of c and t. The relevant parameter is naturally the exciton-phonon coupling constant g defined by

$$g \equiv \frac{E_{LR}}{B} = \frac{c^2/2}{\nu|t|}, \tag{3.5}$$

the ratio of the energy gains associated with the two modes of stabilization in the vanishing limits of other interaction. This problem will be studied in 5.

4. Motion of an Exciton Under Spatially Fluctuating Potential

Let us now study the photo-absorption process under the coexistence of t and c. In order to avoid the effect of subsequent lattice relaxation which takes place in a time of order ω^{-1} or longer, one has to prepare a light pulse with duration time τ shorter than ω^{-1}, which in turn imposes a limitation in spectral resolution: $\Delta E > \omega$. In other words, we can take the adiabatic limit ($\omega \to 0$) ignoring the relaxation as far as we are interested in gross lineshape with width much greater than the phonon energy. The light pulse sees only the momentary configuration coordinates, $Q_n(t_0)$, each of which obeys independently the Boltzmann statistics, $\exp(-Q_n^2/2\kappa_B T)$. Thus, (2.5) reduces to a static potential with only spatial fluctuation.

Denoting the eigenvalues and eigenstates of $H \equiv H_e + H_{ep}$ by E_λ and $|\lambda\rangle$, respectively, and the optically allowed eigenstate $|K=0\rangle$ of H_e simply by $|0\rangle$, one can write the normalized lineshape of the absorption band as [6]

$$F(E) = \sum_\lambda |\langle 0|\lambda\rangle|^2 \delta(E_\lambda - E) \tag{4.1}$$

$$= -\frac{1}{\pi} \mathop{Im}_{\eta\to+0} \langle 0|\frac{1}{H-E+i\eta}|0\rangle = -\frac{1}{\pi} Im \frac{1}{E_0+\Sigma_0(E)-E}$$

$$= \frac{1}{\pi}\frac{\Gamma_0(E)}{[E-E_0-\Delta_0(E)]^2 + [\Gamma_0(E)]^2} \tag{4.1'}$$

where the exciton self-energy $\Sigma_K(E)=\Delta_K(E)+i\Gamma_K(E)$ due to scattering is to be determined self-consistently from

$$\Sigma_K(E) = \sum_{K'} \frac{|(H_{ep})_{KK'}|^2}{E-E_{K'}-\Sigma_{K'}(E)} + \text{(higher order terms)} \quad . \tag{4.2}$$

In the lowest order approximation, one obtains

$$\Gamma_K(E) = \pi D^2 \rho^{(0)}(E) \qquad \text{(golden rule)} \tag{4.3}$$

with the use of (3.4). Here, $\rho^{(0)}(E) \equiv N^{-1}\sum_K \delta(E-E_K)$ is the normalized density of states (DOS) of the unperturbed H_e, which amounts to $(2B)^{-1}$ for a constant DOS. Then one obtains the energy-independent half-width

$$\Gamma_0 = \frac{\pi}{2}\frac{D^2}{B} \propto T \tag{4.4}$$

resulting in a Lorentzian lineshape.

The constant DOS Ansatz is dangerous in the direct gap case since $\rho^{(0)}(E)$ is singular at the band bottom E_0 where the absorption peak appears. As is seen from (4.1'), the half-width Γ_0 on the high energy side of the peak is approximately given by solving $E-E_0=\Gamma_0(E)$ for E. For linear (d=1), square (d=2) and simple cubic (d=3) lattices with nearest neighbor transfer only, one obtains

$$\Gamma_0 = \sqrt{\pi}\,[\Gamma(\tfrac{d}{2})]^{-2/(4-d)}\,[\frac{D}{2\sqrt{\pi}B}]^{d/(4-d)} D \propto T^{2/(4-d)} \tag{4.5}$$

where $\Gamma(\cdots)$ is the Gamma function.

While the above results based on the perturbation theory hold in the weak scattering regime (D<<B), we already have the Gaussian lineshape (3.3) in the strong scattering limit (D>>B). The nature of exciton changes gradually from localized to delocalized type as B, the transfer rate to nearest neighbors, increases and overcomes the potential fluctuation D (see the upper half of Fig.2), with the corresponding change in the physical meaning of the line-width. The width (4.4) or (4.5) governed by the scattering rate of a delocalized exciton can also be considered to result from the motional narrowing of the fluctuation D due to the exciton transfer with rate B. The narrowing factor is given by the dimensionality-dependent exponent of D/B according to (4.5).

Confining ourselves to the weak scattering regime and the direct gap case, we have to note that the quasi-Lorentzian lineshape (4.1') with perturbational expansion (4.2) is valid only where the unperturbed DOS $\rho^{(0)}(E)$ is non-vanishing, namely, only on the high energy side of the absorption peak. The low energy side of the true DOS,

$$\rho(E) \equiv N^{-1}\sum_{\lambda}\delta(E-E_\lambda) \;, \tag{4.6}$$

as well as that of the absorption spectra F(E) is produced only as the result of H_{ep} —— the spatial fluctuation of the potential. Thus the low-energy side is supposed to consist of "bound" or "localized" states, in contrast to the high-energy side consisting of extended states. As a matter of course, the perturbation theory is a very poor approach to the former.

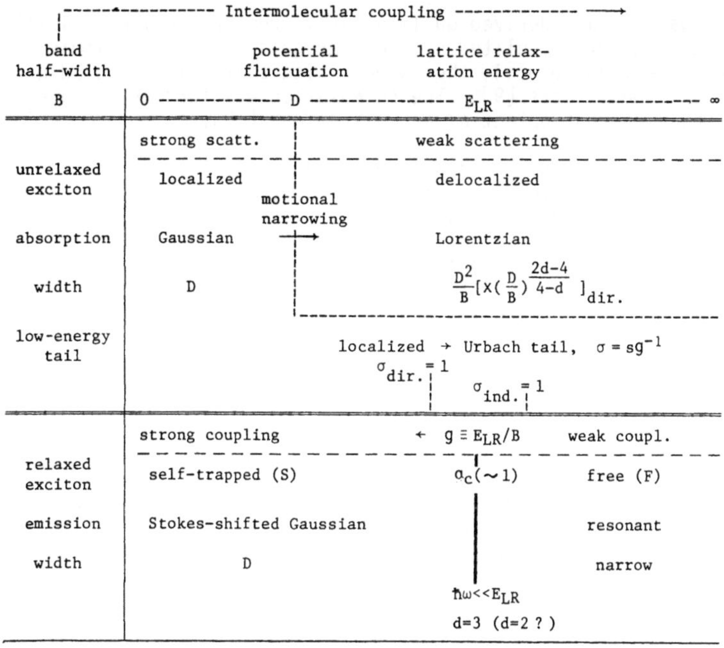

Fig.2 Perspective on localization and delocalization of an exciton in the phonon field as reflected in the optical spectra. B increases towards right

Experimentally, it has been found in a variety of organic and inorganic insulators [7] that the low-energy tail of the exciton (direct or indirect) absoption spectra depends exponentially on the energy, with a decay constant inversely proportional to temperature (at high temperatures):

$$F(E) = A\exp\left[-\sigma\frac{(E_0'-E)}{\kappa_B T}\right] \quad . \tag{4.7}$$

The steepness coefficient σ is a material constant of the order of unity. The universality and simplicity of this empirical rule — now called the Urbach rule — have evoked a number of theoretical and experimental studies to explain it on a microscopic basis [7,8].

If (4.7) is to be derived from the standard Hamiltonian (2.4~6) with the use of the theoretical formula (4.1), σ should be related with g (see(3.5) through

$$\sigma = \frac{s}{g} \tag{4.8}$$

as is obvious from the following scaling argument. When the energy is measured in units of B, the only parameter governing the lineshape is D/B (see (3.4)); the exponent in (4.7) should then be of the form: $[(E_0'-E)/B]/[D^2/B^2]$ (times a dimensionless constant) in order to have the factor T^{-1}. Thus we have (4.8) with the "steepness index" s which no longer depends on the material constant g .

Equation (4.7) was in fact derived qualitatively from (2.4~6) and (4.1') with the use of self-energy calculated by the average t-matrix approximation which allows one to describe exactly the "single site scattering" including bound as well as extended states [9]. The calculation was further improved with the use of the "coherent potential approximation" [10].

More recently, the Monte Carlo calculation of (4.1) and (4.6) was performed for 1-, 2- and 3-dimensional simple lattices by direct diagonalization of the energy matrix $H_e + H_{ep}$ for each configuration $(Q_1, Q_2, \cdots)$ with the Boltzmann distribution [4]. It in fact reproduced (4.7) excellently over four (direct gap) or two (indirect gap) decades for 1-, 2- and 3-dimensional lattices, except for giving $T^{-2/3}$ dependence for d=1 instead of T^{-1} (as for the last point, see the scaling argument presented very recently by IHM and PHILLIPS [11]). The steepness index s was found to be 1.50 (d=3) and 1.24 (d=2) for the direct gap and 0.85 (d=3) and 0.80 (d=2) for the indirect gap.

The numerically calculated lineshape F(E) on the high-energy side of the direct exciton peak fits fairly well to a Lorentzian, whose half-width is in good agreement with (4.5) in the absolute value as well as in the temperature dependence (except that for d=3 a great number of higher order terms in (4.2) have to be considered [6] in order to remove the hundred times discrepancy (too small) of the first-order calculation originating from vanishing $\rho^{(0)}(E)$ at the band bottom).

Another quantity of physical importance is the average oscillator strength per state (AOSPS), defined by

$$f(E) \equiv F(E)/\rho(E) = \underset{E_\lambda = E}{\mathrm{Av}} \left[N |\langle 0|\lambda\rangle|^2 \right] \quad . \tag{4.9}$$

With the use of expansions: $|\lambda\rangle = \sum_n c_n^\lambda |n\rangle$ and $|K=0\rangle = N^{-1/2}\sum_n |n\rangle$, one can write

$$f(E) = \underset{E_\lambda = E}{\mathrm{Av}} \left|\sum_n c_n^\lambda\right|^2 \quad . \tag{4.10}$$

Numerically calculated f(E) has a sharp peak located close to that of F(E), and its high- and low-energy sides behave quite differently in their energy, temperature and dimensionality dependences. The high-energy side fits to a Lorentzian with the above-mentioned T-dependent half-width even better than F(E) does, as is expected because the numerator $\Gamma_0(E)$ in (4.1') is proportional to the true DOS $\rho(E)$.

In contrast, the low-energy tail is almost T-independent if f(E) curves at various T are energy-shifted so as to have a common peak position E_0'', and decay with energy as $(E_0''-E)^{-d/2}$. This is in accordance with the notion that the low-energy tail is due to the localized states. If the localized wavefunction c_n^λ extends over approximately M sites without changing its sign, (4.10) is of the order of M since $c_n^\lambda \sim M^{-1/2}$. This is nothing but the giant oscillator strength first pointed out by RASHBA [12] for an impurity-bound exciton. The binding energy $E_0''-E$ of the localized exciton is of the order of kinetic energy which is proportional to the inverse square of the linear extension: $M^{1/d}$. This establishes the relation $f(E) \propto (E_0''-E)^{-d/2}$.

Thus the remarkable difference between low- and high-energy sides of F(E) and f(E) spectra in the weak scattering regime is the reflection of localized and delocalized states (see Fig.2). In the strong scattering regime, almost all of the exciton states are localized as is reflected in the Gaussian shape of F(E).

5. The Relaxed Exciton in Deformable Lattice

Let us imagine a lattice distortion at a particular site n which causes a local potential $V_n=-cQ_n$ (we assume $V_n < 0$). A bound state will appear at E if the equation $V_n G_0(E)=1$ has a solution below the exciton band, where

$$G_0(E) \equiv \lim_{\eta\to+0} N^{-1}\sum_K (E - i\eta - E_K)^{-1} \tag{5.1}$$

is a lattice Green's function whose imaginary part gives π times the normalized DOS $\rho^{(0)}(E)$ (see 4). Making use of the dispersion relation, $\mathit{Re}G_0(E) = \pi^{-1}\int dE'\,\mathit{Im}G_0(E')/(E-E')$, one immediately finds that the binding energy $E_b \equiv E_{K=0}-E$ rises as $|V_n|^2/2B$, $C\exp[-\pi B/|V_n|]$ and $1.71(|V_n|-V_c)^2/B$ (a bound state exists only when $|V_n| > V_c = 0.65B$) for d=1, 2 and 3 simple lattices, respectively. Making use of the asymptotic form, $E_b(Q_n) = |V_n|-\nu|t|+\nu|t|^2/|V_n|+O(\nu^2|t|^4/|V_n|^3)$, valid for $|V_n|>>B=\nu|t|$, one finds that the adiabatic potential for this bound state, $W^{(en)}(Q) = E_{K=0} - E_b(Q_n) + Q_n^2/2$ takes its minimum value $W^{(S)} = E_{K=0}+B-Bg[1+(2\nu g^2)^{-1}+O(\nu^{-2}g^{-4})]$ at $Q_n^{(S)} = c[1-(4\nu g^2)^{-1}+O(\nu^{-2}g^{-4})]$, where g was defined by (3.5). The adiabatic potential $W^{(en)}(Q)$ is shown schematically in Fig.1(b) for a three-dimensional lattice where a finite distortion Q_n is needed before the bound state is split off from the continuum of extended states. The stable state of the exciton in deformable lattice changes distinctly from the free state $W^{(F)} = E_{K=0}$ without lattice distortion to the localized state $W^{(S)}$ with large lattice distortion as g exceeds $g_c = 1-(2\nu)^{-1}+O(\nu^{-2})$. The latter state (S) is called the self-trapped state.

The discontinuity of the F-S transition is almost synonymous with the existence of a potential barrier between the F and S states. The latter problem has to be studied more carefully by allowing the lattice distortion to extend over many sites since there may be energetically more favorable paths connecting the two states in such multidimensional C.C. space. Let us employ the elastic continuum model [13] for this purpose, and assume that the dilation $\Delta(\vec{r}) \equiv \mathit{div}\vec{u}(\vec{r})$ ($\vec{u}(\vec{r})$ is the displacement of the medium at $\vec{r}$) causes a contact potential $\Xi\Delta(\vec{r})$ for the exciton where Ξ is the so-called deformation potential constant. Denoting by $E[\Delta(\vec{r})]$ the energy of the lowest state of the exciton as a functional of $\Delta(\vec{r})$, we want to calculate the adiabatic energy $W[\Delta(\vec{r})]$ by adding the elastic energy $U = (C/2)\int\Delta(\vec{r})^2 dr$ where C is an appropriate elastic constant.

For our purpose, it is enough to characterize $\Delta(\vec{r})$ by its magnitude Δ and spatial (linear) extension a, namely, we put $\Delta(\vec{r})=\Delta$ for $r < a$ and $\Delta(\vec{r})=0$ for $r > a$. The multi-dimensional C.C. space of $\Delta(\vec{r})$ is thus reduced to a two-dimensional one. The Schrödinger equation for the exciton with effective mass m can then be written, by a suitable scaling, as

$$[-\lambda^{-1}\nabla_\rho^{\,2} - u(\vec{\rho})]\Psi(\vec{\rho}) = \frac{E(\Delta,a)-E_0}{-\Xi\Delta}\,\Psi(\vec{\rho}) \tag{5.2}$$

where $\vec{\rho} = \vec{r}/a$, $u(\vec{\rho}) = 1$ or 0 according as $\rho \lessgtr 1$, and

$$\lambda \equiv \frac{-\Xi\Delta}{K}\ , \quad K \equiv \frac{\hbar^2}{2ma^2} \ . \tag{5.3}$$

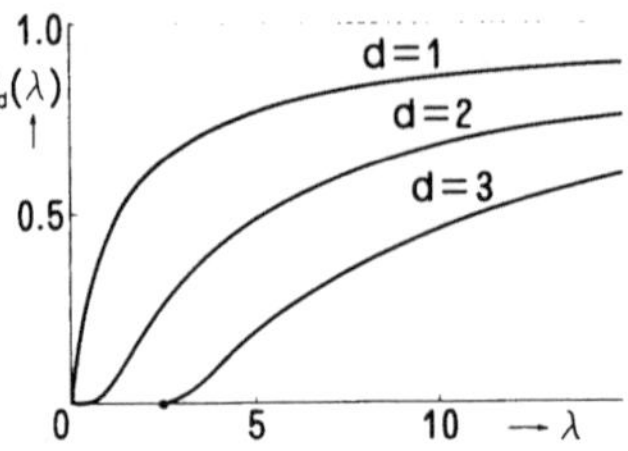

Fig.3 The ratio of the binding energy and the depth of square well potential as a function of the latter, in 1-, 2- and 3-dimensional spaces

For negative $\Xi\Delta$ (square well potential), one can write the binding energy $E_b \equiv E_0 - E(\Delta,a)$ as

$$E_b = (-\Xi\Delta) f_d(\lambda) \quad . \tag{5.4}$$

$f_d(\lambda)$ calculated for dimensionalities d=1, 2 and 3 are shown in Fig.3; their features are essentially the same as described above for the discrete lattices.

The adiabatic energy referred to the free exciton state is then given by

$$W(\Delta,a) - W^{(F)} = (\gamma_d C/2) a^d \Delta^2 - (-\Xi\Delta) f_d(\lambda) \tag{5.5}$$

where γ_d is the volume of the d-dimensional sphere of unit radius. For 1- and 3-dimensional cases, one can bring (5.4) into parameterless form

$$w(x,y) - w^{(F)} = \lambda^{d/2} x^{2-d/2} - f_d(\lambda) x \ , \quad (\lambda = x/y) \quad (d=1,3) \tag{5.5'}$$

by rescaling the energy in unit of W_0:

$$-\Xi\Delta = W_0 x, \quad K = W_0 y, \quad W = W_0 w, \tag{5.6}$$

$$\left(\frac{W_0}{\hbar^2/2m} \right)^{d/2-1} \equiv \frac{\gamma_d}{2} \frac{\hbar^2 C}{2m\Xi^2} \ , \quad (d=1,3) \quad . \tag{5.7}$$

Such is impossible with d=2; for any choice of W one obtains

$$w(x,y) - w^{(F)} = G^{-1}\lambda x - f_2(\lambda) x \ , \quad (d=2) \tag{5.5''}$$

with unremovable coupling constant:

$$G \equiv \frac{4}{\pi} \frac{m\Xi^2}{\hbar^2 C} \ , \quad (d=2) \quad . \tag{5.8}$$

The adiabatic potentials for d=1, 2 and 3 systems calculated with the use of (5.5'), (5.5") and $f_d(\lambda)$ (Fig.3) are shown in Fig.4(a), (b,b') and (c), respectively. Noting that the volume $\gamma_d a^d$ of the distorted region should not be smaller than the atomic volume $a_0{}^d$ of the discrete simple lattice, one finds the upper bound

$$y_d^{(u)} = \gamma_d^{+\frac{2}{d}+\frac{2}{2-d}} (8dg)^{-\frac{2}{2-d}} \ , \quad (d=1,3) \tag{5.9}$$

for physically meaningful y range where we have made use of (5.6), (5.7), (3.5), the relations $E_{LR} = \Xi^2/2Ca_0{}^d$ and

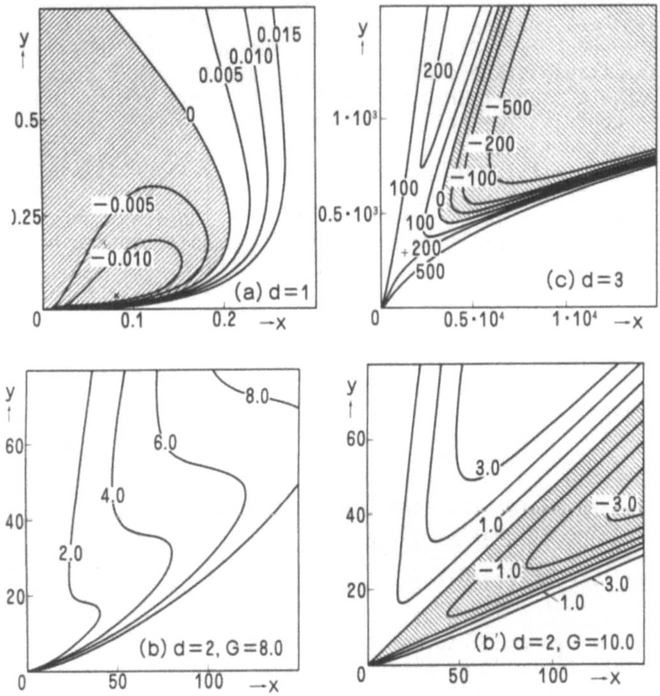

Fig.4 Adiabatic potentials for an exciton in elastic medium with contact-type interaction. x and $y^{-1/2}$ are proportional to the strength of distortion and the radius of distorted region, respectively. The energy is zero along the ordinate ($x=\Delta=0$) and negative in the hatched region

$$\frac{\hbar^2}{2ma_0^2} = |t| = \frac{B}{\nu} \quad . \tag{5.10}$$

With increasing g, the critical line $y=y_d^{(u)}$ moves downward for d=1 and upward for d=3.

Note in Fig.4 that there is the only minimum for d=1 as indicated by x. This means that as g increases the optimum radius of the distorted region and the electron orbital decreases as g^{-1} while the optimum distortion Δ and the energy lowering of the exciton increase respectively as $g^{3/2}$ and g^2 (see (5.3), (5.6) and (5.7)), always continuously. The latter increase still continuously even after $y=y_1^{(u)}$ traverses this minimum point.

In contrast, the stable state in a 3-dimensional system changes discontinuously from the free state F ($\Delta=0$) to the self-trapped state S ($\Delta\neq 0$, $a=(\gamma_3)^{1/3}a_0$ being the minimum possible value) when g exceeds $g_c=2.30$ (namely, when $y=y_3^{(u)}$ reaches the negative energy region in Fig.4(c)). The height of the potential barrier from F to S state (the saddle point in Fig.4(c)) is given by $\varepsilon_a=0.79g^{-2}B$. It should also be noted that the F state remains to be metastable for $g_c < g < \infty$ and that the S state is metastable for $g_c' < g < g_c$ where $g_c'=1.86$ corresponds to $y_3^{(u)}=y$(saddle point).

The situation in a 2-dimensional system is variable even within the continuum model: as $g=(\pi/16)G$ exceeds $g_c=1.83$, the negative energy region sets in like an opening fan. The stable state changes discontinuously from F to S state at $g=g_c$; however, there is neither a potential barrier nor the F-S coexistence range of g, in contrast to the 3-dimensional system.

The critical value g_c for self-trapping estimated above by the continuum model is twice or more greater than that by the discrete lattice model, for 2- and 3-dimensional systems. The first origin for this is the step-function Ansatz for the dilation field $u(\vec{\rho})$ —— somewhat unfavorable constraint for the S state. The second origin is the use of the parabolic band with local effective mass m fitted at k=0 to that of the $(1-\cos(k_x a_0))$ type band of the discrete lattice by (5.10), which results in significant overestimation (~1.5 times) of the electron kinetic energy of strongly localized states, thus tending to suppress self-trapping. For the same reason, the adiabatic potentials in Fig.4(b) and (b') based on the continuum model will be lowered towards increasing y in the case of a discrete lattice, possibly resulting in the appearance of the potential barrier as well as the F-S coexistence region of g.

The adiabatic potential (5.5) for the lowest exciton state of (5.1) is nothing but the minimum of the adiabatic energy

$$E[\Psi(\vec{r}),\Delta(\vec{r})] = \frac{\hbar^2}{2m}\int(\vec{\nabla}\Psi)^2 d\vec{r} + \Xi\int\Delta(\vec{r})\Psi(\vec{r})^2 d\vec{r} + \frac{C}{2}\int\Delta(\vec{r})^2 d\vec{r} \qquad (5.11)$$

with respect to $\Psi(\vec{r})$ [1]. If we are interested only in the extrema of thus obtained adiabatic potential $E[\Delta(\vec{r})]$, an alternative way of obtaining them is first to minimize (5.11) with respect to $\Delta(\vec{r})$ (which gives $\Delta(\vec{r})=-(\Xi/C)\cdot\Psi(\vec{r})^2$) and then extremize the functional thus obtained

$$E[\Psi(\vec{r})] = \frac{h^2}{2m}\int(\vec{\nabla}\Psi)^2 d\vec{r} - \frac{\Xi^2}{2C}\int\Psi(\vec{r})^4 d\vec{r} \qquad (5.12)$$

with respect to Ψ. Choose a trial function which is isotropic, nodeless and specified by a single parameter α representing the linear degree of spatial localization; α is so normalized that it varies between 0 (extended state) and 1 (completely localized state). As an example, one may imagine $\Psi(\vec{r}) \propto \exp[-A\alpha^2 r^2]$ with A being on the order of inverse square of lattice constant a_0. For a d-dimensional lattice, (5.12) immediately gives

$$E(\alpha) = B\alpha^2 - E_{LR}\,\alpha^d \qquad (0 \le \alpha \le 1) \qquad (5.13)$$

with B and E_{LR} representing respectively the exciton kinetic energy and the lattice relaxation energy for complete localization, as were introduced in 3 for a discrete lattice. With d=3, (5.13) has two minima, F (α=0) and S (α=1), separated by a potential barrier $\varepsilon_a=(4/27)g^{-2}B$, as far as $g(\equiv E_{LR}/B) > g_c'=2/3$. S becomes stable and F metastable when g exceeds $g_c=1$. With d=2, the S state suddenly becomes stable as g exceeds $g_c=1$, without any coexistence range. With d=1, α_m and $E(\alpha_m)$ corresponding to the only minimum varies continuously and monotonously as g increases, as was pointed out by EMIN and HOLSTEIN [14] in a more general scaling argument.

RÖSSLER, YERSIN and PERTZSCH studied theoretically [15] and experimentally [16] the effect, on the self-trapping barrier, of the anisotropy in a three-dimensional system, with particular attention to the limiting situations of nearly 1- and nearly 2-dimensional systems.

6. Weakly Nonadiabatic Situation

So far we have confined ourselves to the adiabatic limit. What happens if we consider a finite phonon energy $h\omega$ in (2.6) ? With vanishing transfer t, the lowest eigenstates of the total Hamiltonian $H = H_e + H_{ep} + H_p$ are given by the localized exciton accompanied by localized distortion:

$$\Phi_n(c) = \chi_0(Q_1)\cdots\chi_0(Q_n-c)\cdots\chi_0(Q_N)|n> \tag{6.1}$$

where $\chi_0(Q)$ is the wavefunction for the zero-point vibrational state of each harmonic oscillator in (2.6). The degeneracy of these localized states with n=1, 2, ···, N is removed with introduction of transfer t, whereby the true eigenstates should have the form of a propagating wave:

$$\Psi_K(\delta) = N^{-1/2}\sum_n \exp(ikR_n)\Phi_n(\delta) \quad . \tag{6.2}$$

We left δ as a variational parameter to be chosen so as to minimize the total energy $E_K(\delta)$. Subtracting the zero-point vibrational energy $Nh\omega/2$ of the whole lattice, one obtains

$$E_{K=0}(\delta) = E_a - B\exp[-S(\delta/c)^2] - E_{LR}[2(\delta/c)-(\delta/c)^2] \tag{6.3}$$

where

$$S \equiv E_{LR}/\hbar\omega \quad . \tag{6.4}$$

The exponential factor in the second term of (6.3) is nothing but the square of the overlap integral $\int dQ\,\chi(Q-\delta)\chi(Q)$ between distorted and undistorted states of the harmonic oscillator.

Equation (6.3) has two minima (see Fig.5)

$$F : (\delta/c)_F \sim \gamma \qquad E_F \sim E_a - B - g\hbar\omega \tag{6.5}$$

$$S : (\delta/c)_S = 1 - O(S\,e^{-S}), \quad E_S = E_a - E_{LR}(1 + O(e^{-S})) \tag{6.6}$$

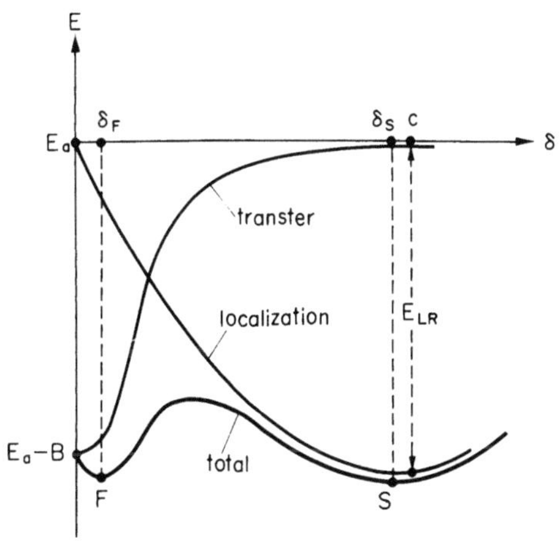

Fig.5 Energy of an exciton moving (with wave vector K=0) in the phonon field, as a function of the accompanying distortion δ chosen as variational parameter

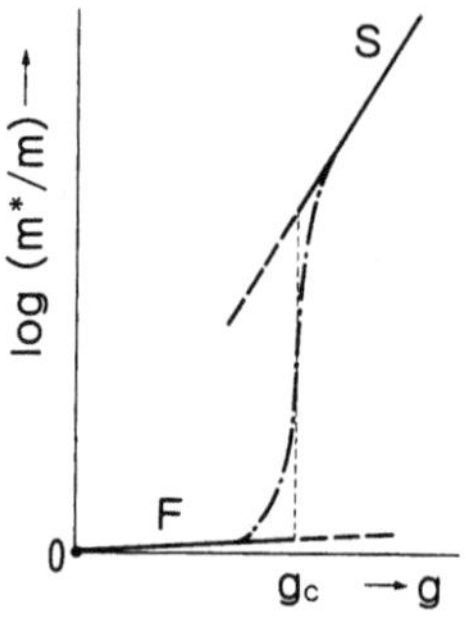

Fig.6 Effective mass m* of the phonon-dressed exciton as a function of coupling constant g

if the non-adiabaticity parameter defined by

$$\gamma \equiv \hbar\omega/B \tag{6.7}$$

is small enough compared to unity (weakly nonadiabatic case). The local distortion δ of the stable state changes discontinuously from (6.5) to (6.6) as $g = E_{LR}/B$ exceeds $g_c \sim 1$, and correspondingly, the effective mass m* of the phonon-dressed exciton, which is proportional to $(d^2E_K/dK^2)^{-1} \propto \exp[S(\delta/c)^2]$, increases discontinuously from $m^* \sim m(1+g\gamma)$ to an enormous value $\sim m\exp(S)$ (note that $S=g\gamma^{-1}$ is much greater than unity in the weakly nonadiabatic situation), as shown schematically in Fig.6.

The next improvement of the wave function is to take the linear combination

$$\Psi_K = a\,\Psi_K(\delta_F) + b\,\Psi_K(\delta_S) \tag{6.8}$$

so as to minimize further the total energy. It has significant effect only within a small region of g around g_c ($|g-g_c| \lesssim \exp[-(1-2\gamma)S]$, the overlap integral between the two states in (6.8)), resulting in a continuous (though rapid) increase of m* as g exceeds g_c (see Fig.6). In the case that γ is not small enough compared to unity, it is no longer meaningful to distinguish between the F and S states.

Let us briefly describe the phonon structures in the absorption and emission spectra which have so far been neglected. We confine ourselves to absolute zero of temperature where the structures show up most clearly. Because of the absence of thermal phonons, the absorption and the emission spectra extend respectively on the high- and the low-energy sides of the common zero-phonon line which corresponds to the optical transition between the lowest vibrational states of the ground and the excited electronic states. The intensity ratio of zero-phonon line and successive phonon sidebands in the direct absorption spectra and in the disorder-allowed indirect emission spectra is approximately given by the successive terms of the power series expansion of $\exp[-S(\delta/c)^2]$ in (6.3). The ratio of one-to-zero phonon bands turns out to be $g\gamma$ and $g\gamma^{-1}$ for the F and S states, respectively. With fixed $\gamma(\ll 1)$ and varying g, we expect a significant difference of the direct absorption spectra below and above g_c [17], as shown schematically in Fig.7 with phonon dispersion taken into account. In the strong coupling case ($g > g_c$), a very weak zero-phonon line is followed by multiphonon sidebands of increasing intensities, with their sum culminating in the Franck-Condon state which is nothing but the F state. Consequently, the decay

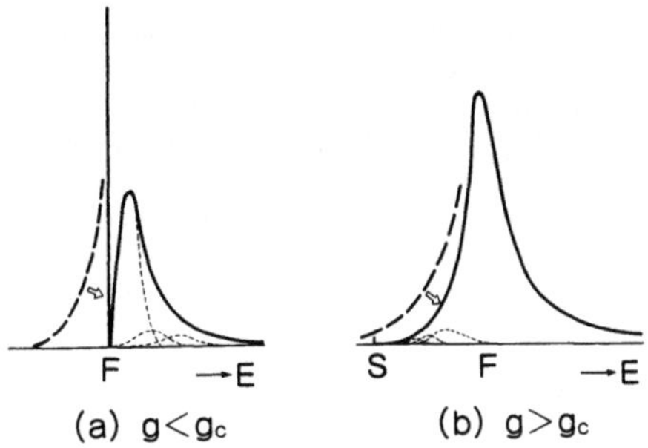

Fig.7 Schematic lineshapes of zero-temperature absorption spectra (solid line) starting with a zero-phonon line followed by phonon sidebands (dotted line) in the weak (a) and strong (b) coupling cases. Expected convergence of the Urbach tail (broken line) is also indicated

constant of the low-energy tail ($\sigma/\kappa_B T$ when the Urbach rule holds) should increase infinitely as $T \to 0$ for $g < g_c$, but should tend to a finite value for $g > g_c$. The statement, often found in literature , that the reciprocal decay constant should be replaced by $\sigma^{-1}\kappa_B T' \equiv \sigma^{-1}(\hbar\omega/2)\coth(\hbar\omega/2\kappa_B T)$ due to the zero-point vibrations, is not justified. Even in the strong coupling case, the finite value to which $\kappa_B T'$ tends as $T \to 0$ should be $\hbar\omega/2$ multiplied by some quantity which vanishes with $(g-g_c)$.

7. Comparison with Experiments

The nature of the relaxed exciton is reflected in the feature of the emission spectra. The luminescence from the F state exciton is a narrow band resonant to the absorption peak (the absorption edge in the indirect gap case), while that from the S state is a Stokes-shifted broad band (see Fig.1(b)). The relaxed excitons in most inorganic semiconductors have been shown to be of F type ($g < g_c$). In contrast, the excitons in alkali halides [18], rare gas solids [19] and some aromatic molecular crystals are relaxed into the S state ($g > g_c$). In some of the latter cases, a weak F-type luminescence from the metastable F state has also been observed. In pyrene and α-perylene, for instance, the temperature dependence of the intensity ratio of the two types of luminescence indicates that the populations of F and S states are in thermal equilibrium, wherefrom the energy difference between the two states has been determined [20]. In contrast, the F-S population in alkali halides is governed not by thermal equilibrium but by the rate of F→S transition (only the F excitons are produced initially in the optical excitation) because of higher potential barrier; the thermally activated process (at high T) and the quantal tunneling process (at low T) have been analysed from the temperature dependence of the F-S intensity ratio [18]. Direct measurements of the F→S transition rates have also been made for KI [21] and solid Xe [22].

The luminescence in $AgBr_{1-x}Cl_x$ alloys at very low temperature changes its feature from F to S type rather abruptly as x exceeds $x_c \sim 0.45$ [23], the fact which has been taken as the most direct evidence of the F-S discontinuity. The F-S coexistence within a narrow concentration range around x_c $(0.4 < x < 0.5)$ might partly be ascribed to the alloy inhomogeneity. We think, however, that the F-S coexistence in the emission spectra is at least partly due to the formation of the F-S resonance state: (6.8). The evidence is found in the F component of the emission spectra which consists of several optical phonon sidebands (starting from the disorder-allowed zero-phonon line). The effective $\bar{S}$ value of their Poisson-like intensity distribution $(\exp(-\bar{S})\bar{S}^n/n!)$ shows remarkable increase from ~ 0.7 to $\gtrsim 10$ within the narrow concentration range $0.4 < x < 0.5$. As seen from (6.5) and (6.3), S value of the F state is given by $S(\delta/c)_F^2 = E_{LR}\cdot\hbar\omega/B^2$, which is nothing but the squared amplitude of virtually mixed one-phonon states as calculated by perturbation theory. B in the denominator comes from the recoil kinetic energy $\hbar^2K^2/2m^*$ due to the virtual emission of a K phonon. In the F-S resonance state, however, the effective mass m^* of the exciton increases rapidly as g increases, and so does $\bar{S}$.

Recently the observations of the S-F transition of an exciton by applying pressure have been reported for a quasi-1-dimensional material [16] as well as for the alloy system mentioned above [24]. It would be interesting to see whether and how much the F-S resonance effect is playing a role there.

As for the absorption spectra, we know that excitons in most inorganic and some [25] organic crystals have typical Lorentzian shapes (usually on the high-energy side) characteristic of weak scattering. Thus, the excitons in alkali halides belong to the intermediate case of Fig.2 — weak scattering and strong coupling. The strong scattering situation may be realized in triplet excitons or dipole forbidden singlet excitons of organic molecular crystals as well as in core excitons of insulators in general.

The Urbach rule for the low-energy tail of the absorption spectra has been found in a variety of crystals: some of them are shown in Table 1 in the increasing order of the steepness coefficient σ measured. As mentioned in [4], σ is related with g by $\sigma=s/g$. Considering the simple cubic lattice for which $g_c=1-(2\nu)^{-1}=0.917$ and s=1.50 (direct) or 0.85 (indirect), we find the critical values $\sigma_c=1.64$ (direct) or 0.93 (indirect) below which the exciton is expected to be self-trapped according to the theory. In the third column we indicated the nature of the relaxed excitons found from luminescence studies. Thus, the two kinds of experimental data, one from absorption and the other from luminescence, can be correlated with each other in satisfactory agreement with the theory.

Table 1 Correlation between the observed steepness coefficient σ and the observed nature (F or S) of the relaxed exciton

predicted:	direct : $\sigma_c=1.64$		indirect : $\sigma_c=0.93$		
Observed:					
Se	0.6	?			
NaCl	0.76	S			
KBr	0.79	S			
KI	0.82	S			
α-perylene	0.93	S			
β-perylene	1.38	S			
pyrene	1.38	S			
GeS	1.45	?	AgCl	0.78	S
PbI_2	1.48	?	$AgCl_{.45}Br_{.55}$	0.89	← S=F
anthracene	1.7 ~1.5	F	AgBr	0.98	F
CdS	2.2	F	TlCl	1.16	F
ZeSe	2.4	F			
CdSe	2.5	F			
ZnTe	2.8	F			
CdTe	3.0	F			
Te	3.1	?			

The experimental situation is not very clear with the low-temperature behavior of the low-energy tail, possibly because of residual imperfections and low-lying exciton states (such as triplet) which might affect the otherwise sharp spectral structures expected in the weak coupling case. Only two typical examples will be mentioned. The low-energy tail of exciton absorption in CdTe ($g < g_c$ according to Table 1) reveals with decreasing temperature the structures due to the optical phonon absorption, the intensities being governed by the thermal population of the phonons [26]. This means that the low-energy tail tends to vanish as $T \to 0$, being consistent with Fig. 7(a), although the real situation is not so simple. In alkali halides ($g > g_c$), it has been empirically established that as $T \to 0$ the reciprocal decay constant $\sigma^{-1}\kappa_B T'$ tends to a finite value as shown in Fig.7(b).

It seems to be rather difficult to separate the true zero-phonon line in nearly adiabatic situations because of the following reason. With $g < g_c$, the acoustic phonon sideband comes very close to the zero-phonon line due to the motional effect of exciton: the separation is given by $ms^2/2$ (s: the sound velocity) which is two orders of magnitude smaller than the average phonon energy [27]. With $g > g_c$, the zero-phonon line is usually too weak to be observed. The situation is subtle in the critical region $g \sim g_c$ possibly with the F-S resonance, and much better in the nonadiabatic situation. In this respect, the zero-phonon line and phonon sidebands of absorption and emission spectra in some organic charge transfer compounds, observed and analysed by HAARER [28] and TOKURA and KODA [29], seem to show important physics of exciton-phonon system yet to be clarified.

The author wishes to express his gratitude to Drs. M. Schreiber, A. Sumi, H. Sumi and K. Cho whose collaborations have been essential for the development of the study surveyed here.

References

1 Y. Toyozawa: Proc. 4th Intern. Conf. Vacuum Ultraviolet Radiation Physics (Vieweg-Pergamon, Braunschweig, 1974) p. 317
2 Y. Toyozawa: in Relaxation of Elementary Excitations, ed. R. Kubo and E. Hanamura (Springer, Berlin-Heidelberg-New York, 1980)
3 Y. Toyozawa: Physica 117B & 118B, 23 (1983)
4 M. Schreiber and Y. Toyozawa: J. Phys. Soc. Jpn. 51, 1528, 1538, 1544 (1982); 52, 318 (1983)
5 See the Appendix to Part I of reference 4
6 Y. Toyozawa: Progr. Theor. Phys. 27, 89 (1962); J. Phys. Chem. Solids 25, 59 (1964)
7 M.V. Kurik: Phys. Stat. Solidi (a) 8, 9 (1971)
8 See the references in Part III of reference 4
9 H. Sumi and Y. Toyozawa: J. Phys. Soc. Jpn. 31, 342 (1971)
10 H. Sumi: J. Phys. Soc. Jpn. 32, 616 (1972)
11 J. Ihm and J.C. Phillips: Phys. Rev. B 27, 7803 (1983)
12 E.I. Rashba and G.E. Gurgenishvilli: Sov. Phys.-Solid State 4, 759 (1962)
13 Y. Toyozawa: Progr. Theor. Phys. 26, 29 (1961)
14 D. Emin and T. Holstein: Phys. Rev. Lett. 36, 323 (1976)
15 B. Pertzsch and U. Rössler: Solid State Commun. 37, 931 (1981)
16 U. Rössler and H. Yersin: Phys. Rev. B 26, 3187 (1982)
17 K. Cho and Y. Toyozawa: J. Phys. Soc. Jpn. 30, 1555 (1971)

18 Ch.B. Lushchik: in Excitons, ed. E.I. Rashba and M.D. Sturge (North=Holland, 1982) p. 505 and references therein
19 I.Ya. Fugol: Advances in Physics 27, 1 (1978) and references therein
20 A. Matsui, M. Iemura and H. Nishimura: J. Luminescence 24/25, 445 (1981); H. Nishimura, A. Matsui and M. Iemura: J. Phys. Soc. Jpn. 51, 1341 (1982)
21 Y. Masumoto, Y. Unuma and S. Shionoya: J. Phys. Soc. Jpn. 51, 3915 (1982)
22 E. Roick, R. Gaethke, G. Zimmerer and P. Gürtler: Solid State Commun. 47, 333 (1983)
23 H. Kanzaki, S. Sakuragi and K. Sakamoto: Solid State Commun. 9, 999 (1971)
24 M. Kobayashi, T. Kawamura, S. Endo, K. Cho and S. Narita: Solid State Commun., to be published
25 D.M. Burland, U. Konzelmann and R.M. Macfarlane: J. Chem. Phys. 67, 1926 (1977); D.M. Burland, D.E. Cooper, M.D. Fayer and C.R. Gochanour: Chem. Phys. Lett. 52, 279 (1977)
26 D.T.F. Marple: Phys. Rev. 150, 728 (1966); B. Segall: Phys. Rev. 150, 734 (1966)
27 Y. Toyozawa: J. Luminescence 1, 798 (1969)
28 D. Haarer: J. Chem. Phys. 67, 4076 (1977); Festkörperprobleme 20, 341 (1980)
29 Y. Tokura and T. Koda: Solid State Commun. 40, 299 (1981)

Vibronic Excitons in the Intermediate Coupling Regime

Sighart F. Fischer
Technische Universität München, D-8046 Garching, Fed. Rep. of Germany

Abstract

Electronic excitations of a linear molecular crystal are investigated. The theory applies to arbitrary coupling of the exciton to molecular vibrations as well as arbitrary strength of the exciton transfer. As an example the spectrum of a J aggregate, the dye 1,1'-diethyl-NN'-isocyaninchlorid (PIC), is discussed. Furthermore, the spectra and the dynamics of excimer formation for α-perylene are analyzed within the spirit of this model.

1. Introduction

The problem of excitons interacting strongly with a vibrating lattice has attracted the attention of theorists for many years [1-6]. If one considers a linear chain of interacting molecules the simplest system may be classified by three parameters:

(i) a frequency ω of the molecule that couples most strongly,
(ii) the coupling parameter λ, or the localization energy $E_L=\lambda^2\omega$ and
(iii) the matrix element V for nearest-neighbour exciton transfer.

If the localization energy E_L is small compared to the transfer term V the excitations can be described within a band model as delocalized states which are scattered by the vibrations. If E_L is large compared to V the eigenstates are best described as localized states. If both E_L and V are large compared to ω one may obtain both types of solutions, the delocalized state also referred to as the F state, and the localized one referred to as self trapped S state [4].

In this paper we shall use the term intermediate coupling for all cases in which perturbation theory with respect to λ as well as V fails. We shall see that the PIC aggregate [7-8] is a good example for a case in which all three parameters are of the same magnitude. In this case the eigenstates are largely delocalized. We derive critical parameters which separate the different regimes in which only delocalized, coexisting localized and delocalized, and only localized states exist. In a second application we try to interpret the dynamics of excimer formation for α-perylene [9-12] within the frame of this model. In this case we include intramolecular vibrations as well as intermolecular phonon modes in our model. A critical discussion of these results indicates, however, that a proper interpretation of experiments on α-perylene requires a full threedimensional treatment of the system characterized by the symmetry of the four molecules within the elementary crystal cell.

2. The Hamiltonian

We consider N identical molecules with one vibrational mode each having the frequency ω. For an electronic excitation this vibrational mode may change its equilibrium position due to a local coupling proportional to λ. Finally, there will be an electronic transfer matrix element between nearest neighbours, denoted by V which is given in units of ω. Thus we get in these units ($\hbar$=1):

$$H = \omega \sum_{n=0}^{N-1} b_n^+ b_n + \lambda\omega \sum_{n=0}^{N-1} (b_n + b_n^+)\, a_n^+ a_n$$

$$+ \sum_{n=0}^{N-1} V\omega (a_n^+ a_{n+1} + a_{n+1}^+ a_n) \,. \tag{1}$$

The local phonon operators b_n and b_n^+ fulfill Bose commutation relations and the electronic operators a_n and a_n^+ are Fermi operators. We assume cyclic boundary conditions, that means the site N coincides with the site 0.

In a first step we decouple the system with the help of a displacement operator G

$$G = \exp\left\{ i \sum_q q b_q^+ b_q \right\} \; ; \; q = 2\pi l \cdot N^{-1}, \quad l = 0,1 \ldots N-1 \tag{2}$$

where we introduced the Fourier transform of b_n:

$$b_q = N^{-1/2} \sum_n e^{iqn} b_n \,. \tag{3}$$

This operator G has the following property

$$G^+ b_n G = b_{n-1} \tag{4}$$

which allows us to relate the excitation of site n to the one at site 0. We do this with a unitary transformation

$$H = S_E^+ H S_E \tag{5}$$

where the unitary operator is defined as

$$S_E = \sum_n G^n a_n^+ a_n \,. \tag{6}$$

The transformed Hamiltonian has the form

$$H = \omega \sum_n b_n^+ b_n + \lambda\omega (b_o + b_o^+) \sum_n a_n^+ a_n$$

$$+ V\omega \sum_n (G\, a_n^+ a_{n+1} + G^+ a_{n+1}^+ a_n) \,. \tag{7}$$

Introducing also the Fourier transform with regard to the electronic operator a_n we obtain the electronically decoupled Hamiltonian

$$H = \omega\sum_k h_k a_k^+ a_k \tag{8}$$

with the vibronic operator

$$h_k = \sum_q b_q^+ b_q + \lambda N^{-1/2} \sum_q (b_q + b_q^+) + V\,(Ge^{ik} + G^+ e^{-ik})\,. \tag{9}$$

3. A Self consistent Lattice Distortion

In order to get approximate eigenvalues of our Hamiltonian h_k we make a further canonical transformation with a variable distortion parameter $\lambda_q(k)$ for every phonon mode q, which depends upon the electronic quantum number k. In order to get real displacements we require $\lambda_q = \lambda^*_{-q} = \lambda_{-q}$ and set

$$\tilde{h}_k = S_\lambda^+(k) h_k S_\lambda(k) \tag{10}$$

with

$$S_\lambda(k) = \exp\left\{ N^{-1/2} \sum_q \lambda_q(k)(b_q - b_q^+)\right\}\,. \tag{11}$$

The transformed Hamiltonian reads:

$$\begin{aligned} \tilde{h}_k = {} & \sum_q b_q^+ b_q + N^{-1/2} \sum_q (\lambda - \lambda_q(k))(b_q + b_q^+) + N^{-1} \sum_q (\lambda_q^2(k) - 2\lambda\lambda_q(k)) \\ & + VGe^{ik} \exp\left\{ -N^{-1/2} \sum_q \lambda_q(k)(b_q e^{iq} - b_q^+ e^{-iq})\right\} \exp\left\{ N^{-1/2} \sum_q \lambda_q(k)(b_q - b_q^+)\right\} \\ & + Ve^{-ik} \exp\left\{ -N^{-1/2} \sum_q \lambda_q(k)(b_q - b_q^+)\right\} \exp\left\{ N^{-1/2} \sum_q \lambda_q(k)(b_q e^{iq} - b_q^+ e^{-iq})\right\} G^+\,. \end{aligned} \tag{12}$$

We now determine the distortion parameters $\lambda_q(k)$ by minimizing the ground state with respect to these parameters.

As approximate ground state we take the normalized state $|\Omega\rangle$ with the property $b_k\,|\Omega\rangle = 0$. This gives us the phononless energy spectrum E_k

$$\begin{aligned} E_k = \omega\langle\Omega|\tilde{h}_k|\Omega\rangle = {} & \omega N^{-1} \sum_q \left(\lambda_q^2(k) - 2\lambda\lambda_q(k)\right) \\ & + 2\omega\, V \cos k \exp\left\{ N^{-1} \sum_q \lambda_q^2(k)\,(\cos q - 1)\right\}\,. \end{aligned} \tag{13}$$

The variation of $\lambda_q(k)$ gives us the set of equations which can be expressed in terms of a generalized Debye-Waller factor R(k):

$$\frac{\partial}{\partial\lambda_q(k)} \langle\Omega h_k \Omega\rangle = N^{-1}(\lambda_q(k) - \lambda) + 2N^{-1} V \lambda_q \cos k\,(\cos q - 1) R_k \quad \text{with} \tag{14}$$

$$R_k = \exp\left\{N^{-1} \sum_q \lambda_q^2(k)(\cos q - 1)\right\}. \tag{15}$$

Setting Eq. (14) to zero we can express $\lambda_q(k)$ in terms of R_k

$$\lambda_q(k) = \lambda(1 + 2R_k V \cos k\,(\cos q - 1))^{-1} \tag{16}$$

and the self consistency equation for R_k follows from substituting (16) in (15)

$$R_k = \exp\left\{N^{-1} \lambda^2 \sum_q (1 + 2 R_k V \cos k\,(\cos q - 1))^{-2} (\cos q - 1)\right\}. \tag{17}$$

The integrals can be performed in the limit $N\to\infty$ and we obtain our final results for R_k and E_k:

$$R_k = \exp\left\{-\lambda^2 (1 - 4R_k V \cos k)^{-3/2}\right\} \tag{18}$$

and

$$E_k = 2R_k V \cos k - 2\lambda^2\omega\left[1 - 4R_k V \cos k\right]^{-1/2}$$
$$+ \lambda^2\omega(1 - 2R_k V \cos k)\left[1 - 4R_k V \cos k\right]^{-3/2}. \tag{19}$$

Finally, we want to evaluate the equilibrium displacements $\delta_n(k)$ for the different sites n which are due to excitation into the state E_k. If we use the dimensionless coordinates
$r_n = 2^{-1/2}(b_n + b_n^+)$ we obtain

$$\delta_n(k) = \langle \tilde{r}_n \rangle - \langle r_n \rangle = (2N)^{-1/2} \sum_q \lambda_q \cos qn. \tag{20}$$

Substituting λ_q from eq. (16) we find

$$\delta_n(k) = \delta_o(k) \cdot Z_k^n \tag{21}$$

with the central displacement

$$\delta_o(k) = 2^{1/2} \lambda \left[1 - 4R_k V \cos k\right]^{-1/2} \tag{22}$$

and

$$Z_k = \frac{2^{1/2}\lambda - \delta_o(k)}{2^{1/2}\lambda + \delta_o(k)} \quad . \tag{23}$$

This result shows that the range of displacements is given by $\ln|Z_k|$. Since the interaction V is negative for most cases of interest, Z_k will be smaller than one for $k = 0$. It becomes zero for $k = \pi/2$ and negative for $k = \pi$. The latter case refers to the upper band edge. In this case the displacements would oscillate with an amplitude at zero, $\delta_0(\pi)$, which is larger than the displacement of the isolated molecule. This prediction may be unphysical and could be changed if we allow for more flexible relaxations including a variation in the momentum and frequency changes. MERRIFIELD [3] did not make the assumption $\lambda_q=\lambda_{-q}$. As a result he got a phase factor in R_k which led to more complex self consistency equations. Our interest here is mainly to concentrate on the $k = 0$ state or the lower part of the exciton band. We want to do this in particular for the intermediate coupling regime where we may find coexisting localized and delocalized states. Our relatively simple equations (18) and (19) make the detailed model study of the so-called self-trapping process possible.

4. Self trapping and Delocalization

We want to study the equations (18) and (19) in the intermediate coupling regime, where we may get more than one solution for E_k for given values of λ and V. Since R_k can not be expressed explicitly in terms of V and λ we find it convenient to treat R_k as a variable and we express λ^2 and E_k in terms of R_k with V as a parameter. The corresponding equations follow directly from (18) and (19) to read

$$\lambda^2(R_k,V_k) = -\left[1+2R_kV_k\right]^{3/2} \ln R_k \tag{24}$$

and

$$E_k(R_k,V_k) = \ln R_k - R_kV_k(1-3\ln R_k) \tag{25}$$

with the positive interaction

$$V_k = -2\,V\cos k \;. \tag{26}$$

In Fig. 2 we plotted $\lambda^2(R,V_k)$ and $-E_k(R,V_k)$ as function of R which varies from zero to one, and treated V_k as parameter. In order to find the energies $E_k(\lambda^2,V)$ for given values of λ^2 and V or V_k we have to invert the equation $\lambda^2(R,V_k) = \lambda^2$ graphically to get $R_k(\lambda,V_k)$. We can then use these solutions to get $E_k(R_k(\lambda,V_k), V_k) = E_k(\lambda^2,V)$. We see that $\lambda^2(R,V_k)$ and $E_k(R,V_k)$ are monotonic functions of R as long as V_k is smaller than a critical value V_c. For $V_k \to V_c$ we find three regimes separated by $\lambda_m^2(V_k)$ and $\lambda_1^2(V_k)$ which are the maxima and minima respectively of the function $\lambda^2(R,V_k)$. For $\lambda^2 < \lambda_m^2(V_k)$ there exists only one solution $\lambda^2(R_k,V_k)$

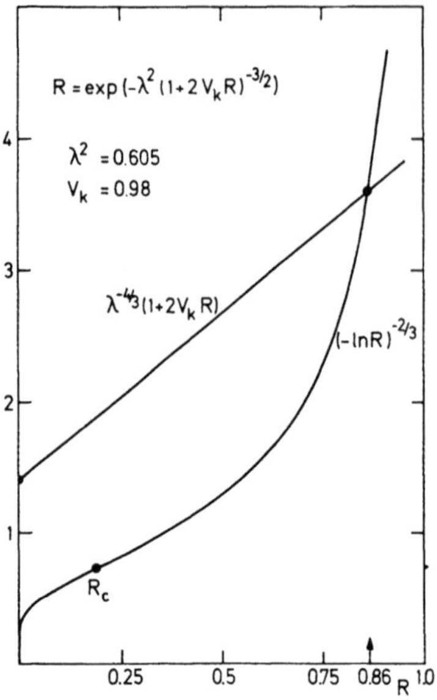

Fig. 1

Graphical solution of the self-consistent range parameter R. The values $\lambda^2 = 0.605$ and $V_k = 0.92$ refer to PIC which is discussed in section 5. The critical range parameter R_c is also shown. As can be seen, large values of λ and large values of V_k may lead to three intersection points, one very close to zero. For $V_k < 0$ only one intersection can occur

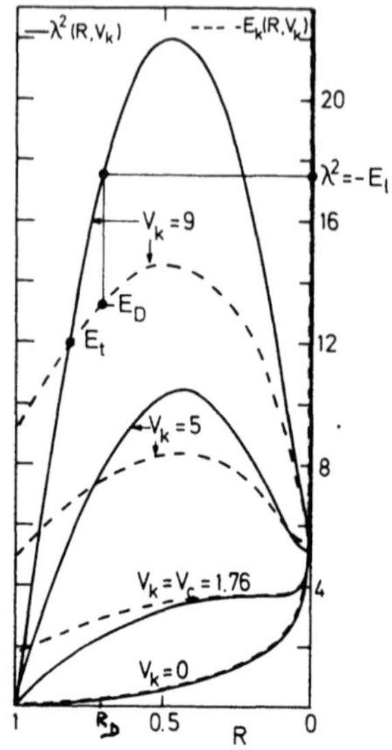

Fig. 2

The functions $\lambda^2(R,V_k)$ as well as $-E_k(R,V_k)$ are plotted versus the range parameter R, with V_k as parameter. For $V_k = 0$ the two curves coincide. Furthermore the functions for $V_k = V_c$ and for $V_k > V_c$ are shown.
The point at $-E_D$ is a delocalized solution related to the localized solution at $-E_L$ by the same parameter λ^2. E_t describes the point of a crossover from a stable localized to a stable delocalized state

and consequently also one energy $E_k(\lambda,V)$. This solution occurs at values of R which are relatively close to one. We call these a delocalized state solution. For

$\lambda_1^2(V_k)<\lambda^2 < \lambda_m^2(V_k)$ there are three solutions. One is found at very small values of $R = R_L$. It corresponds to the localized state. The corresponding energy is given approximately by

$$E_k(R_L,\lambda,V) = E_L = -\lambda^2\omega . \tag{27}$$

Then there exists a solution still at relatively small values of R. The third at R_D can be identified with the delocalized state. Its energy is somewhat lower than the energy of the entirely delocalized state ($\lambda = 0$):

$$E_k(R_D,\lambda^2,V_k) < 2\,V\cos k . \tag{28}$$

It should be noted that the delocalized state may be higher or lower in energy than the localized state, depending upon the size of λ^2 for given V_k. However, the range of λ^2 values, for which the lowest state is the delocalized one, is rather small. As λ^2 increases it will clearly separate from the delocalized state until it reaches the maximum value $\lambda_m^2(V_k)$. For $\lambda^2 > \lambda_m^2$ only the localized state exists and its energy is now given by E_L with very high accuracy.

Now we want to determine the parameters which characterize the different regimes for the existence of localized and delocalized states. The extrema $\lambda_m^2(V_k)$ and $\lambda_1^2(V_k)$ follow from the equation

$$\frac{\partial\lambda^2(R,V_k)}{\partial R} = 0 \tag{29}$$

which can be brought in the form

$$1 + 2RV_k + 3RV_k \ln R = 0 \, . \tag{30}$$

Since V_k has to be much larger than one in order to produce extrema, we can in lowest order neglect the one to find the maximum value R_m and iterate to get to lowest order

$$R_m \approx \exp\left\{-\frac{2}{3} - \frac{1}{3} e^{+2/3} V_k^{-1}\right\} \approx e^{-2/3} - (3V_k)^{-1} \, . \tag{31}$$

For the minimum the value of R_ℓ is quite small. We can start the iteration with $\ln R = -\lambda_c^2$ where $\lambda_c^2 = 3.6$ as is shown below. Thus we get to first order

$$R_\ell \approx (3\lambda_c^2-2)^{-1} V_k^{-1} \approx 0.1 \; V_k^{-1} \, . \tag{32}$$

The corresponding values for $\lambda_m^2(V_k)$, $\lambda_1^2(V_k)$ as well as $E_m^2(V_k)$ and $E_1^2(V_k)$ read

$$\lambda_m^2(V_k) \approx 2^{5/2}(3e)^{-1} V_k^{3/2} \approx 0.7 \; V_k^{3/2} \tag{33}$$

$$\lambda_1^2(V_k) \approx 2.2(1+0.2 \; V_k)^{3/2} \tag{34}$$

and

$$E_m(V_k) \approx -\frac{2}{3} - 3e^{-2/3} V_k \tag{35}$$

$$E_1(V_k) \approx -(3\lambda_c^2-2)^{-1}\left[1+3\ln\left(V_k(3\lambda^2-2)\right)\right] - \ln\left(V_k(3\lambda^2-2)\right) \approx -1.3 \ln V_k - 3 \, . \tag{36}$$

The critical parameters λ_c^2, V_c, R_c and E_c follow from the condition that $\lambda^2(R,V_k)$ must have a turning point at R_c. So that the two conditions must hold:

$$\frac{\partial\lambda^2(R_c,V_k)}{\partial R} = 0 \text{ and } \frac{\partial^2\lambda^2(R_c,V_k)}{\partial R^2} = 0 \,. \tag{37}$$

These equations can be solved analytically for R_c and we get:

$$\begin{aligned} R_c &= e^{-5/3} = 0.19 \\ \lambda_c^2 &= (\tfrac{5}{3})^{5/2} = 3.59 \\ V_c &= \tfrac{1}{3}\, e^{5/3} = 1.76 \\ E_c &= \tfrac{11}{3} = 3.67 \,. \end{aligned} \tag{38}$$

Since V_k varies with cosk, the critical value V_c will always be reached within the band for any interaction V with $|2V|>V_c$. For k values such that V_k becomes smaller than V_c only the localized solution persists. We mentioned earlier that within the intermediate coupling regime, which provides several solutions for $E_k(R,\lambda^2,V_k)$, the localized or the delocalized state may be the lowest, depending upon the relative magnitude of λ^2 and V_k. We can estimate the relation between λ^2 and V_k if we assume that the transition of the lowest energy state from the delocalized state to the localized is caused by changing the coupling λ^2 and both λ^2 and V_k are so large that the transition takes place at a value R_t relatively close to one. We simply equate $E_k(R,V_k)$ and $\lambda^2(R,V_k)$, expand R at one to get $R_t(V_k)$ and set $E_k(R_t,V_k) = \lambda_t^2$ to get

$$\lambda_t^2 = -E_t(V_k) \approx V_k + (\frac{V_k}{2})^{1/2} \,. \tag{39}$$

The transition from the delocalized state to the localized one is accompanied by a change in the relative effective mass from

$$(\frac{m^*}{m})_{del.} = 1 + \frac{1}{2}(2V_k)^{-1/2} \text{ to } (\frac{m^*}{m})_{loc.} = \exp\left\{+\lambda^2\right\} \,. \tag{40}$$

5. Application to PIC

As a first application we consider the J aggregate (PIC). This system is well suited to test the theory because the relevant parameters are known and the aggregate seems to exist of linear chains. The frequency ω, the coupling constant λ^2, and the transfer integral V have been determined by KOPAINSKY, HALLERMEIER and KAISER [13], from a detailed spectroscopy study of the monomer and the dimer. The resulting parameters are:

$$\begin{aligned} \omega &= 1368 \text{ cm}^{-1} \\ V_{k=0} &= 0.92 \\ \lambda^2 &= 0.605 \,. \end{aligned} \tag{41}$$

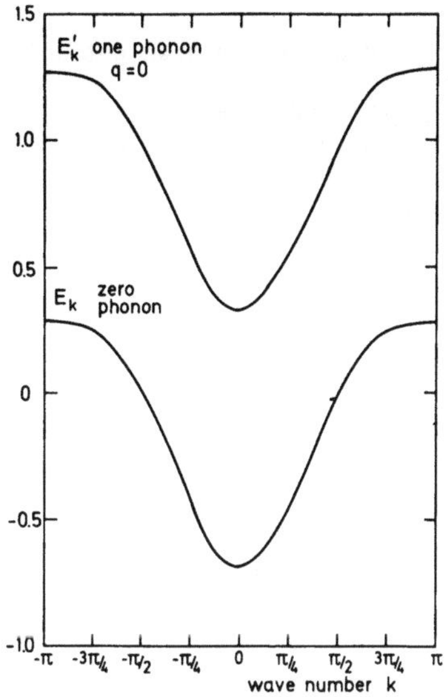

Fig. 3

Energy dispersion curves for the zero and the one-phonon states. It should be noted that there are strong deviations from the cosk dependence. The upper edge of the band is pushed towards lower energies. The Stokes shift, λ^2, is added to the energy, so that the centre of E_k for $k = \pi/2$ lies at zero

It is clear that both V_k and λ^2 are smaller than the critical parameters V_c and λ_c^2 so there exists only one solution. Nevertheless, we may regard the system as a representative of intermediate coupling since ω, $\lambda^2\omega$ and $V_0\omega$ are all of the same magnitude and low-order perturbation theory is not directly applicable. Experimentally the absorption spectrum is known. There exists also some information on the polarisation [7]. From (18) we obtain R_k as is illustrated for k=0 in Fig. 1. The resulting value is $R_{k=0} = 0.87$. Given R_k we calculate from Eq. (19) the energy E_k corresponding to the phononless energy band. The one-phonon energy $E'_{k,q}$ can be evaluated as

$$E'_{k,q} = \langle\Omega b_q \tilde{h}_k b_q^+ \Omega\rangle = E_k + 1 - R_k \,(V_{k+q} - V_k) \,. \qquad (42)$$

The two dispersion curves for E_k and $E'_{k,q=0}$ are shown in Fig. 3.

The transition probabilities into the corresponding energy states are

$$F_k = |\langle\Omega|S_\lambda(k)|\Omega\rangle|^2 = \exp\left\{-\sum_q \lambda_q^2\right\} = \exp\left\{\lambda^2 \,(1 + V_k R_k)(1 + 2V_k R_k)^{-3/2}\right\} \qquad (43)$$

and

$$F'_{k,q} = |\langle\Omega S_\lambda(k)\, b_q^+ \Omega\rangle|^2 = F_k\, N^{-1} \lambda^2 \Big(1 - V_k R_k(\cos q - 1)\Big)^{-2}$$

$$= F_k\, N^{-1}\, \lambda^2\, (E'_{k,q} - E_k)^{-2} . \qquad (44)$$

The density of states $(\frac{\partial E'_{k,q}}{\partial q})^{-1}$ diverges at $q = 0$ and $q = \pi$. Thus we get only contributions to the spectrum at band edges $k=0$, $q=0$; $k=0$, $q=\pi$; $k=\pi$, $q=0$ and $k=\pi$ $q=\pi$. The values for the range parameters, the energies, and the intensities are given in Table 1. The corresponding line spectrum is shown in Fig. 4 together with the experimental spectrum.

Table 1

Energies: $E^{0,1}_{k,q}+\lambda^2$		Intensities $I^{0,1}_{k,q}$	
$E^0_{00}+\lambda^2 = -0,68$	$E^0_{\pi 0}+\lambda^2 = 0.29$	$I^0_{00} = 0.77$	$I^0_{\pi 0} = 0.33$
$E^1_{00}+\lambda^2 = 0.32$	$E^1_{\pi 0}+\lambda^2 = 1.29$	$I^1_{00} = 0.18$	$I^1_{\pi 0} = 0.09$
$E^1_{0\pi}+\lambda^2 = 1.91$	$E^1_{\pi\pi}+\lambda^2 = 0.91$	$I^1_{0\pi} = 0.03$	$I^1_{\pi\pi} = 0.28$
$R_0 = 0.87$	$R_\pi = 0.24$	range parameters	

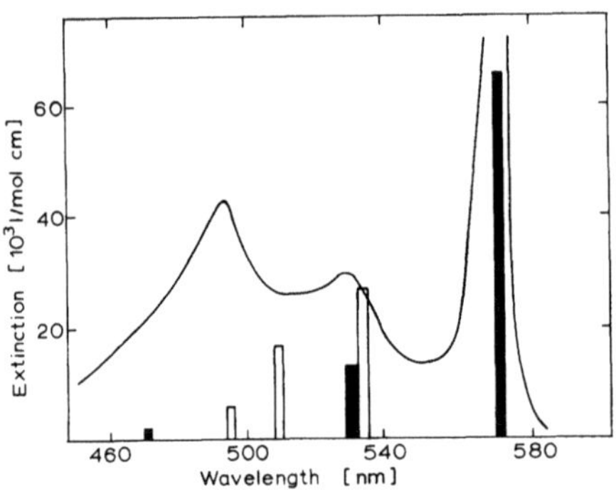

Fig. 4

Absorption spectrum of PIC. Experimental curve from Ref. 8. The filled stick diagram gives the intensity for the zero and one-phonon transitions polarized parallel to the chain. The open sticks refer to polarisation perpendicular to the chain

SCHERER [14] has performed numerical calculations for a chain of 12 units including all two-phonon excitations. His results show that this theory gives very good agreement with the k=0 states, but the agreement of the intensity distribution of the upper band edge transitions is less satisfying, as can be seen from the comparison with the experimental results as well. In particular the $k=\pi$, $q=0$ state carries not enough intensity.

6. Application to α-perylene

It would be interesting to find a system which fulfills the conditions

$$\lambda_c < \lambda < \lambda_m \text{ and } V > V_c . \qquad (45)$$

In this case one could hope to observe the localized and also the delocalized state. For purely intramolecular modes it is difficult to find such a strong coupling. But if the chain consists of pairs of molecules the optical

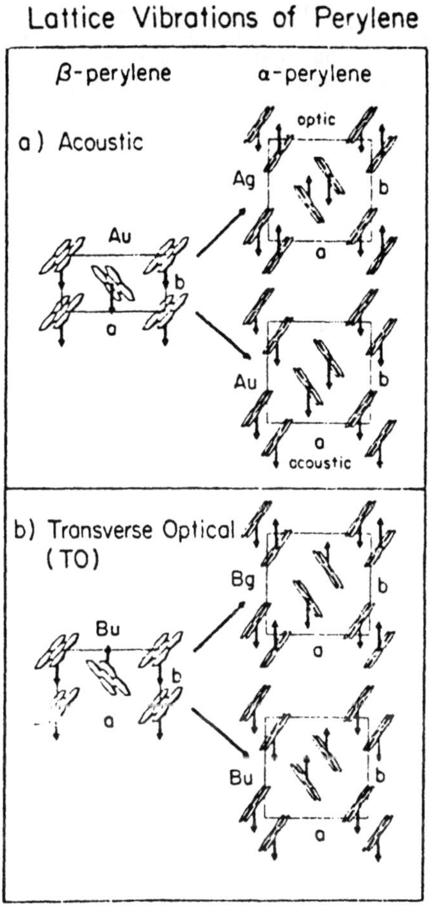

Fig. 5

Lattice vibrations of perylene from Ref. 15. In the crystal there are always three modes of the symmetry type shown. The frequencies of the Ag modes are 200 cm^{-1}, 161 cm^{-1} and 104 cm^{-1} respectively. Also shown is ß-perylene and the symmetry relations of the Au(Bu) state to the states Ag, Au (Bg, Bu) in α-perylene. This correlation holds also for the electronic states

mode of the pair can couple very strongly. Such modes are typically in the range between 50 cm^{-1} and 200 cm^{-1}. For α-perylene a mode that changes the distance of a pair of molecules is shown in Fig. 5. It is an Ag mode with a frequency of 200 cm^{-1} in the ground state. Molecular crystals which form stacks of such pairs often show excimer emission. This is characterized by a broad unstructured band, not seen in absorption, and a large Stokes shift. For α-perylene it is found at 15873 cm^{-1} and it has a half-width of 1600 cm^{-1} [9]. After excitation into a higher delocalized state at room temperature, emission is seen from this excimeric E state only. At low temperatures $T < 70$ K emission appears also from a higher state at 18868 cm^{-1} denoted as the Y state.

Recent experiments by AUWETER et al. [12] have shown that the Y state is a precurser of the E state and that the transition from the Y to the E state is governed by a temperature-dependent rate

$$k = k_0 + k_1 \exp\left\{ - \frac{\Delta E}{kT} \right\} \text{ with}$$

$$k_0 = 2.4 \cdot 10^7 s^{-1};\ k_1 = 6.6 \cdot 10^{11} s^{-1} \text{ and } \Delta E = 310\ cm^{-1}. \qquad (46)$$

There exists further experimental evidence that the Y state is not an impurity state [9]. It has been described as a partially trapped state which, however, does not show exciton annihilation [9]. The fluorescence is similar to the fluorescence from ß-perylene, the so-called monomeric form, but it does not show the detailed structure as the ß-perylene does.

In order to simulate these results within our model we used the frequency of $\omega_1 = 200\ cm^{-1}$ and a coupling $\lambda_1^2 = 17.5$, which creates a Stokes shift of $3500\ cm^{-1}$. In addition we included an intramolecular mode with frequency $\omega_2 = 1300\ cm^{-1}$ and a coupling parameter $\lambda_2^2 = 1.5$ which accounts for the most important intramolecular couplings. Our theoretical model Hamiltonian did not include several coupling modes. In Fig. 2 the energies E_L and E_D are shown for $\lambda_1^2 = 17.5$. The extension of the theory to several modes causes no difficulties. The inclusion of the high-frequency mode ω_2 is essential for the comparison with the experiment. Due to this mode the selftrapped E_L state (E state) shifts by $\lambda_2^2\omega_2$ but the delocalized state E_D (Y state) is less affected. As a result, these high-frequency modes widen the energy difference between the localized and the delocalized state. The transfer term is not available from the experiments, but one may get a proper estimate by choosing the corresponding value from the Davydov splitting of the excited singlet, which is reported to be $820\ cm^{-1}$ [10]. This is of course the effective bandwidth given by $2V_oR$. The Y state may have a somewhat smaller width since it is not symmetry allowed. In order to simulate the observed energy difference between the Y and the E state we needed a rather strong coupling of $V_o = 1800\ cm^{-1}$. Interestingly we obtain at the same time a novel interpretation of the activation energy ΔE. Varying the qunatum number k corresponds to a reduction of our potential parameter V_k relative to V_o. The maximum value at which there still exists a delocalized state is reached at $V_k = V_m = 1560 cm^{-1}$ and the energy differences between the corresponding states are

$$E_m - E_D = \Delta E = 4020\ cm^{-1} - 3750\ cm^{-1} = 270\ cm^{-1}\ . \qquad (47)$$

The interpretation of this activation energy is slightly different from the one used by TANAKA [10] or AUWETER et al. [12], who assume an activation barrier along a configuration coordinate. Within our model we would get an excitonic band for the Y state, which does not extend to the upper band edge. It becomes unstable for a k value $V_k = V_m$. Thermal excitation within the band leads to transitions into the selftrapped state. The special instability of the band could also explain the absence of annihilation within this band. For band states below E_m nonradiative transitions from the partially delocalized Y state to the E state exist. These are only weakly temperature dependent and can account for the rate k_o in Eq. (46). The faster rate k_1 would correspond to the transfer from the highest k value of the Y state to the E state.

Even though the general concept might be attractive to interpret such systems which show excimer states as well as partially delocalized states, we want to add some critical remarks and offer an alternative interpretation. The actual system is not one-dimensional. There are four molecules in the unit cell and as mentioned already by HOCHSTRASSER and NYI [11] there are four factor group states belonging to the C_{2h}^5 symmetry. Two are of Au and Bu symmetry respectively. They carry oscillator strength and can be observed in absorption. The remaining two states of Ag and Bg symmetry are not identified. HOCHSTRASSER and NYI [11] suggested that the lower one of these might be the Y state. We would argue that only the Ag state can couple strongly with the totally symmetric optical mode, shown in Fig. 5. We would therefore like to correlate this state with the E state. The fact that the electronic nature of this state may vary as a function of the Ag normal mode as the relaxed position of the excimer is reached could be included in the model. This leads us to correlate the Y state with the Bg state. The Bg state can not couple so strongly to the Ag mode since the electronic excitation would tend to shorten the distance of one pair and widen the distance of the other. Comparing the two models, there is a basic difference. If there are two electronic states due to Davydov splitting they exist also when the transfer energy to the other neighbours is relatively small, smaller than V_c. Our interpretation would require a very strong transfer term.

In conclusion we like to stress that the model interpretation developed here may be helpful to understand the excitations of molecular onedimensional systems. The coexistence of a localized and a partially delocalized state is often observed. It is still an open question if the Y state in α-perylene is well interpreted as the partially delocalized state, the F state, in the nomenclature of TOYOZAWA [4]. However, a detailed analysis of all electronic states within the unit cell is needed to arrive at a more quantitative picture.

Acknowledgements: The author wishes to thank Dipl. Phys. Scherer for discussions and detailed investigations related to PIC. Also the help of Dr. E. W. Knapp in the simulation of the α-perylene system has added to this paper.

References

1. A. S. Davydov, Theory of molecular excitons (Plenum Press, N. Y. 1971)
2. E. G. McRae, Austr. J. Chem. 14,329 (1961)
3. R. E. Merrifield, J. Chem. Phys. 40,445 (1964)
4. Y. Toyozawa, Prog. Theor. Phys. 26,29 (1961)
5. S. F. Fischer and S. A. Rice, J. Chem. Phys. 52,2089 (1970)
6. M. Grover and R. Silbey, J. Chem. Phys. 52,2099 (1970)
7. G. Scheibe, Angew. Chem. 50,212 (1937)
8. B. Kopainsky, J. K Hallermeier, and W. Kaiser, Chem. Phys. Letters 87,7 (1982)
9. E. von Freydorf, J. Kinder, and M. E. Michel-Beyerle, Chem. Phys. 27,199 (1978)
   M. B. Cohen, R. Haberkorn, E. Huber, Z. Ludmer, M. E. Michel-Beyerle, D. Rabinovich, R. Sharon, A. Warshel, and V. Yakhot, Chem. Phys. 27,211 (1978)
10. J. Tanaka, Bull. Chem. Soc. Japan 36,1237 (1963)
11. R. M. Hochstrasser and C. A. Nyi, J. Chem. Phys. 72,2591 (1980)
12. H. Auweter, D. Ramer, B. Kunde, and H. C. Wolf, Chem. Phys. Letters 85, 325 (1982)
13. B. Kopainsky, J. K. Hallermeier, and W. Kaiser, Chem. Phys. Letters 83,498 (1981)
14. P. Scherer, unpublished results
15. T. J. Kosic, C. L. Schosser, and D. Dlott, Chem. Phys. Letters 96,57 (1983)

Photo-Induced Electron or Excitation Transfer Enhanced During Vibrational Relaxation and Generalized Förster's Formula

H. Sumi

Institute of Materials Science, University of Tsukuba
Sakura-mura, Ibaraki 305, Japan

Electron or excitation transfer can greatly be enhanced during vibrational relaxation (VR) after optical excitation of a donor. When it occurs after VR with a thermal activation energy E_A, enhanced transfer during VR is also governed by an activation energy E_H which, however, is smaller than E_A or even vanishes. After optical excitation the phonon system evolves drawing a trajectory on a potential energy surface and E_H is determined by the minimum-energy trajectory contributing to transfer.

Much interest has recently been aroused in dynamic behaviors of a localized electron-phonon system occurring during vibrational relaxation (VR) after a pulse excitation. To detect them optically, we must observe the hot luminescence emitted during VR, which is different from the ordinary one emitted after VR [1]. In this case, however, it is necessary to use a pulse width Δt much smaller than the time τ_V for VR which is usually of the order of 1 psec. Then, one would need femtosecond laser techniques, now being developed [2], which might, however, be premature for application. Under steady excitation, the intensity ratio of hot to ordinary luminescence is very small, being of the order of τ_V/τ_0, where τ_0 is the radiative lifetime which is usually of the order of nanoseconds. On the other hand, photo-induced electron or excitation transfer from a donor to an acceptor can be a sensitive probe of the dynamic behaviors of the donor, since its rate can greatly be enhanced only during VR [3]. The enhanced transfer has been called the hot transfer [4].

Excitation transfer occurring after VR is called the ordinary transfer, and its rate is given by the well-known Förster formula

$$k_0 = (2\pi/\hbar)\, J^2 \int dE'\, D(E')A(E') \quad , \tag{1}$$

where J represents the transfer integral, D(E') the spectrum of ordinary luminescence from the donor, and A(E') the absorption spectrum of the acceptor, with normalization $\int D(E')dE' = \int A(E')dE' = 1$. In the case of electron transfer, D(E') represents the energy spectrum of the outer sphere electron of the donor, while A(E') is the energy spectrum of the unoccupied level of the acceptor to which the electron is to be transferred. In the following, terminologies appropriate to excitation transfer are used for simplicity. Hot transfer during VR, on the other hand, is described by the generalized Förster formula

$$k(t;E) = (2\pi/\hbar)\, J^2 \int dE'\, D(E';t;E)A(E') \quad , \tag{2}$$

where D(E';t;E) with normalization $\int D(E';t;E)dE' = 1$ represents the spectrum of hot luminescence from the donor, which depends on the time t elapsed after pulse excitation and the excitation energy E. (Here, t and E are regarded as having uncertainties satisfying the Heisenberg's relation). Thus, with the hot-transfer rate k(t;E) of (2), we detect the hot luminescence from the donor through a window of the absorption spectrum A(E') of the acceptor. When t is much larger than the time τ_v for VR, $D(E';t;E)$ approaches the ordinary luminescence spectrum $D(E')$, and hence $k(t;E)$ approaches the ordinary transfer rate k_0 of (1). The hot-luminescence spectrum $D(E';t;E)$ changes with time t. Therefore, if its overlap with $A(E')$ in (2) is greater at some t during VR than the ordinary luminescence spectrum does in (1), the hot-transfer rate $k(t;E)$ becomes much larger than the ordinary transfer rate k_0.

Usually, however, the pulse width Δt of the exciting photon is much larger than the time τ_v for VR. Therefore, hot transfer has already been terminated when observation starts. Then, observable by time-resolved experiments are only the ordinary transfer rate k_0 and the luminescence-decay time τ_F. However, if the quantum yield of hot transfer Y is nonvanishing, it reduces the quantum yield of luminescence Y_F from the value τ_F/τ_0 obtained without consideration of hot transfer to $Y_F = (1-Y)\tau_F/\tau_0$. On the other hand, it increases the quantum yield of excitation transfer Y_R from the value $k_0\tau_F$ obtained without consideration of hot transfer to $Y_R = Y+(1-Y)k_0\tau_F$. Each of these two equations enables us to determine Y by measuring (Y_F, τ_F, τ_0) or (Y_R, k_0, τ_F). The usual analysis, however, does not recognize the presence of Y, and assumes *a priori* $Y_F = \tau_F/\tau_0$ and $Y_R = k_0\tau_F$.

In the following, let us investigate a case that the ordinary transfer rate k_0 is governed by a thermal activation energy E_A. Such a case has often been found in solid-state physics, chemistry and biophysics. In this case it can be shown that the quantum yield of hot transfer Y is also governed by an activation energy E_H which, however, is smaller than E_A or even vanishes [3]. The thermally activated temperature dependence of k_0 is obtained when both the donor spectrum $D(E')$ and the acceptor one $A(E')$ have a Gaussian shape as

$$D(E') \propto \exp[-\tfrac{1}{2}(E'-E_d)^2/\Delta_d^2], \quad \text{and} \quad A(E') \propto \exp[-\tfrac{1}{2}(E'-E_a)^2/\Delta_a^2],$$

where $D(E')$ has a peak at energy E_d with width Δ_d, while $A(E')$ has a peak at energy E_a with width Δ_a. Both Δ_d and Δ_a are regarded as much larger than the average phonon energy and proportional to the square root of the temperature T. Optical processes in this case can be understood in the adiabatic potential energy curve shown in Fig.1a along the configuration coordinate Q_d of the donor. When optical excitation occurs at the vertical transition with energy E shown therein, hot luminescence is emitted during VR which is represented also therein by a dotted trace. Its spectrum has also a Gaussian shape [1] as

$$D(E';t;E) \propto \exp[-\tfrac{1}{2}(E'-E_d(t;E))^2/\Delta_d(t)^2] \,, \tag{3}$$

where the time-dependent peak energy and width are respectively given by

$$E_d(t;E) = [1-r(t)]E_d + r(t)E, \quad \text{and} \quad \Delta_d(t) = [1-r(t)^2]^{\frac{1}{2}}\,\Delta_d \,,$$

whose time dependences are determined by the relaxation function

$$r(t) = \sum_j V_j^2\cos(\omega_j t)/\sum_j V_j^2 \,.$$

Now, r(t) shows a damped oscillation due to dephasing among normal-mode components contributing to r(t) with different frequencies ω_j and coupling constants V_j, and it is directly proportional to the damped oscillation of Q_d after optical excitation shown in Fig. 1a by the dotted trace. Especially, at the instant of excitation at t=0, r(t) equals unity, and hence $E_d(t;E)$ equals the excitation energy E and $\Delta_d(t)$ vanishes, since the emission occurs immediately after optical excitation without VR. After VR for $t \gg \tau_v$, on the other hand, r(t) dampens to zero, and hence $E_d(t;E)$ and $\Delta_d(t)$ approach respectively the corresponding values E_d and Δ_d of ordinary luminescence from the donor, emitted around the relaxed configuration at $Q_d \simeq 0$. During VR for $t \lesssim \tau_v$, however, r(t) sweeps approximately from about -1 to 1 during its damped oscillation, and hence $E_d(t;E)$ oscillates too with magnitude given by a vertical energy difference during VR in Fig.1a. The total energy of phonons liberated

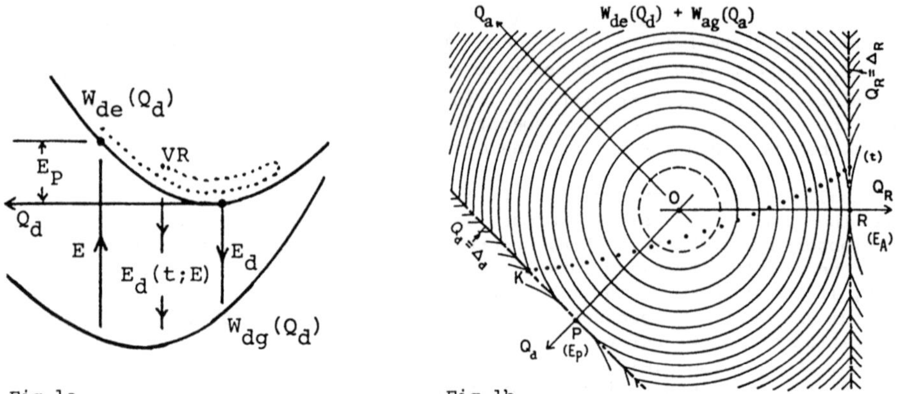

Fig.1a   Fig.1b

during VR is given by E_P in Fig.1a, and it drives the hot transfer, as will be shown below. Therefore, E_P is called the phonon-kick energy hereafter.

Inserting the hot-luminescence spectrum (3) into the generalized Förster formula (2), we see that the hot-transfer rate $k(t;E)$ is governed by a time-dependent activation energy $E_A(t;E)$ as,

$$k(t;E) \propto \exp[-E_A(t;E)/k_BT], \text{ with } E_A(t;E) = [\sqrt{E_A}+gr(t)\sqrt{E_P}]^2/[1-g^2r(t)^2], \quad (4)$$

where the absolute value of g is given by $|g| = \Delta_d/(\Delta_d{}^2+\Delta_a{}^2)^{1/2} < 1$. The physical meaning of g will be explained later. The minimum value, E_H, of $E_A(t;E)$ can be obtained by changing time t, that is, changing $r(t)$ from -1 to 1, as

$$E_H = E_A - E_P \text{ for } E_P < g^2E_A, \quad = [\sqrt{E_A}-|g|\sqrt{E_P}]^2/(1-g^2) \text{ for } g^2E_A < E_P < E_A/g^2,$$
$$\text{and } = 0 \text{ for } E_P > E_A/g^2. \quad (5)$$

The quantum yield Y of hot transfer is given by the integral of $k(t;E)$ during the time τ_v for VR. Therefore, E_H obtained here should give the activation energy of Y as $Y \propto \exp(-E_H/k_BT)$. Note that E_H of (5) decreases, starting from the activation energy E_A of ordinary transfer, with increasing the phonon-kick energy E_P. Especially in the third region of (5) where E_P exceeds E_A/g^2, E_H vanishes and the hot transfer occurs without thermal activation. In this third region, the peak energy $E_d(t;E)$ of hot luminescence from the donor crosses the absorption-peak energy E_a of the acceptor during its damped oscillation, while in the first and the second region, $E_d(t;E)$ does not reach E_a.

Let us interpret these results in terms of the configuration coordinates Q_d and Q_a of phonons describing respectively optical properties of the donor and the acceptor. Adiabatic potentials associated with the ground and the excited states are written respectively as $W_{dg}(Q_d)$ and $W_{de}(Q_d)$ for the donor, and as $W_{ag}(Q_a)$ and $W_{ae}(Q_a)$ for the acceptor. The optical excitation of the donor occurs at a configuration where $W_{de}-W_{dg} = E$, and this configuration is written as $Q_d = \Delta_d$. After optical excitation, the adiabatic potential is given by $W_{de}+W_{ag}$, and it changes to $W_{dg}+W_{ae}$ after the excitation has transferred. Then, the excitation transfer occurs at a configuration where $W_{de}+W_{ag} = W_{dg}+W_{ae}$, and this configuration is written as $Q_R = \Delta_R$ with the use of a reaction coordinate Q_R which is given by a linear combination of Q_d and Q_a in a harmonic-potential model [3]. Energy contours shown in Fig.1b represent the potential energy surface $W_{de}+W_{ag}$ obtained after optical excitation of the donor. Figure 1a corresponds to the intersection of Fig.1b cut along the axis of the configuration coordinate Q_d of the donor. The optical excitation occurs at a straight line $Q_d = \Delta_d$ shown therein. This line is parallel to the axis of the configuration coordinate Q_a of the acceptor also shown therein.

After optical excitation of the donor, the phonon system relaxes at the origin O where the potential energy becomes minimum with value 0. The excitation transfer occurs at another straight line $Q_R = \Delta_R$. The minimum potential energy on the first line, obtained at point P, gives the phonon-kick energy E_P, while that on the second line, obtained at point R, gives the activation energy E_A of ordinary transfer. Associated with VR from P to O, the phonon kick occurs in a direction $\overrightarrow{PO}$, while distortions effective to excitation transfer must occur in a direction $\overrightarrow{OR}$. We can show as in Ref.3 that g appearing in (4) and (5) gives just the direction cosine between these vectors $\overrightarrow{PO}$ and $\overrightarrow{OR}$. Therefore, g represents the effectiveness of the phonon kick with respect to direction. At point P, phonons are excited only at the donor.

Now, let us set a starting point of the phonon system at point K, different in general from P, on the line of $Q_d = \Delta_d$. The dotted trace in Fig.1b shows an example of trajectories of the phonon system, determined under the influence of the potential energy surface, starting from point K. In this example, the phonon system has a nonvanishing initial velocity at point K. If these trajectories successfully reach the goal line of $Q_R = \Delta_R$, the excitation transfer occurs subsequently to optical excitation of the donor. The time when they reach the goal is denoted by t. When the time t is fixed, the minimum energy of the phonon system obtained by changing both the point K and the initial velocity thereat is written as $E_P + E_A(t)$, using a positive energy $E_A(t)$. The phonon-kick energy E_P was supplied by optical excitation of the donor to the phonon system. Therefore, this result shows that in order for hot transfer to occur at time t, the phonon system must have had at least an energy $E_A(t)$ already before optical excitation of the donor. Thus, $E_A(t)$ should give the activation energy of the hot-transfer rate k(t;E) obtained at time t. When the potential energy surface is quadratic in Q_d and Q_a, we can easily calculate $E_A(t)$, and $E_A(t)$ thus obtained agrees just with $E_A^a(t;E)$ determined before in (4) with the use of the generalized Förster formula. The minimim value E_H of $E_A(t)$ is meaningful for hot transfer, since it is not necessary to specify t.

When the exciting photon energy E is changed, the phonon-kick energy E_P is also changed as $E_P \propto (E-E_d)^2$. Then, at a certain value E_t of E, E_P should exceed E_A/g^2 and the activation energy E_H of hot transfer vanishes. Then, the quantum yield of hot transfer Y and hence the reaction yield Y_R should rise around E_t with increasing E. The excitation-energy dependence of the reaction yield as noted here has often been found in various fields. Then, the present work predicts that Y and Y_R in the tail part of their E dependence for $E \lesssim E_t$ should be governed by an activation energy E_H smaller than its thermal-equilibrium value E_A. The reduction of the activation energy under the presence of nonthermal-equilibrium phonons has in fact been observed in recombination-enhanced defect reactions in semiconductors, where electron-hole nonradiative recombination at a deep-level defect emits many phonons which accelerate various defect reactions such as movement or production of a defect [5,6]. In this case, the phonon-kick energy E_P is determined by the energy of the deep level in the forbidden gap. The hot transfer mentioned in the present work is an example of more general phenomena of chemical reactions enhanced by nonthermal-equilibrium phonons in a very short time during their relaxation.

References

1 Y. Toyozawa: J. Phys. Soc. Jpn. 41, 400 (1976); and
Y. Toyozawa, A. Kotani and A. Sumi: *ibid.* 42, 1495 (1977).
2 *Picosecond Phenomena III*, ed. K.B. Eisenthal *et. al.* (Springer, 1982).
3 H. Sumi: Phys. Rev. Lett. 50, 1709 (1983).
4 I. Tehver and V. Hizhnyakov: Sov. Phys. - JETP 42, 305 (1975).
5 L.C. Kimerling: Solid State Electron. 21, 1391 (1978).
6 H. Sumi: Physica B+C 117&118, 197 (1983) and Phys. Rev. B (submitted, 1983).

Molecular Aggregates in Liquids Resolved by a Novel Raman Spectroscopy

W. Zinth and W. Kaiser

Physik Department der Technischen Universität München
D-8000 München, Fed. Rep. of Germany

1 Introduction

Raman spectroscopists are frequently faced with the problem of broad bands which are made up of closely spaced overlapping transitions. In some cases, the broad band might even result from a continuous distribution of frequencies. These inhomogeneously broadened Raman bands give little spectroscopic information.

In recent years, time-resolved picosecond techniques have been devised which allow one to measure directly two important dynamic properties of a vibrational transition: The dephasing time T_2 and the population lifetime T_1 /1,2/. More recently we have shown that an extension of the coherent probing technique allows us to observe individual Raman transitions in highly congested spectral regions, where spontaneous Raman techniques show featureless broad inhomogeneous bands /3/. In this paper we apply the new technique to the hydrogen-bonded system of pyridine and methanol. We find clear evidence of distinct H-bonded species which are not seen in the spontaneous Raman spectrum.

2 Theory

The molecular transitions investigated here are treated as two-level systems with resonance frequencies ω. Using a density matrix formalism we may describe the ensemble of molecular transitions in the limit of small excitation by one macroscopic quantity, the collective amplitude $q(\omega,t)$ /1,4/. The inhomogeneous broadening of the transition is introduced in our model by a distribution $g(\omega)$ of the transition frequencies. For simplicity, we assume the inhomogeneous distribution to be static for the duration of the excitation and probing process of approximately 10^{-11} s. The dephasing time T_2 is taken to be constant over the inhomogeneous distribution. Under these conditions the spontaneous Raman band (for a polarized Raman transition) results from the convolution of an inhomogeneous distribution $g(\omega)$ and a Lorentzian function due to a dephasing time T_2.

As an example we present in Fig. 1 the spontaneous Raman band of a simple situation. The inhomogeneous distribution $g(\omega)$ consists of two equal transitions separated by $\delta\omega = 1/T_2$. The homogeneous broadening of both transitions (dephasing process with

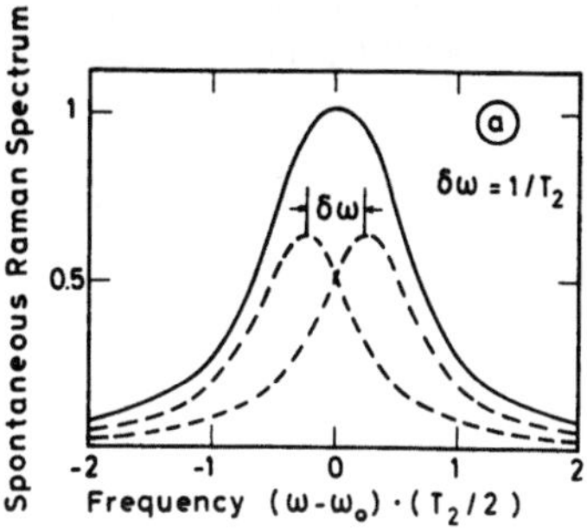

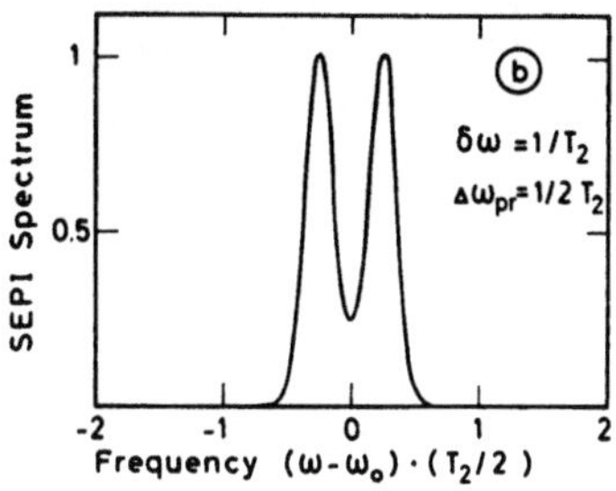

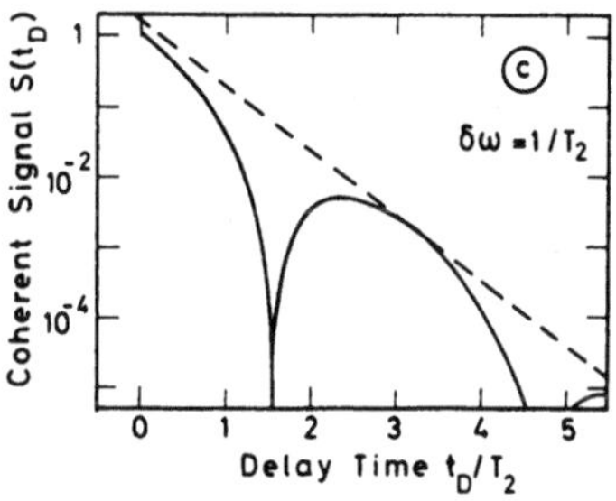

Fig. 1
(a) Calculated spontaneous Raman spectrum for a pair of homogeneously broadened lines separated by $\delta\omega$

(b) Calculated SEPI spectrum for the same molecular system and a width of the Gaussian shaped probing pulse $\Delta\omega_{pr} = 1/2T_2$

(c) Coherent scattering signal $S(t_D)$ calculated using δ-function shaped exciting and probing pulses

time constant T_2) leads to a broad spontaneous Raman spectrum with a single band of width $\Delta\omega \simeq 2.7/T_2$.

The excitation of the coherent vibrational amplitude q at a given frequency ω by the driving force $\tilde{A}(t,z) = A(t,z)\cos(\omega_0 t)$ is determined by the equation of motion /1,5/:

$$\frac{\partial^2}{\partial t^2} q + \frac{2}{T_2}\frac{\partial}{\partial t} q + \omega^2 q \propto \tilde{A}(t,z). \tag{1}$$

In a Raman excitation process the driving force is proportional to the product of two light fields $\tilde{A} \propto E_1 E_2^*$. The driving frequency ω_0 equals the frequency difference $\omega_0 = \omega_1 - \omega_2$. A solution of (1) is as follows:

$$q(\omega,t,z) \propto \mathrm{Re}\left\{\frac{ie^{-i\omega t - t/T_2}}{2\omega}\int_{-\infty}^{t} dt' \left[e^{t'/T_2} A(t',z)\, e^{i\Delta\omega t'}\right]\right\}. \tag{2}$$

$\Delta\omega$ is the frequency mismatch with the driving force $\Delta\omega = \omega - \omega_0$. The essential features of the coherent amplitude may be deduced from (2): The coherent amplitude rises with the driving force and decays exponentially with the time constant T_2. The molecular excitation q in the sample may be monitored by a delayed coherent Raman scattering process. A probing laser field $\tilde{E}_{L2}(t-t_D) = E_{L2}(t-t_D,z)\cos(\omega_{L2}t)$ interacts with the coherently vibrating molecules. Maxwell's equations containing a nonlinear polarisation $P^{NL} \propto q$ give (3) which allows to calculate the scattered coherent Stokes field E_{S2} /1/:

$$\frac{\partial E_{S2}}{\partial z} \propto \int d\omega \; g(\omega) \; \tilde{E}_{L2}(t-t_D, z) \; q(\omega, t, z) \; . \tag{3}$$

According to (3) each frequency component of the coherent excitation oscillating at ω produces a frequency component $\omega_{S2} = \omega_{L2} - \omega$ of the scattered light. In general, the coherently scattered light is detected at the end of the sample of length L. In the following we assume that the material excitation is constant over the entire length of the sample.

Now we discuss a probe-scattering experiment where the frequency spectrum of the coherently scattered signal is detected /6/. The excitation is made by a short driving force in order to excite most of the inhomogeneous distribution $g(\omega)$ (see (2)). A Gaussian-shaped pulse of long duration ($t_p \gtrsim T_2$) probes the coherent material excitation at a late delay time after the excitation process. The spectrum of the scattered light is detected by a spectroscopic device of high resolution. We call this technique a short excitation and prolonged interrogation (SEPI) spectroscopy /3/. Using a δ-function shaped driving force, $A(t) \propto \delta(t)$, we obtain at late delay times t_D the spectral intensity $S(\Omega)$:

$$S(\Omega) = |E_{S2}(\Omega)|^2 = \left| \int_{-\infty}^{+\infty} dt \; E_{S2}(t) \; e^{i\Omega t} \right|^2 \tag{4}$$

$$E_{S2}(\Omega) \propto \int_{-\infty}^{+\infty} d\omega \; g(\omega) \; \underline{\int_{-\infty}^{+\infty} dt \; e^{i(\Omega - \omega_{L2} + \omega)t} \; E_{L2}(t) \; e^{-t/T_2}} \; . \tag{5}$$

The underlined expression in (5) represents the Fourier transform of the product, probing field E_{L2} times exponential decay $\exp(-t/T_2)$. With a Gaussian-shaped probing field $E_{L2}(t)$ the Fourier transformation leads to a Gaussian frequency dependence, independent of the dephasing time T_2. The width of this Gaussian function is determined by the duration of the probing pulse. When the Raman transition consists of a single frequency component at $\tilde{\omega}$, the measured coherent spectrum shows a Gaussian-shaped peak at the frequency $\omega_{L2} - \tilde{\omega}$. The width of this spectrum can be narrowed by using longer probing pulses. The effect of the dephasing time on the Raman line is removed in a SEPI experiment /3,6/.

Figure 1b shows the calculated results of a SEPI experiment for a distribution $g(\omega)$ consisting of two transitions (Fig. 1a). The duration of the probing pulse was $t_p = 5.6\; T_2$, which narrows each spectral component by a factor of four. Under these conditions the two components of the broad spontaneous spectrum in Fig. 1a are clearly resolved. The SEPI spectrum gives the resonance frequencies directly without further manipulation.

The application of the SEPI technique to a general inhomogeneity $g(\omega)$ shows the following result. The measured coherent spectrum reproduces the convolution of the inhomogeneous distribution with the spectral shape of the probing laser pulse (independent of the homogeneous width). The application of the SEPI technique for the study of inhomogeneously broadened lines seems promising for the case of moderate inhomogeneities, where the spontaneous Raman spectra are strongly influenced by the dephasing process. In this case, the SEPI technique gives new information on the distribution function $g(\omega)$.

Until recently, most coherent probe scattering experiments of inhomogeneously broadened transitions have been performed in the time domain /1,2/: The time dependence of the coherently scattered pulse energy is detected without considering the frequency spectrum. The quantity measured is the time-integrated coherent signal $S(t_D)$. The time dependence of $S(t_D)$ is obtained by a number of measurements with various delay times t_D:

$$S(t_D) \propto \int_{-\infty}^{+\infty} dt\, |E_{S2}(t,t_D,L)|^2 \propto$$

$$\propto \int_{-\infty}^{+\infty} dt \left| \int_{-\infty}^{+\infty} d\omega\; g(\omega)\; \tilde{E}_{L2}(t-t_D,z)\; q(\omega,t,z) \right|^2 \tag{6}$$

$$S(t_D) \propto \int_{-\infty}^{+\infty} dt\, |E_{L2}(t-t_D)|^2 \times |Q_{tot}(t)|^2 . \tag{7}$$

The coherent signal is the convolution of the probing intensity and the absolute square of the frequency integrated coherent excitation $Q_{tot}(t,z) = \int d\omega\; g(\omega)\; q(\omega,t,z)$. Optimum information on the material excitation is obtained when the probing pulses and the exciting force are very short. In this limit the coherent signal reproduces the absolute square of the total coherent excitation $S(t_D) \propto |Q_{tot}(t_D)|^2$.

For a short driving force Q_{tot} becomes the Fourier transform of the spontaneous Raman spectrum.

$$Q_{tot}(t) \propto \int_{-\infty}^{+\infty} d\omega \int_{-\infty}^{+\infty} d\Omega\; g(\Omega) \frac{1}{(\Omega-\omega)^2+(1/T_2)^2}\, e^{i\omega t} . \tag{8}$$

In Fig. 1c we have calculated the coherent signal for the pair of transitions shown in Fig. 1a. We used δ-function shaped driving and probing pulses. The envelope of the signal decays with the dephasing time T_2 whereas the interference between the two modes leads to the beating process allowing to deduce the frequency difference $\delta\omega$. In this especially simple case of two closely spaced transitions the information on $g(\omega)$ is more easily obtained from $S(t_D)$ than from the spontaneous Raman spectrum in Fig. 1a.

The time-dependent data of the coherent probe-scattering experiments allow us to measure the absolute square of the Fourier transform of the spontaneous Raman spectrum. On the other hand, the time and frequency resolved SEPI technique allows us to gain new information in congested spectral regions where the spontaneous and time-dependent coherent techniques fail.

3 Experimental

The short dephasing times in molecular liquids require picosecond pulses for the coherent Raman measurements discussed here.
Figure 2 shows a schematic of part of our experimental system to measure SEPI spectra. At the top, three pulses from a mode-locked Nd:glass laser system enter the figure. These pulses of frequency ν_1 = 18,990 cm^{-1} originate from one picosecond pulse using three beam splitters. The pulse in beam 1 passes through the polarizer P1 and the sample, but is blocked by polarizer P2 in its straight path. In the center beam 2 of Fig. 2 a new frequency ν_2 is produced in the generator by a stimulated Raman process. Changing the medium of the generator one readily obtains pulses of different frequencies ν_2. These pulses are blocked by the polarizer P2 and are spectrally monitored by spectrograph SP2. On account of the transient generation process the pulses of frequency ν_2 are shorter in duration than the incident pulses ν_1 by a factor of approximately three /5/. The two pulses of the beams 1 and 2 simultaneously enter the sample, coherently exciting molecular vibrations via transient stimulated Raman scattering at the frequency

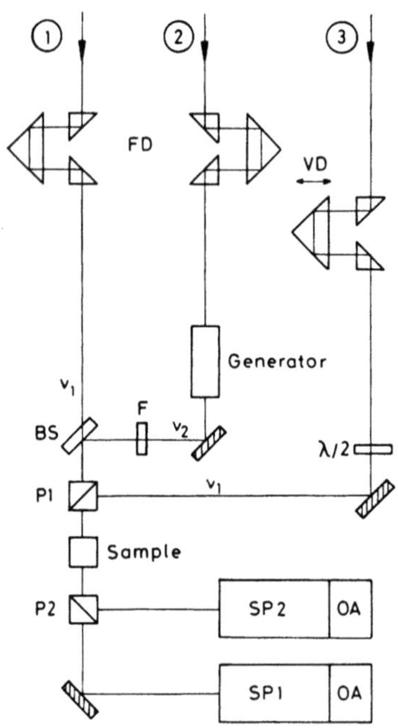

Fig. 2. Schematic of the experimental system used for the study of SEPI spectra of liquids. Beam splitters BS, variable and fixed delay VD and FD, polarizers P1 and P2, blocking filter F, spectragraphs SP1 and SP2, used in conjunction with optical spectrum analysers OA

$\nu_0 = \nu_1 - \nu_2$. In the optical path, 3, a delayed pulse with polarization perpendicular to the pulses of 1 and 2 is produced. This third pulse interacts with the coherently excited molecules of the sample producing a Raman-shifted signal pulse. For Stokes scattering the three pulses travel collinearly through the sample. The spectrum of the coherently scattered light is studied by a 2 m spectrograph SP1 and a cooled optical spectrum analyser OA. The experimental system has a resolution of 0.3 cm^{-1} per channel.

4 Experimental Results

We have performed SEPI measurments using a number of organic and inorganic liquids. First we present data on CCl_4 with a Raman shift of 460 cm^{-1}. Second, we show - for the first time - SEPI spectra of a hydrogen-bonded system. In pyridine-methanol mixtures new Raman lines are observed at Raman frequencies close to 990 cm^{-1}.

a. Carbon Tetrachloride

The natural abundance of chlorine, $^{35}Cl:^{37}Cl$ = 75:25, leads to different isotope species of CCl_4. As a result, the symmetric vibrational mode $\nu_1(a_1)$ at 460 cm^{-1} consists of several components.

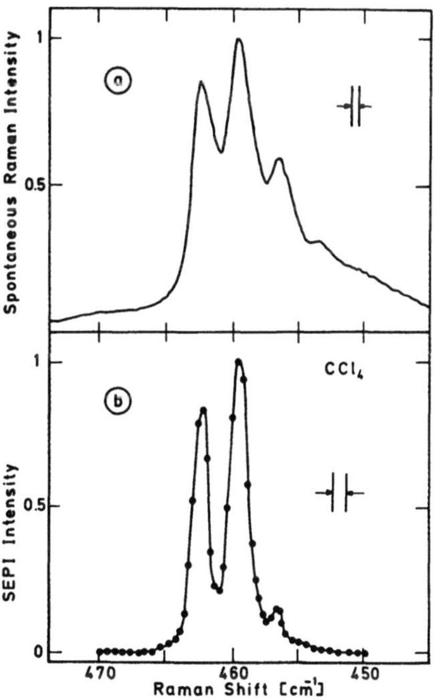

Fig. 3 Raman spectra of the ν_1 mode of liquid CCl_4. (a) Polarized spontaneous Raman spectrum, instrumental resolution 0.5 cm^{-1}. (b) SEPI spectrum of CCl_4. Instrumental resolution 1 cm^{-1}

The spontaneous Raman spectrum of Fig. 3a indeed shows four peaks the intensity of which is in good agreement with the distribution of the four major molecular species. The different Raman lines in Fig. 3 overlap strongly in spite of the high resolution of the Raman spectrometer of 0.5 cm^{-1}. The individual Raman lines are broadened by the dephasing time T_2 = 6.0 ps which gives rise to the observed width $\Delta\nu_{spont} \simeq 1/\pi c T_2$ = 1.8 cm^{-1}.

The SEPI spectrum of the same vibrational mode of CCl_4 is depicted in Fig. 3b. In this case the sample was excited for approximately 7 ps by a laser pulse at ν_1 together with a Stokes shifted pulse at ν_2 with $\nu_1 - \nu_2 \simeq 458$ cm^{-1}. Under these conditions the two neighboring major molecular components are strongly excited and give rise to the observed scattering spectrum. The data of Fig. 3b were obtained with long probing pulses of t_p = 20 ps and at a late delay time of 40 ps. We point to the sharp lines in the SEPI spectrum which are narrower than the line widths $\Delta\nu_{spont}$ in the spontaneous spectrum of Fig. 3a. The linewidths depicted in Fig. 3b are determined by the limited spectral resolution of the spectrometer. Averaging over a number of laser shots we can determine the interval between the two isotopic components to be 3.0±0.08 cm^{-1}.

b. Pyridine:Methanol

As another example for the short excitation and prolonged interrogation (SEPI) technique we present Raman data of the system pyridine:methanol in the range of the ν_1 mode of pyridine close to 990 cm^{-1}. This system has recently been investigated via infrared absorption measurements and by an advanced spontaneous Raman technique /7,8,9/. The following interesting features have been reported: (i) The ν_1 mode of pyridine shifts slightly upon dilution. (ii) A hydrogen-bonded complex is formed, shifted from the ν_1 mode by approximately 7 cm^{-1} at a concentration of pyridine in methanol of χ=0.5 mole fraction. This complex shifts with concentration by 3 cm^{-1} to higher energies. Simultaneously, the band width of the hydrogen complex changes.

Figure 4 shows the results for pure pyridine. The spontaneous Raman spectrum (Fig. 4a) demonstrates one strong line close to 989 cm^{-1} /8/. Using the SEPI technique we obtain a single line at the same frequency. The width of the SEPI line is 1.5 cm^{-1}, considerably narrower than the width of the spontaneous Raman line. There is no indication of an additional Raman transition between 990 cm^{-1} and 1000 cm^{-1}.

The experimental results for the pyridine-methanol mixtures are shown in Fig. 5. The spontaneous Raman spectrum (Fig. 5a) represents the pyridine line at 989 cm^{-1} and a new, broader band close to 997 cm^{-1} /8/. The concentration for the spontaneous Raman spectrum was χ=0.5.

In Fig. 5b we present four SEPI spectra taken at a concentration of χ=0.3. The spectrum at the r.h.s. is the line corresponding to the pyridine molecules not involved in strong interaction with the methanol molecules. Averaging over a number of experiments we find that this line is slightly shifted by 0.3 cm^{-1}

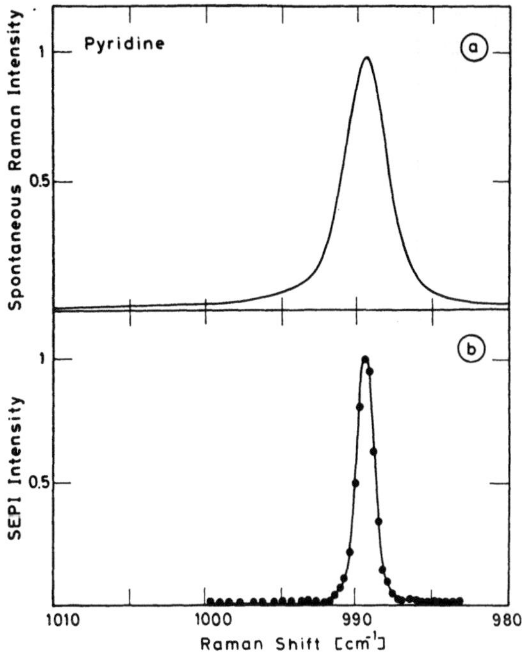

Fig. 4 Raman spectra of the ν_1 mode of liquid pyridine. (a) Polarized spontaneous Raman spectrum, (b) SEPI spectrum (short excitation and prolonged interrogation of the coherent excitation)

to smaller Raman shifts. This finding is in full agreement with the data from spontaneous Raman measurements. The three additional SEPI spectra result from H-bonded aggregates. Under opitmal conditions we were able to record all three lines in a single laser shot. The transition frequencies located at the high-frequency side of the ν_1 mode of pyridine are shifted by 3.6 cm^{-1}, 6.3 cm^{-1}, and 8.8 cm^{-1}. The position of the SEPI lines are marked by arrows in Fig. 5a. Varying the concentration of pyridine in the sample between $\chi=0.1$ and $\chi=0.5$ the positions of the spectral lines remain fixed within 0.5 cm^{-1}. At a concentration of $\chi=0.1$ we could record only the two lines at 8.8 cm^{-1} and 6.3 cm^{-1}.

In Fig. 5b the four SEPI bands are normalized for clarity. It should be noted that in the present system the intensity of the bands gives only qualitative information on the concentration of the individual species.

The SEPI spectra of Fig. 5b quite convincingly show that three distinct species appear when methanol is added to pyridine. The band shifted by 3.6 cm^{-1} from the pyridine line is always small, suggesting a small concentration of this complex. The varying intensity of the two other species with concentration is responsible for the previously reported shift of the broad spontaneous Raman band at approximately 998 cm^{-1}.

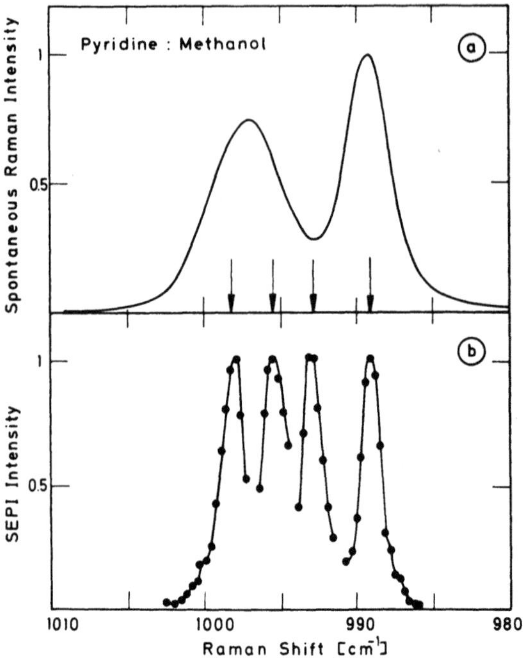

Fig. 5 Raman spectra of the solution pyridine:methanol. (a) Polarized spontaneous Raman spectrum. The line at 989 cm^{-1} is due to the free pyridine molecules; the line at 997 cm^{-1} is caused by H-bonded complexes. The arrows mark the average frequency positions found in SEPI spectroscopy. (b) SEPI spectra showing four transitions. They were obtained by four exciting laser shots of different frequencies. The broad spontaneous Raman band at 997 cm^{-1} is made up of the three high-frequency components

The present data do not yet allow a final assignment of the observed Raman lines to specific aggregates. Several possibilities have to be considered: (i) The H-bonding of pyridine to one methanol molecule and to aggregates containing methanol, and (ii) the existence of one or two hydrogen bridges to the nitrogen atom of pyridine.

5 Conclusion

We have presented new results on coherent Raman probe scattering of inhomogeneously broadened transitions. The data of this article convincingly show the usefulness of the short excitation and prolonged interrogation spectroscopy for the study of molecular transitions in pure and diluted liquids, especially for the case of hydrogen-bonded systems.

Acknowledgement:

The authors wish to thank Thomas Robl for valuable experimental assistance.

References

1 A. Laubereau, W. Kaiser: Rev. Mod. Phys. 50, 607 (1978)
A. Penzkofer, A. Laubereau, W. Kaiser: Prog. Quant. Electron. 6, 55 (1979)
2 W. Zinth, H.J. Polland, A. Laubereau, W. Kaiser: Appl. Phys. B26, 77 (1981)
3 W. Zinth, M.C. Nuss, W. Kaiser: Chem. Phys. Lett. 88, 257 (1982)
4 J.A. Giordmaine, W. Kaiser: Phys. Rev. 144, 676 (1966)
5 R.L. Carman, F. Shimizu, C.S. Wang, N. Bloembergen, Phys. Rev. A2, 60 (1970)
S.A. Akhmanov, K.N. Brabovich, A.P. Sukhorukov, A.S. Chirikin: Sov. Phys. JETP 32, 266 (1971)
6 W. Zinth: Optics Commun. 34, 479 (1980)
W. Zinth, M.C. Nuss, W. Kaiser: Optics Commun. 44, 262 (1983)
7 H. Takahashi, K. Mamola, E.K. Plyler: J. Mol. Spectrosc. 21, 217 (1966)
B.P. Asthana, H. Takahashi, W. Kiefer: Chem. Phys. Lett. 94, 41 (1983)
H. Takahashi, B.P. Asthana, W. Häfner, W. Kiefer: J. Raman Spectrosc. 14, 102 (1983)

Part V

Excimers, Charge Transfer Excitons, and Exciton Fission

Charge Transfer Spectra of Aromatic Hydrocarbon Crystals*

Willem Siebrand and Marek Z. Zgierski

Division of Chemistry, National Research Council of Canada
Ottawa, Ontario, Canada K1A 0R6

1. Introduction

Charge-transfer (CT) spectra are commonplace in the class of molecular crystals known as CT crystals [1]. They refer to electronic transitions in which an electron is transferred between two different molecules, one behaving as the donor, the other as the acceptor. Although there is no obvious reason why the donor and acceptor molecules must be different, CT spectra in molecular crystals built from one type of molecule have long escaped detection. Presumably they are masked by the more intense spectra associated with intramolecular transitions ("Frenkel excitons"). Their prominence in CT crystals, one might argue, is simply the result of their being the lowest-energy absorption bands in the crystal, and hence not subject to masking.

In the absence of direct spectroscopic information on CT states in one-component molecular crystals, our understanding of these states was based on indirect experimental evidence and on theoretical estimates. Much of the early work concerned anthracene. It was demonstrated [2,3] that pure crystals of this material are intrinsically photoconductive, with charge carriers being produced via the ONSAGER mechanism [4]. This mechanism describes carrier formation as the result of dissociation of CT states by thermal diffusion of the component ions, i.e., of bound electron-hole pairs. CT states were also invoked to explain the mutual annihilation of pairs of triplet excitons in these crystals [5].

The first self-consistent calculation of CT energies in molecular solids was reported by JURGIS and SILINSH [6,7] who realized the importance of induced dipole moments as sources of polarization. Their method was cast in a more efficient form by BOUNDS and her coworkers [8-12] who showed that in anthracene the CT potential is both Coulombic and isotropic, or nearly so [10]. Unfortunately, these calculations relate CT potentials to conduction bands and do not locate the correponding CT states relative to known Frenkel exciton states, unless additional, drastic assumptions are introduced [10]. One possibility is to adopt a one-dimensional model developed by MERRIFIELD [13], in which the kinetic energy of the CT excitons is neglected altogether. This model relates the CT energy levels directly to the energy of the "parent" Frenkel exciton (i.e., the unseparated electron-hole pair involving the same singly-occupied molecular orbitals). In a still older treatment, due to WANNIER [14], the CT excitons are treated as hydrogen-like atoms in a homogeneous dielectric. Both treatments indicate that the CT levels form a hydrogenic series with the conduction band as ionization

*Issued as NRCC No. 21234

limit. These ideas have been used in an attempt to explain the observed wavelength and temperature dependence of the intrinsic photoconductivity of anthracene [10] and pentacene [11].

Great progress was made when SEBASTIAN et al. [15], following earlier work by HANSON on dichloroanthracene [16], observed a number of CT states in tetracene and pentacene. These states show up as absorption bands that shift when an electric field is applied in a manner characteristic of CT transitions. It appears therefore that CT transitions in these crystals are not weak but at least as strong as the vibronic Frenkel exciton bands of the same energy.

In a later paper, SEBASTIAN et al. [17] reported results for anthracene where the CT transitions appear to be much weaker. No direct evidence for CT bands was observed in the zero-field absorption spectrum, but their location could be derived from field-dependent band shifts.

To assign the observed CT bands, SEBASTIAN et al. [15,17] used an ad hoc generalization of Merrifield's model to three dimensions. They associated a CT level with each lattice vector (molecule-molecule separation) and neglected the vibronic structure of the CT states. Associating the lowest-energy CT band with the shortest lattice vector, the next higher band with the next longer vector, etc., they plotted the CT energies against the corresponding inverse vectors and obtained reasonably linear relationships, indicating the applicability of Coulomb's law. The slope and intercept of these plots yielded the average dielectric constant and the photoconductivity band gap, respectively, in good agreement with other estimates.

In this contribution, we report on attempts to reconstruct the observed CT spectra from theoretical models. Among the questions that remain to be answered, we consider the following: (i) what is the reason for the observed intensity distributions; (ii) do the CT bands show vibronic structure; (iii) why are CT transitions so much weaker in anthracene than in tetracene and pentacene; and (iv) can one recognize Davydov components among the observed CT bands?

2. Model

Our model calculations are based on a crystal with two molecules per unit cell. We consider two molecular excited states, one labeled S_1, corresponding to the lowest excited singlet state, and the other labeled S_3 corresponding to a very strong singlet transition ($f \approx 1$). In anthracene S_1 and S_3 correspond to the states ^1L_a and ^1B_b in PLATT's notation [18]. In addition, we consider one molecular vibration, namely the 1430 cm^{-1} CC-stretching mode prominent in the S_1 spectrum. In the crystal, S_1 and S_3 form Frenkel exciton states, each with two Davydov components because of the two molecules per unit cell; moreover, each of the S_1 components exhibits a CC-stretch progression. To this model system, we add a third excited state, namely a CT state corresponding to a positive and negative molecular ion on neighbouring sites. The restriction to a single CT state (which may have two Davydov components and a CC-stretch progression) is introduced for practical reasons; it is consistent, however, with the Merrifield model, as extended by POLLANS and CHOI [19] and BOUNDS et al. [10], which indicates that the lowest CT transition will be much stronger than the others. In the present model this transition will borrow its intensity, and possibly some other properties as well, from the S_1 and S_3 excitons.

This borrowing is described by a system of six coupled equations, corresponding to the three zeroth-order excited states, each with two components. The "exact" excited states, built from these six components, are written as

$$\Psi_j = C_{1j}|A^*B^0\rangle + C_{2j}|A^0B^*\rangle + C_{3j}|A^+B^-\rangle + C_{4j}|A^-B^+\rangle + C_{5j}|A^{**}B^0\rangle + C_{6j}|A^0B^{**}\rangle, \quad (1)$$

where A and B label the two molecules in the unit cell, and the superscripts 0,*,**,+ and - denote the ground state, S_1,S_3, and the positive and negative ion, respectively. The energies of these components (and hence the coefficients C_{ij}) will depend on the CC-stretch coordinate Q:

$$\begin{aligned}
E_{11} &= E^* + \tfrac{1}{2}\omega^2(Q_A^2+Q_B^2) + b^*Q_A \\
E_{22} &= E^* + \tfrac{1}{2}\omega^2(Q_A^2+Q_B^2) + b^*Q_B \\
E_{33} &= E^{\pm} + \tfrac{1}{2}\omega^2(Q_A^2+Q_B^2) + b^+Q_A + b^-Q_B \\
E_{44} &= E^{\pm} + \tfrac{1}{2}\omega^2(Q_A^2+Q_B^2) + b^-Q_A + b^+Q_B \\
E_{55} &= E^{**} + \tfrac{1}{2}\omega^2(Q_A^2+Q_B^2) + b^{**}Q_A \\
E_{66} &= E^{**} + \tfrac{1}{2}\omega^2(Q_A^2+Q_B^2) + b^{**}Q_B, \qquad (2)
\end{aligned}$$

where the displacement parameters b measure the shift in CC-stretch equilibrium position relative to the ground state. On account of the pairing relations for pi-electron orbitals in alternant systems, we set $b^+ = b^- \equiv b$.

These molecular components are subject to intermolecular (off-diagonal) interactions which we represent by

$$\begin{aligned}
&\langle A^*B^0|H|A^0B^*\rangle = V^*; \quad \langle A^{**}B^0|H|A^0B^{**} = V^{**}; \quad \langle A^+B^-|H|A^-B^+\rangle = V^{\pm} \\
&\langle A^+B^-|H|A^*B^0\rangle = \langle A^-B^+|H|A^0B^*\rangle = V_+^* \\
&\langle A^-B^+|H|A^*B^0\rangle = \langle A^+B^-|H|A^0B^*\rangle = V_-^* \\
&\langle A^-B^+|H|A^{**}B^0\rangle = \langle A^+B^-|H|A^0B^{**}\rangle = V_-^{**} \\
&\langle A^+B^-|H|A^{**}B^0\rangle = \langle A^-B^+|H|A^0B^{**}\rangle = V_+^{**} \\
&\langle A^*B^0|H|A^0B^{**}\rangle = \langle A^0B^*|H|A^{**}B^0\rangle = V_*^{**}. \qquad (3)
\end{aligned}$$

The coupled equations are solved numerically with the help of a program developed by PAWLIKOWSKI [20]. However, as shown by SCHLOSSER and PHILPOTT [21], this will not produce the observed Davydov splittings unless higher excited states are included. To avoid this complication, we assume that the relevant Frenkel exciton Hamiltonian has been prediagonalized. We further assume that delocalization of the Frenkel excitons will reduce their displacement parameters, since these parameters should vanish in the limit of strong intermolecular coupling. To incorporate these assumptions, we introduce the following transformation of parameters:

$$V_*^{**} \to 0,\ V^* \to V_1,\ V^{**} \to V_3,\ V_\pm^* \to V_{1,\pm},\ V_\pm^{**} \to V_{3,\pm},\ E^* \to E_1,\ E^{**} \to E_3,\ b^* \to b_1(\propto V_1^{-1}),$$
$$b^{**} \to 0.$$

The parameters E_1,E_3,V_1 and V_3 are directly taken from experimental spectra; $E^{\pm}$ is adjusted to yield the strongest of the observed CT bands and $V^{\pm}$ is neglected. All other energy parameters are adjusted to give the best representation of the observed absorption spectra. The same holds for the displacement parameters, except that the range of b values is restricted by the condition $b=\frac{1}{2}b^*$ [10] and b_1 is assumed to be inversely proportional to V_1. To reproduce the spectra we also require transition moments of S_1 and S_3 as well as linewidths; these are taken directly from the observed spectra. The spectra of so-called amorphous solids are calculated by averaging over all orientations, those of oriented microcrystals by choosing the most suitable orientation. The parameter values used are summarized in Table 1.

Table 1. Energy parameters (in cm^{-1}) and displacement parameters (in dimensionless units $\Delta Q\omega^{\frac{1}{2}}$) used to reproduce the observed CT transitions in the crystals listed.

	$E^{\pm}-E_1$	$V_{1,+}$	$V_{1,-}$	$V_{3,+}$	$V_{3,-}$	B_1	$B_{\pm}$
pentacene	2430	430	290	1150	-1150	0.65	1.1
tetracene	3360	430	215	750	-750	0.9	1.1
anthracene	2930	215	145	640	430	1.25	1.1

3. Results

Absorption spectra for pentacene calculated with the parameters listed in Table 1 are compared with experimental spectra in Fig. 1. The top and bottom parts refer to the "amorphous" and "microcrystalline" spectra, respectively, observed by SHIROTANI et al. [22], the former being very similar to the absorption spectrum of Ref. [15]. Figure 1 also displays calculated stick spectra in which the transitions are depicted as sharp lines, solid and broken lines denoting different polarizations. The lines marked X have predominant CT character. The dot-dash curve depicts the CT component of the absorption spectrum given in Ref. [15].

Inspection of Fig. 1 shows that the model can reproduce the observed absorption spectra very well if the bands above 17000 cm^{-1} are assigned to a vibrational progression with the lowest CT state as origin. As pointed out in Ref. [15], these bands do not coincide with the Frenkel exciton vibronic bands. The present calculations confirm this and indicate that their intensity is also too high for this "conventional" assignment. Although the calculated absorption spectrum compares favourably with the observed spectrum for both microcrystalline and amorphous pentacene, the calculated CT transition energies do not correlate very well with CT components deduced by SEBASTIAN et al. [15] and indicated by upward pointing arrows. To test their assumption that the higher energy CT bands should be assigned to non-nearest-neighbour CT states, we have indicated the corresponding transitions by downward pointing arrows; their relative positions have been deduced from CT energy calculations of BOUNDS, EISENSTEIN and MUNN [12]. The method used to relate the CT potential energies to the CT state energies is the same as that used in [15]. It is seen that the spacings of the upward and downward pointing arrows are quite different; this supports the idea that nonnearest-neighbour CT states contribute only modestly to the observed CT spectrum.

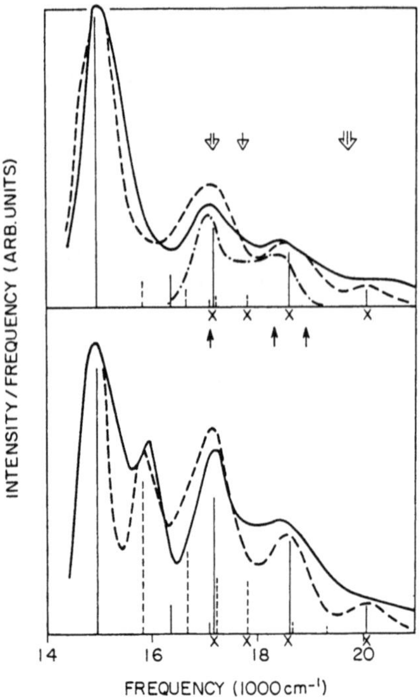

Fig. 1. Observed (solid curves [22]) and calculated (dashed curves) absorption spectra of pentacene in amorphous (top) and microcrystalline (bottom) form. The calculated spectrum is also presented as a stick spectrum with solid and broken lines indicating different polarizations and lines marked X having predominant CT character. The dot-dash curve represents the observed [15] CT component, the arrows pointing up indicating its principal band positions. Arrows pointing down represent calculated CT bands corresponding to increasing electron-hole separations (see text and Fig. 2 caption)

In addition to the CT bands, marked X, the calculated stick spectra show a number of weaker bands to be assigned as vibronic components of S_1; however, these bands are not resolved in the observed spectra. They mix with CT bands close in energy, thus leading to the calculated splittings of these bands.

The results calculated for tetracene are displayed in Fig. 2. The experimental spectra are again taken from SHIROTANI et al. [22]; the absorption spectrum reported by SEBASTIAN et al. [15] is close to the "amorphous" spectrum of [22], but with the first four bands all resolved as peaks and shoulders. According to our analysis the bands above 22000 cm^{-1} should be assigned to a vibrational progression of CT bands with the nearest-neighbour CT band as origin. The calculated CT bands do not agree well with the CT bands deduced by Sebastian et al. (dot-dash curve); especially disturbing is the absence of an observed CT band near 24100 cm^{-1} which corresponds to an absorption maximum that cannot be easily explained as a vibronic component of S_1. The calculated spacing of the series of CT

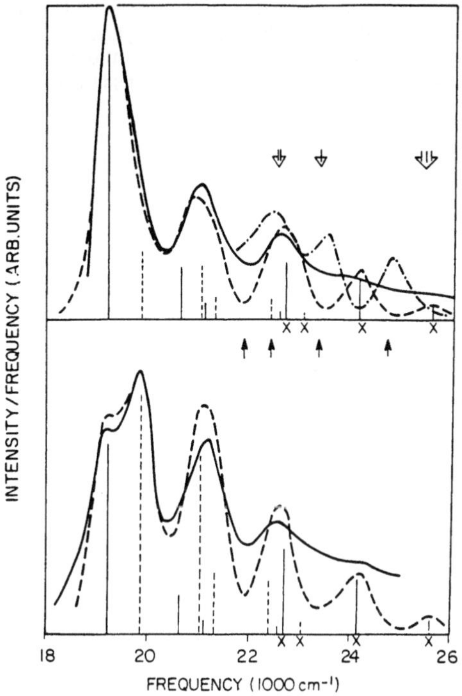

Fig. 2. Same as Fig. 1 but for tetracene. In both figures the arrows pointing down correspond, from left to right, to the following sets of lattice vectors [12]: (i) [0-½½] and [0½½]; (ii) [010]; and (iii) [011], [0-11], and [100]. The calculations in both figures are based on the parameter values listed in Table 1 together with information directly obtained from the absorption spectra

states, deduced from the CT energies calculated by BOUNDS et al. [12], (down arrows) are somewhat closer to the CT bands deduced by Sebastian et al. (up arrows) than for pentacene, but not close enough to make their identification compelling.

As seen from Table 1, the parameter values used to match the pentacene and tetracene spectra are very similar as expected for these closely related hydrocarbons. However, this similarity breaks down for anthracene, where the CT bands were found to be much weaker than in the other molecules [17]. Since they are too weak to show up in the absorption spectrum, we cannot follow the same procedure as for tetracene and pentacene: instead of trying to match the observed structure of the absorption spectrum, we can only try to match the overall intensity of the CT component. This intensity can be reproduced without significantly changing the absolute values of the parameters used for tetracene and pentacene, provided the sign of one exchange integral is changed. In tetracene and pentacene, $V_{3,+}$ and $V_{3,-}$ were found to have opposite sign, but in anthracene, the model indicates that they have the same sign (see Table 1). A possible explanation for this sign change is the presence in tetracene and pentacene of other strong transitions which mix strongly with the Davydov components of S_3. It has been shown [21] that these higher-energy transitions are responsible for the large Davydov splitting of the S_1 exciton in these crystals. The absence of such interfering transitions in anthracene may be the reason for the lower CT intensities.

Table 2. Comparison of calculated and observed [17] CT band energies (in cm^{-1}) and relative intensities in anthracene

Band nr	1	2	3	4	5
calc. energy	(28950)		30380	31810	33240
obs. energy	28550	29120	30330	31940	33470
calc. intensity	(1.00)		0.78	0.28	0.06
obs. intensity	0.44	1.00	0.56	0.20	0.07

4. Discussion

We are now in a position to answer the questions posed in the Introduction. The calculations indicate that the observed intensity distributions of CT bands represent vibrational structure of the lowest and most intense CT band, at least in part. The structure calculated on the basis of a reasonable assumption about CC-stretch displacements accounts very well for the observed absorption spectrum of tetracene and pentacene. The fact that it accounts less well for the CT spectra deduced from field-dependent absorption measurements could be due to approximations introduced in the analysis of these measurements. Our calculations indicate that the vibronic CT levels mix with vibronic Frenkel levels of the same energy. The mixing may give rise to splitting of the CT bands into two components of different polarization, analogous to Davydov splitting; the magnitude of this splitting depends on the presence of accidental resonance and is therefore hard to predict. The presence of such split levels may well interfere with the analysis of the field-dependent absorption spectra.

The low CT intensity in anthracene compared to that in tetracene and pentacene has been related to the sign of a CT-Frenkel exchange integral and indirectly to the structure of the higher excited Frenkel exciton states. If this argument is valid, it may apply to all aromatic hydrocarbons of the size of anthracene and smaller, since all of them will have their higher excited states too far removed to be effective in the charge-transfer region. The conclusion should then be that CT states are readily observable in larger aromatic hydrocarbon crystals but tend to remain hidden in smaller ones. Unfortunately, much of the earlier search for these states has focussed on the latter crystals. It may be timely to reverse this trend and pay attention to crystals such as tetracene and pentacene where CT bands provide the dominant structure in readily accessible regions of the absorption spectrum. It would be of great interest to study these bands in absorption or reflection for various crystal directions.

It has been known for some time that the quantum yield of optical carrier generation approaches unity in tetracene and pentacene for the right excitation wavelengths and high temperatures [7]. In anthracene, on the other hand, this yield remains of the order of 10^{-4} [23,24]. Hence there is a direct relation between the absorption coefficient of CT bands and this yield, suggesting that direct optical population of CT states is the first step in optical carrier generation. For tetracene and pentacene this mechanism may be considered proved: the Frenkel exciton absorption coefficient in the photoconductivity region is too weak to provide an alternative. In anthracene, light absorption predominantly populates Frenkel exciton states and the carrier yield is low. This low yield may be due either to the smallness of the CT absorption coefficient or to the inefficiency of the transition from Frenkel to CT excitons. Whatever the answer may be, the basic message is clear: efficient photoconductors are found only among those molecular crystals that have absorption regions where CT transitions dominate.

Acknowledgement. The work reported here has benefitted greatly from contributions by Pamela Bounds, Ithakar Eisenstein, Robert Munn, Marek Pawlikowski and Piotr Petelenz.

References

1. D. Haarer and M.R. Philpott in: Modern Problems in Solid State Physics eds. V.M. Agranovich and R.M. Hochstrasser (North-Holland, 1982).

2. R.H. Batt, C.L. Braun, and J.F. Hornig, J. Chem. Phys. 49, 1967 (1968).

3. R.R. Chance and C.L. Braun, J. Chem. Phys. 59, 2269 (1973); 64, 3573 (1976).

4. L. Onsager, Phys. Rev. 54, 554 (1938).

5. N. Wakayama and D.F. Williams, J. Chem. Phys. 57, 1770 (1972); M. Wittmer and I. Zschokke-Gränacher, ibid 63, 4187 (1975).

6. A. Jurgis and E.A. Silinsh, Phys. Stat. Sol. 53b, 735 (1972).

7. E.A. Silinsh, Organic Molecular Crystals: Their Electronic States (Springer, Berlin, 1980).

8. P.J. Bounds and R.W. Munn, Chem. Phys. 44, 103 (1979); 59, 41,47 (1981).

9. P.J. Bounds and W. Siebrand, Chem. Phys. Lett. 75, 414 (1980).

10. P.J. Bounds, P. Petelenz and W. Siebrand, Chem. Phys. 63, 303 (1981).

11. P.J. Bounds and W. Siebrand, Chem. Phys. Lett. 85, 496 (1982).

12. P.J. Bounds, I. Eisenstein and R.W. Munn, unpublished results.

13. R.E. Merrifield, J. Chem. Phys. 34, 1835 (1961).

14. G.H. Wannier, Phys. Rev. 52, 191 (1937).

15. L. Sebastian, G. Weiser and H. Bässler, Chem. Phys. 61, 125 (1981).

16. S.C. Abbi and D.M. Hanson, J. Chem. Phys. 60, 319 (1974).

17. L. Sebastian, G. Weiser, G. Peter and H. Bässler, Chem. Phys. 75, 103 (1983).

18. J.R. Platt, J. Chem. Phys. 17, 484 (1949).

19. W.L. Pollans and S. Choi, J. Chem. Phys. 52, 3691 (1970).

20. M. Pawlikowski, unpublished results.

21. D.W. Schlosser and M.R. Philpott, Chem. Phys. 49, 181 (1980).

22 I. Shirotani, Y. Kamura and H. Inokuchi, Mol. Cryst. Liq. Cryst. 28, 345 (1975).

23. L.E. Lyons and K.A. Milne, J. Chem. Phys. 65, 1474 (1976).

24. K. Kato and C.L. Braun, J. Chem. Phys. 72, 172 (1980).

Exciton Band Structure and Excimer Formation

L.A. Dissado

Physics Department, Chelsea College London, Pulton Place
London SW6 5PR, United Kingdom

S.H. Walmsley

Chemistry Department, University College London, 20 Gordon Street
London WC1H 0AJ, United Kingdom

1 Introduction

Molecular crystals exhibit a complicated range of behaviour following ultraviolet absorption. Fluorescence, chemical reaction, degradation into heat and excimer formation are all possible secondary processes. Which, if any, of these is dominant depends on the detailed nature of the conditions. Previous papers have dealt with trapping at chemical impurities and self-trapping [1], exciton-phonon coupling [2] and scattering and vibrational relaxation in relation to excimer formation [3]. The most characteristic feature of the excimer is that the equilibrium separation of the two molecules concerned is significantly different from their separation in the crystal and this causes difficulties in formulating a complete theory. Here attention is concentrated on a simple model of the incorporation of lattice distortion into the pure electronic exciton and the way in which this distortion can favour excimer formation.

2 Excitons and Lattice Distortion

In the simplest form of the theory, the exciton band, which includes the states responsible for photon absorption, is formulated assuming a static lattice. The energy of a single level in the band is given by

$$E(\underset{\sim}{k}) = \Delta\varepsilon + D(R_0) + \sum_{m\neq 0} \exp(i\underset{\sim}{k}.\underset{\sim}{m})\, M(m0,R_0) \,. \tag{1}$$

Following the usual notation, $\Delta\varepsilon$ is the excitation energy of the parent molecule electronic excited state, D is the shift of the one molecule energy due to the presence of the surrounding molecules and M(m0) is the resonance energy for transfer of excitation from the origin molecule 0 to m. The additional label R_0 indicates that all terms are evaluated at the static lattice structure. Attention is confined to the case of one molecule per unit cell and the state is uniquely classified by the wave vector $\underset{\sim}{k}$.

When the lattice molecules are allowed to move, the most important effect for the present treatment is the inclusion of lattice distortion. During the time the excitation is present at a particular site, a partial relaxation towards equilibrium is possible giving a change in vibrational origins. The exciton is modified so that this total distortion is transferred from site to site with the electronic excitation. The Hamiltonian describing the distortion is symmetry breaking since it refers to a particular site. It takes the form of a sum over phonons, the oscillation periods of which are short enough for them to be effective during the site residence time. The new band has a coherence based on the equivalence of the whole set of distorted sites. Full details are given in an earlier paper [2] in which the particle annihilation operator is denoted by A_k.

For the present purpose, the important point is that the modified band energies retain the same form as (1). In particular, the resonance interaction M(mn) is replaced by $\bar{M}$(mn) as follows:

$$\bar{M}(mn,R) = M(mn,R)\, \exp[-\sum_{\bar{s}\bar{q}}\sum |\Lambda_{\bar{s}\bar{q}}|^2 \{1-\exp(i\underset{\sim}{q}.(\underset{\sim}{m}-\underset{\sim}{n}))\}^2] \ . \tag{2}$$

Two effects are incorporated in this way. First, R_0 is replaced by R representing the distorted configuration. Second, the exponential factor gives the overlap between vibrational wave functions when the distortion is centred at m and n respectively. The sum over $\bar{s}\bar{q}$ implies a restriction to those phonons which contribute to the distortion.

$$\Lambda_{\bar{s}\bar{q}} = \chi_{\bar{s}\bar{q}}/h\Omega_{\bar{s}\bar{q}} \tag{3}$$

for a phonon of frequency Ω. χ is the first derivative of D with respect to the corresponding phonon coordinate, evaluated at R_0.

An important point is that the vibrational overlap factor (the exponential factor in (2)) departs from unity when the sites m and n have different environments as they will have in general in the distorted configuration. Equally, it may be said that the sites m and n are in resonance when the overlap factor is unity. It will be seen that a characteristic of excimer formation is that the pair of molecules forming the incipient excimer remain in resonance.

These ideas are now applied to special examples of distortions.

3 Symmetric Distortion

For simplicity, attention is confined to a one-dimensional crystal with one molecule per unit cell. Only nearest-neighbour interactions are taken into account. In the static lattice case, the exciton band has extrema

$$E(\pi) = \Delta\varepsilon + D(R_0) - 2M(01,R_0) \tag{4}$$

$$E(0) = \Delta\varepsilon + D(R_0) + 2M(01,R_0) \tag{5}$$

in which R_0 is the equilibrium value of the nearest-neighbour separation.

In the symmetric distortion, the nearest neighbours relax to a new distance R, so that the band edge energies become

$$E(\pi) = \Delta\varepsilon + D(R) - 2\bar{M}(01,R) \tag{6}$$

$$E(0) = \Delta\varepsilon + D(R) + 2\bar{M}(01,R) \ . \tag{7}$$

M is replaced by $\bar{M}$ since the excited molecule is no longer in resonance with the other molecules. For the case of R less than R_0, the qualitative changes can be deduced. D is negative and increases in magnitude with decrease in R. M, the electronic part of $\bar{M}$, increases in magnitude with decrease in R but this effect tends to be outweighed by the exponential vibrational overlap factor.

The total effect is a narrowing of the band width and a displacement to lower energies. This is a familiar result in many problems in solid state physics, discussed in general terms for excitons by Toyozawa [4].

4 Excimeric Distortion

In the excimeric distortion, molecules 0 and 1 are symmetrically displaced so that their mid point is stationary. The remaining molecules are not altered in position. The band extrema as a function of the distortion are now given by

$$E(\pi) = \Delta\varepsilon + D(R) - M(01,R) - \bar{M}(0\text{-}1,S) \tag{8}$$

$$E(0) = \Delta\varepsilon + D(R) + M(01,R) + \bar{M}(0\text{-}1,S) \tag{9}$$

in which R is the distance between the displaced molecules 0 and 1 and S given by $\frac{1}{2}(3R_0-R)$ is the distance between 0 and its undisplaced neighbour -1.

The new feature is that the pair 01 remain in resonance during the distortion and hence the vibrational overlap factor is unity. On the other hand, the pair 0-1 is not in resonance and their interaction is represented by an $\bar{M}$ term.

Moreover, closer examination shows that there is a bifurcation of the band. From the discussion for the symmetric distortion it can be deduced that the magnitude of M(01,R) is greater than that of M(0-1,S). The inequality is increased by the exponential factor which ensures that M is greater than $\bar{M}$. The separation of the two bands is given by the difference of the two energies

$$E(\tfrac{1}{2}\pi,1) = \Delta\varepsilon + D(R) - M(01,R) + \bar{M}(0\text{-}1,S) \tag{10}$$

$$E(\tfrac{1}{2}\pi,2) = \Delta\varepsilon + D(R) + M(01,R) - \bar{M}(0\text{-}1,S)\,. \tag{11}$$

At zero distortion, R=S and the two levels are degenerate. Within the very simple model used here an infinitesimal change is sufficient to give a band gap.

The bifurcation into two sub-bands reflects the fact that it is easier to transfer the excitation to the nearer than to the further neighbour. In this way, each sub-band has its own relative phase, positive or negative, between the pair of molecules, which can therefore be regarded as an incipient excimer. The true excimer requires a much larger displacement and its electronic stability will arise largely from other terms not taken into account in this treatment. The mechanism described here is one which points the system towards excimer formation. The excimer state undoubtedly exists, it is considerably more stable than the exciton and the two are linked by a common potential surface. The problem is to make this potential surface dynamically available.

5 Discussion

The equations given for the two types of distortion are exact for the simplified model considered: one-dimensional crystal, nearest-neighbour interaction, exciton-phonon coupling linear in molecular displacements and restriction to one free molecule electronic excited state. The general qualitative picture, however, transcends all these approximations.

The important conclusion is that incorporation of a component of the kind suggested in section 4 into the travelling distortion gives access to the possibility of stabilising the excimeric electronic state. The magnitudes of all the energy terms suggest that the formation of the excimer becomes progressively easier the larger the excimeric distortion that can be incorporated. The symmetric distortion, on the other hand, tends to oppose excimer formation since it maintains the one site location of the electronic excitation and travels relatively easily through the lattice. It may be speculated that a total distortion which contained significant contribution from both symmetric and excimeric components could provide easy transfer for an excimeric electronic state. Fuller discussion requires a more complete treatment and this is given in a paper in preparation [5].

References

1. D.P.Craig, L.A.Dissado and D.H.Walmsley: Chem. Phys. Letters, 46, 87 (1977)
2. D.P.Craig, L.A.Dissado and S.H.Walmsley: Chem. Phys. 46, 87 (1980)
3. L.A.Dissado and S.H.Walmsley: Chem. Phys. Letters, 87, 74 (1982)
4. Y.Toyozawa: J.Luminescence 1/2, 732 (1970)
5. L.A.Dissado and S.H.Walmsley: in preparation

Fission and Radiationless Transitions in Organic Molecular Crystals in Highly Excited States

G. Klein

Centre de Recherches Nucléaires, 23, rue du Loess
F-67037 Strasbourg Cedex, France

Introduction

Direct study of radiationless transitions in organic molecular crystals is difficult or impossible because the rate constants ($> 10^{12}s^{-1}$) of these processes are too large compared with those of radiative transitions and they are generally not accessible in time-resolved experiments. However, it is sometimes possible to get more information from excitation spectra of various processes that compete with one another during the relaxation of a highly excited state.

The excitation spectrum of the yield of pairs of triplet excitons from one initially excited singlet state can be obtained by measuring the magnetic field effect on the fluorescence yield [1-3]. The yield of pairs of singlet excitons from one high energy exciton can be measured by detecting coincidences between two fluorescence photons with steady light excitation [4;5].

Excitation spectra of the yield of photoconductivity [6-9] and of photoelectrons [10] were widely studied.

By measuring coincidences between fluorescence photons and photoelectrons in anthracene crystals excited with photons of more than 10 eV energy it is possible to get more information on the slowing down of electrons in the condensed phase [11].

In the present paper we are most concerned with fission processes, or more generally with production of pairs of excitons. A convincing study of these radiationless transitions cannot ignore the other processes that compete with it.

Radiationless transitions in the first excited singlet state S_1

The excitation spectrum of singlet excitons fission into two triplets η_{2T} in anthracene crystals is shown in Fig. 1; it is taken from ref. [2].

S_1 excitons are created with intramolecular vibrational excitation when the incident photon energy is between 3.12 and 4.35 eV. In this energy range two particle states are created [12,13]. This means that there is no binding energy between the electronic excitation and the vibrational excitation, although these two excitations are initially on the same crystal site. The transfer of the electronic excitation from one site to a neighboring one occurs with a transfer of a minimum of vibrational quanta because Franck-Condon factors are maximum in this case as shown schematically in Fig. 2.

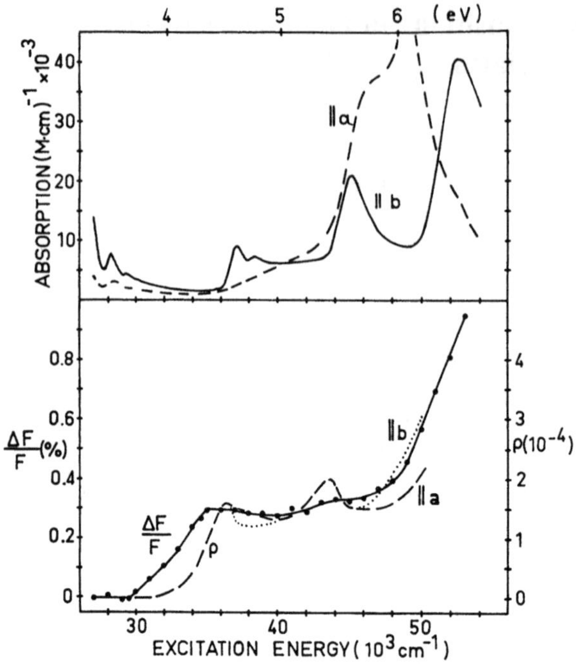

Fig. 1 The upper figure represents the anthracene crystal absorption spectrum taken from ref. [21]. The lower figure shows the relative variations $\Delta F/F_0$ of the prompt fluorescence intensity upon application of a magnetic field in the resonance conditions. In these conditions $\eta_{2T} = 5 \times \Delta F/F_0$. The yield of charge carrier ρ is taken from ref. [22]

Therefore after a few transfers of the electronic excitation from site to site we may consider that the two excitations, electronic and vibrational, are separated in the crystal. The rate constant k_r for this vibronic relaxation is approximately independent of the number of initial vibrational quanta. It is given by the exciton hopping rate constant ($\approx 10^{14} s^{-1}$) corresponding to a dipole-dipole interaction between two neighboring molecules and to the Davydov splitting (300 cm^{-1}). This rate constant is two orders of magnitude larger than that of vibrational relaxation in isolated molecules.

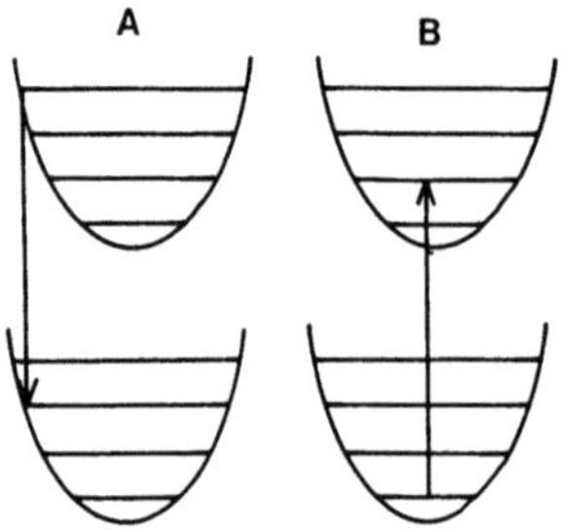

Fig. 2 Electronic excitation transfer from molecule A to molecule B

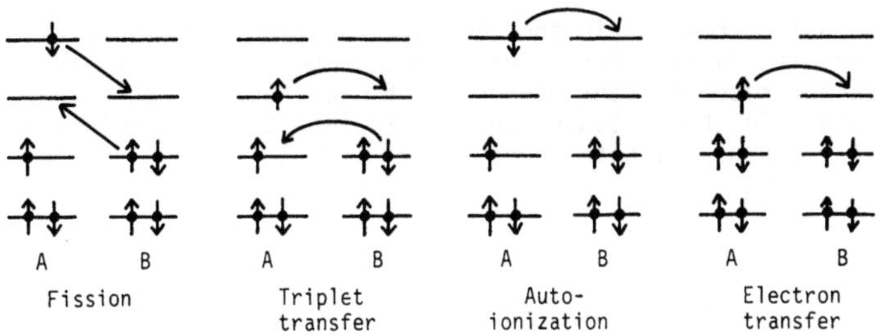

Fig. 3 Intermolecular radiationless transitions between two molecules A and B

In the excitation energy range 3.66 - 4.35 eV the yield of fission η_{2T} is less than 1.5×10^{-2} corresponding to a fission rate constant

$$k_{2T} = \eta_{2T}(k_r + k_{2T}) \approx \eta_{2T}\, k_r = 1.5 \times 10^{12}\, s^{-1} .$$

This value is of the same order of magnitude as the rate of fission measured independently in tetracene crystals [14]. It is also of the same order of magnitude as the triplet exciton hopping time in anthracene crystals [16]. This is not unexpected since the two transitions are very similar as shown in Fig. 3.

The linear increase of η_{2T} in Fig. 1 from $2E(T_1) = 3.66$ eV to 4.35 eV may be explained by the linear increase of the number of possible repartitions of the excess vibrational quanta on the two triplet excitons. If n vibrational quanta are in excess above the threshold energy $2E(T_1)$, there are (n+1) possibilities. If n is not too large (n = 0 - 3), the Franck-Condon factors are approximately the same for each final state and k_{2T} increases linearly with energy above $2E(T_1)$.

ALBRECHT et al. [17] showed that in deuterated anthracene the yield of fission η_{2T} is lower by a factor of 2 or 3 than in protonated anthracene. These experimental results can be interpreted as a lower Franck-Condon factor for the transition in deuterated anthracene. For a given excitation energy the number of vibrational quanta of the initial singlet state is higher in this case, and therefore the overlap with the triplet vibrational wavefunction (in the final state) is smaller.

The steps in the excitation spectra η_{2T} observed by MOLLER and POPE [18] in tetracene and by ARNOLD [19] and ALBRECHT et al. [17] in anthracene crystals were not found by us [2]. This difference remains unexplained but does not change the conclusions of the present discussion.

The possibility of charge transfer states as intermediate states in fission processes in this energy range seems unlikely for two reasons. The first one is that in anthracene at 4.35 eV, $\eta_{2T} = 1.5 \times 10^{-2}$, while the experimental yield of geminate charge pair states is 2×10^{-3} [8]. The second reason is that in tetracene crystals, where the charge transfer states are located between $2E(T_1)$ and $E(S_2)$, one would expect a resonance in the fission excitation spectrum η_{2T}. This is not observed [2].

However, we cannot understand why the yield of geminate charge pair states is so low compared with fission. One would expect that the reverse is true, because fission is a two-electron transition whereas auto-ionization involves only one-electron transfer to a neighboring molecule as is shown in Fig. 3. If the fission into two triplets is similar to the triplet exciton transfer, then auto-ionization may be compared to the electron transfer to a neighboring site as is shown in Fig. 3. It is known that in anthracene crystals, the electron transfer integral in the lowest conduction band is 10 times larger than the triplet T_1 transfer integral [20]. Therefore one would expect an auto-ionization rate constant one order of magnitude larger than the fission rate constant in the S_1 state. This would lead to the conclusion that the experimental yield of geminate charge pairs from S_1 states and the activation energies of charges separation given by KATO and BRAUN [8] are underestimated.

Radiationless transitions in the second singlet excited state S_2

When the anthracene crystal is excited in the second singlet excited state S_2, above 4.35 eV (the peak is at 4.6 eV), the fission yield η_{2T} (Fig. 1) is constant with excitation energy and is restricted to the fission from the vibrationally excited S_1 exciton created from S_2 by vibronic relaxation and internal conversion. Then we conclude that fission is not possible from S_2. The absence of fission from vibrationally excited S_2 states can be explained by a very fast vibronic relaxation, one order of magnitude faster than for S_1 because the dipole-dipole interaction between two neighboring molecules is stronger by a factor of ten than for S_1 excitons.

However the rate constant for the internal conversion $S_2 \rightarrow S_1$ is probably not more than 10^{14} s^{-1} . Therefore it seems necessary to suppose that the rate constant of the fission is significantly lower than for S_1 excitons. This can be explained only partly as follows. The S_2 state is a linear combination of two degenerated states corresponding to the transitions of an electron from the sixth π orbital to the eighth, and a transition from the seventh to the ninth*, $\Psi_{S2} = (1/\sqrt{2})(\Psi_{6\rightarrow 8} \pm \Psi_{7\rightarrow 9})$. A fission into two T_1 excitons from $\Psi_{6\rightarrow 8}$ can occur only with three-electron transitions, and then is slower than the transition $S_1 + S_0 \rightarrow T_1 + T_1$ involving only two electrons. The same effect is expected for the auto-ionization process which is possible with one-electron transition from S_1, and only a two-electron transition from $\Psi_{6\rightarrow 8}$ because the energy corresponding to the lower lying hole is not directly available for the kinetic energy of the upper orbital. However, we see no reason why fission should be impossible from $\Psi_{7\rightarrow 9}$.

This effect may be more important in other crystals. The lack of fission from S_2 is probably the reason why fission was not observed up to 6 eV in several crystals, p-terphenyl, stilbene and pyrene [24,25]. In these crystals $2E(T_1) > E(S_2)$ and then fission from S_1 is energetically impossible.

A strong coupling between S_2 and charge transfer states, as is suggested by BOUNDS et al. [23], would imply a large intermolecular electron orbital overlap and then a high exchange integral for the fission into two triplets. However, this is not observed in the excitation spectrum of η_{2T}.

Radiationless transitions in excited states above the ionization potential ($E > 6$ eV)

The excitation spectra η_{2T} of the yield of fission into a pair of triplet excitons and η_{2S} of fission into a pair of singlet excitons were measured in anthracene crystals for photon excitation energies up to 50 eV [26]. These results are shown in Fig. 4.

---

* In anthracene in the ground state there are 7 π orbitals filled by 2 times 7 electrons.

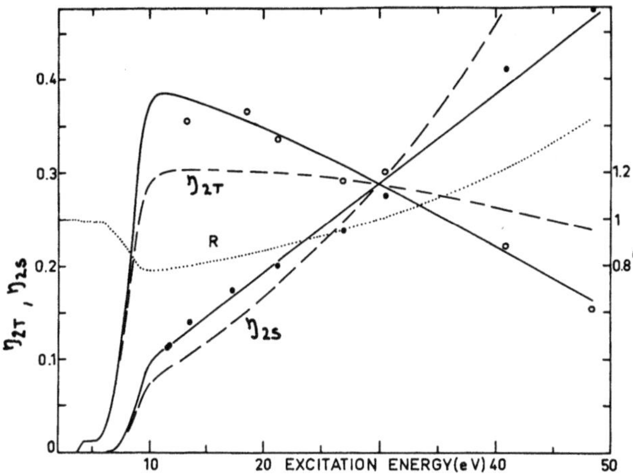

Fig. 4 Broken lines are the yields of fission into a pair of triplet excitons η_{2T} and into a pair of singlet excitons η_{2S}. These curves are calculated from experimental results (open and closed circles) in ref. [26]. R is the fluorescence quantum yield calculated as $R = 1 + \eta_{2S} - \eta_{2T}$

There is a very steep increase of η_{2T} with excitation energy above 6 eV. The same increase is observed in p-terphenyl crystals with a threshold at 8.6 eV [25] and in stilbene at 6.8 eV [24]. In liquid benzene a similar increase of the fluorescence yield is observed at 7 eV, and it was attributed to the onset of auto-ionization [27]. This phenomenon seems to be quite general. We think that it corresponds to the increase of the auto-ionization yield and of transitions to neutral excited states with large intermolecular overlap. There is no evidence for a particular relation between the threshold energy of this effect and the ionization potential, since in anthracene they have approximately the same value whereas in p-terphenyl crystals the ionization potential [10] is 2.5 eV lower.

The smooth and quadratic increase of η_{2S} above threshold $2E(S_1) = 6.3$ eV in anthracene crystals [4,28] can be interpreted by the increase of the number of possible final pair states, and by the sharp increase of the density of ionized states that can be intermediate states in the fission process, since the direct transition is symmetry forbidden near the threshold [4].

The following sequence of transitions in Fig. 5 has been proposed for this high-energy region [26], where electrons (e) and holes (h) are the intermediate states. The star means that there is some kinetic or excitation energy.

The excited hole h* corresponds to a lack of electrons in an energy level lower than the highest filled levels. The transfer of the electronic excitation from a molecule A to a neighboring molecule B without transfer of charge may be written $A^{+*} + B \rightarrow A^{+} + B^{*}$. The matrix elements for such transitions are very similar to those for the transfer of high-energy excited states (S_2, S_3) which were studied experimentally [29,30] and it was shown that this reaction is faster than the other intramolecular radiationless transitions.

The creation of singlet and triplet excitons by energetic electrons (e*) are well established and documented by experiments on photoelec-

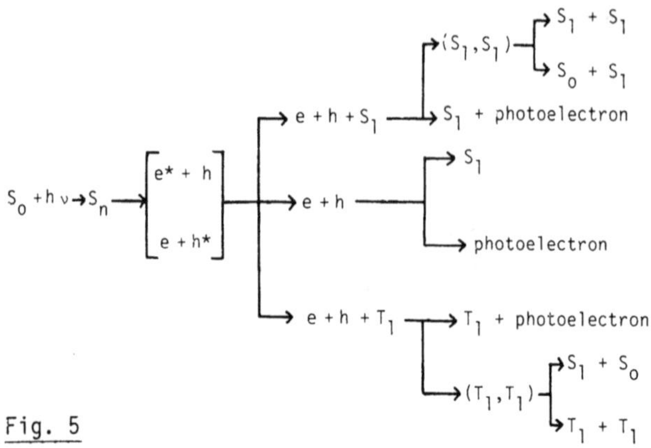

Fig. 5

tron energy distribution curves [10] and on electron energy loss spectra [31]. The channel ending at S_1+ photoelectron (see Fig. 5) has been studied in anthracene crystals by measuring coincidences between fluorescence photons and photoelectrons [11]. This proves the validity of the proposed model of Fig. 5.

The steep increase of η_{2T} and η_{2S} up to 10 eV excitation energy can be interpreted as an increase in the auto-ionization yield and increase of secondary electron and hole energy, and also an increase of the delocalization of electrons in the high-energy excitons. The change of slope of η_{2S} and η_{2T} at 10 eV indicates a saturation of these channels.

The variations of η_{2T} and η_{2S} with excitation energies above 10 eV are governed by the increase of e* and h* excess energies. At the lower energies (up to 30 eV) the low-energy excitons (T_1) are created with higher probabilities by e*, whereas at the higher energies $E(e^*)$ or $E(h^*) \gg E(S_1)$, and the secondary singlet excitons are created (dipole allowed transition) rather than triplet excitons (involving only exchange integrals), then η_{2S} rises with excitation energy, while η_{2T} is approximately constant.

It is necessary to analyze with more detail the relation between the magnetic field effect on the fluorescence yield and η_{2T}. It was shown [26,32] that the fission yield η_{2T} calculated from magnetic field effect measurements on prompt fluorescence is valid if the (T_1,T_1) pair state in fission experiments is the same as in fusion experiments, which means that in fission experiments the two T_1 excitons have to be created on two neighboring sites with a total singlet spin state. The condition of an initial singlet state is certainly fulfilled. The precursors of the T_1 excitons, higher excited triplet states or electron hole pairs undergo some spin evolution, different from that of T_1 excitons and lead to (T_1,T_1) pairs with some triplet and quintet character. But this effect can be neglected because these precursor lifetimes ($\approx 10^{-12}$ s) are short compared with time constants of spin evolution ($\approx 2 \times 10^{-10}$ s) [3]. This first step of the fission is a spin conserving transition and magnetic field independent. The magnetic field dependent step is, as in fusion experiments, the annihilation $(T_1,T_1) \rightarrow S_1 + S_0$.

The other condition that the two T_1 excitons are formed on two neighboring sites is difficult to understand if we adopt the model of Fig. 5. The electron e* with a kinetic energy between $E(T_1)$ and $E(S_1)$ is a quasi free electron and can diffuse a great distance before and after creating a triplet exciton. After the electron hole recombination, the two triplets can be several intermolecular distances apart or in two different (a,b) planes, and the (T_1,T_1) complex partly dissociated. Then the recombination probability

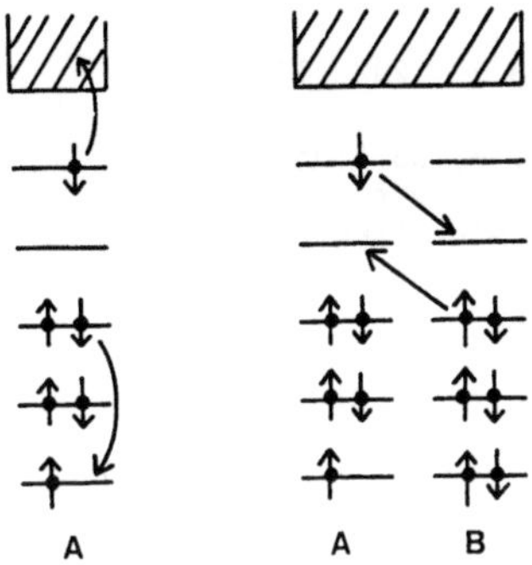

Fig. 6 Auto-ionization and fission

will be smaller and the magnetic field effect smaller too. However, this effect does not seem to be very important since the value of η_{2T} in Fig. 4 is never less than 0.25 between 10 and 50 eV.

Two reasons can explain this result. The first one is that an electron-triplet metastable complex may be formed so that the electron-hole recombination necessarily happens near the triplet and yields two triplets very near one another. The existence of the electron-triplet complex has already been suggested by BADER et al. [33]. The second reason may be that direct ionization is not possible for momentum conservation rules, as the incident photon momentum is $k \approx 0$. If we assume that no important bound orbitals are in the ionization continuum, then the probability of creation of an electron e* of more than $E(T_1)$ kinetic energy by a one-electron auto-ionization is low, and we think that high-energy electrons are produced by a two-electron transition as shown in Fig. 6. In this case, the rate constant of direct fission (see Fig. 6) for high-energy delocalized orbitals may compete with auto-ionization. Thus, the direct fission into two triplets may be compared with indirect triplet pair creation through the auto-ionization channel as shown in Fig. 5.

The conclusion of this discussion is that the initially highly excited state is a one-electron excitation that shares very rapidly its energy with other electrons of the same molecule and neighboring molecules. The rate constants of these transitions are typically $10^{14}\,s^{-1}$. In this time scale vibrational excitations play only a minor part.

References

1. R.C. Johnson and R.E. Merrifield : Phys. Rev. B1, 896 (1970)
2. G. Klein, R. Voltz and M. Schott : Chem. Phys. Letters 19, 391 (1973)
3. C.E. Swenberg and N.E. Geacintov : Organic Molecular Photophysics ; Ed. J.B. Birks (Wiley, New York 1973)
4. G. Klein and R. Voltz : Int. J. Radiat. Phys. Chem. 7, 155 (1975)
5. Ch.B. Lushchik and T.I. Savikhina : Bull. Acad. Sciences USSR, Phys. Ser. 45, 34 (1982)
6. N. Geacintov and M. Pope : J. Chem. Phys. 50, 814 (1969)
7. R.R. Chance and C.L. Braun : J. Chem. Phys. 64, 3573 (1976)
8. K. Kato and C.L. Braun : J. Chem. Phys. 72, 172 (1980)
9. E.A. Silinsh, V.A. Kolesnikov, I.J. Musikante and D.R. Balode : Phys. Stat. Sol. (b) 113, 379 (1982)
10. K. Seki, T. Hirooka, Y. Kamura and H. Inokuchi : Bull. Chem. Soc. Japan. 49, 904 (1976)

11. G. Klein : Chem. Phys. Letters 95, 305 (1983)
12. E.I. Rashba : Sov. Phys. JETP, 23, 708 (1966) and 27, 292 (1968)
13. E.F. Sheka : Sov. Phys. Uspekhi 14, 484 (1972)
14. M. Pope, N. Geacintov and F. Vogel : Mol. Cryst. 6, 83 (1969)
16. P. Avakian and R.E. Merrifield : Mol. Cryst. Liq. Cryst. 5, 37 (1968)
17. W.G. Albrecht, H. Coufal, R. Haberkorn and M.E. Michel-Beyerle : Phys. Stat. Sol. (b) 89, 261 (1978)
18. W.M. Moller and M. Pope : J. Chem. Phys. 59, 2760 (1973)
19. S. Arnold : J. Chem. Phys. 61, 431 (1974)
20. A. Tiberghien and G. Delacote : Le Journal de Physique 31, 637 (1970)
21. L.B. Clark : J. Chem. Phys. 53, 4092 (1970)
22. C.L. Braun and J.F. Hornig : IV th Molecular Crystal Symposium (Enschede, The Netherlands, 1968)
23. P.J. Bounds, P. Petelenz and W. Siebrand : Chem. Phys. 63, 303 (1981)
24. G. Klein and R. Voltz : Molecular Spectroscopy of Dense Phases (Elsevier, Amsterdam, 1976)
25. G. Klein and M.J. Carvalho : Chem. Phys. Letters 51, 409 (1977)
26. G. Klein, Chem. Phys. Letters 97, 114 (1983)
27. C. Fuchs and R. Voltz : Chem. Phys. Letters 18, 394 (1973)
28. G. Klein and R. Voltz : IV International Conference on Vacuum Ultraviolet Radiation Physics (Hamburg, 1974)
29. C. Fuchs, F. Heisel and R. Voltz : J. Phys. Chem. 76, 3867 (1972)
30. V.A. Benderskii, V.Kh. Brikenstein, A.G. Lavrushko, P.G. Filipov and A.V. Yatsenko : Chem. Phys. Letters 56, 443 (1978)
31. M. Rei Vilar, M. Heyman and M. Schott : Chem. Phys. Letters 94, 522 (1983)
32. M. Chabr and D.F. Williams : Phys. Rev. B 16, 1685 (1977)
33. G. Bader, L. Caron and L. Sanche : Sol. State Comm. 38, 849 (1981)

Part VI

Electronic Excitations in Disordered Systems

Transport and Thermodynamics of Physical Systems with Fractal Geometry

R. Orbach
Physics Department, University of California, Los Angeles, CA 90024, USA

Abstract

An analysis of the physical properties of fractals is given. Three dimensionalities are required: d - Euclidean (embedding); $\bar{d}$ - Hausdorff; and $\bar{\bar{d}}$ - spectral (fracton). These three are sufficient to specify the transport and thermodynamic properties of self-similar systems. Application is made to relaxation time measurements on polypeptides; thermodynamic and transport studies of glasses, polymers, and irradiated quartz; and computer simulations of percolating networks.

I. Introduction

It is now clear that many physical systems exhibit fractal geometry [1] over appropriate length scales. Among many are percolating networks [2]; polypeptide structures [3]; glasses, polymers, and irradiated quartz [4]; and gels [5]. One suspects that fractal geometry may be more general than Euclidean. Said another way, nature may more often exhibit dilation symmetry [6] than translation symmetry.

Fractals are self-similar structures embedded in Euclidean spaces of (the usual) dimensionality d. The mass distribution (site density) is specified by the Hausdorff dimensionality $\bar{d}$ [1]. The mass enclosed by a sphere of radius r increases as $r^{\bar{d}}$. Because of the "tortured" path a particle follows when diffusing on a fractal, the diffusion constant D(r) is range dependent, falling off as $D(r) \sim r^{-\theta}$. This leads to a third dimension, $\bar{\bar{d}}$, known as the spectral dimension, found by ALEXANDER and ORBACH [7]. Thus, the vibrational (fracton) density of states on a fractal, $N(\omega) \sim \omega^{\bar{\bar{d}}-1}$. They show $\bar{\bar{d}} = 2\bar{d}/(2+\theta)$.

These three dimensionalities, d, $\bar{d}$, and $\bar{\bar{d}}$, so far (!) appear to be sufficient to specify completely the physical properties of solids possessing fractal geometries. For Euclidean spaces, these three dimensionalities are equal (i.e. "accidentally" degenerate). Quoting [6]: "... the study of physics on self similar spaces is not only an exploration of a new world, it also allows for a deeper understanding of the physics on more traditionally trodden spaces."

II. Relevance to Physical Systems

Self-similar structures may be a common feature of nature. The requirement of self-similarity is that the physical system "looks" identical on a length scale 1) longer than an atomic or other shortest length, a, in a system; and 2) shorter than a characteristic length, L, beyond which the system is Euclidean.

The prototypical fractal is the percolating network for which the bond length is a, and the characteristic length $L = \xi_p$, the percolation correlation length. Thus, fractal behavior of physical systems exhibits short length scale properties (high frequency). Traditionally, condensed matter scientists have focused rather on long length scale properties (low frequency). The length scale L implies a crossover frequency, $\omega_{c.o.}$. For vibrational states on a fractal, [7] shows that

$$\omega_{c.o.} \sim L^{-[1 + (\theta/?)]} \quad . \tag{1}$$

Thus, for percolating networks [4,7], the vibrational density of states crosses over from

$$N(\omega) \sim \omega^{d-1} \quad , \quad \omega < \omega_{c.o.} \tag{2a}$$

to

$$N(\omega) \sim \omega^{\bar{\bar{d}}-1} \quad , \quad \omega_{c.o.} < \omega < \omega_{fD} \quad , \tag{2b}$$

where the fracton Debye frequency, ω_{fD}, is given by

$$\omega_{fD} = \omega_D (L/a)^{\theta/2} \quad , \tag{3}$$

where ω_D is the extrapolated Euclidean Debye frequency.

The remarkable property of many "disordered" systems in nature is that $\omega_{c.o.}$, or L, may cover the majority of the phase space. Thus, in glasses [4] claims that L~60 Å, or $\omega_{c.o.}$~8 K. For gels, or percolating networks, the $\omega_{c.o.}$ depends on the percolation correlation length:

$$\omega_{c.o.}(p) \sim \xi_p^{-[1+(\theta/2)]} \tag{1a}$$

where $\xi_p \sim (p-p_c)^{-\nu}$, p being the occupation probability, and p_c being the critical percolation probability (p_c = 1/2 for (d = 2) bond percolation, for example, on a square network). By varying p, one can thereby change $\omega_{c.o.}$ at will.

An even more remarkable property of percolating networks is the very recent conjecture [7] that $\bar{\bar{d}}$ = 4/3, 1 < d ⩽ 6. This provides the much sought after relationship between the transport exponent t (the conductivity scales like $(p - p_c)^t$) with the "single particle" exponents ν and β (the probability of belonging to the infinite cluster scaling as $(p - p_c)^\beta$). The basis for this conjecture is as follows.

It is easy to show [7] that, in general,

$$\bar{\bar{d}} = 2(d\nu - \beta)/(t - \beta + 2\nu). \tag{4}$$

For d = 6, it is generally accepted [8] that β = 1, ν = 1/2, and t = 3. This leads immediately from (4) to $\bar{\bar{d}}$ = 4/3. Taking "accepted" values for β, ν, and t, for d = 2,3,4, and 5 one finds [7] $\bar{\bar{d}}$ = 1.36, 1.42, 1.39, and 1.44, respectively. Hence <u>if</u> one conjectures $\bar{\bar{d}}$ = 4/3, independent of d between 1 < d ≤ 6, one finds from (4) that

$$t = (1/2)\,[\nu(3d - 4) - \beta]. \tag{5}$$

This allows us to use, in d = 2, the den Nijs' conducture (now proven), [9]: ν = 4/3 and β = 5/36. Using (5), we find t = 91/72 = 1.264. This value is very close to what is now generally regarded as the best estimate of t at d = 2 by DERRIDA and VANNIMENUS [10], t = 1.28.

A Monte Carlo simulation of diffusion on (d = 2) percolating networks by BEN AVRAHAM and HAVLIN [2] also provides evidence in favor of the scaling

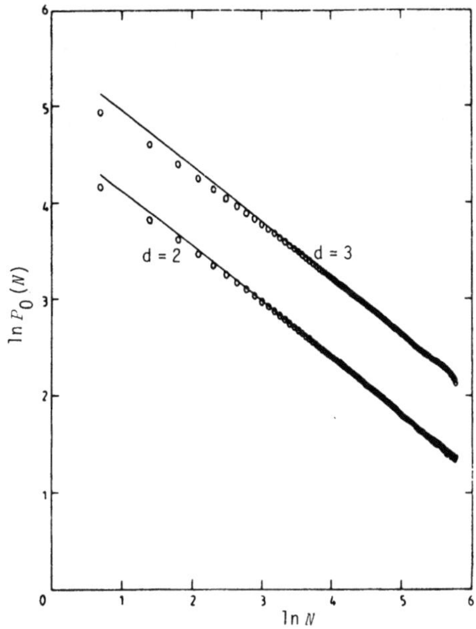

Fig. 1. Plot of ln $<P_0(t)>$ against lnN. The open circles are the numerical data and the straight line is the best fit (from [2], p. L-695)

theory results of [7]. They simulated critical clusters on square (d = 2) and cubic (d = 3) lattices. From Sec. I, the diffusion constant on a fractal was stated to scale with distance r as $D(r) \sim r^{-\theta}$. Thus, the mean square distance traversed by a diffusing particle after time t equals

$$< r^2(t) > \sim t^{2/(2+\theta)} \quad . \tag{6}$$

The probability of a particle located at the origin at t = 0 being found at the origin at time t is therefore

$$\begin{aligned} < P_0(t) > &\sim < r^2(t) > ^{-\bar{d}/2} \\ &\sim t^{-\bar{d}/(2+\theta)} \quad . \end{aligned} \tag{7}$$

BEN-AVRAHAM and HAVLIN [2] compute this quantitiy as a function of the number of Monte Carlo steps N.

We identify lnN with lnt, so that the slopes in Fig. 1 give a direct measure of $\bar{\bar{d}}/2$. Notice that the straight lines for d = 2 and 3 in Fig. 1 are parallel. Thus, $\bar{\bar{d}}$ appears to be independent of d for d = 2 and 3 as claimed. Ben-Avraham and Havlin actually find $\bar{\bar{d}} = 1.26 \pm 0.10$, "somewhat lower but in overall good agreement with the predictions (4/3) of Alexander and Orbach ...."

The percolating network's fractal geometry, at least, appears to allow one to relate directly the "transport" exponent t to the "thermodynamic" exponents ν and β. RAMMAL and TOULOUSE [6] have taken this concept further with regards to the conductance on a fractal, and YU, CHAIKIN, and ORBACH [5] have computed the scaling property of the elastic compliance on a fractal.

Taken together, the famous DE GENNES relation [5] can easily be proven: the conductance/conductivity scales with the force constant/bulk modules on a fractal.

It has been claimed by STAPELTON *et al.* [3] that fractional exponents observed in the spin-lattice relaxation time of low spin iron in polypeptides can be traced to their fractal vibrational properties. For temperatures generally greater than about 4 K (to at least 20 K, beyond which measurements become difficult), COLVIN *et al.* [3] report $1/T_1 \sim T^x$, with x varying from 5.86 to 6.32. Assuming (this may not be correct) the local density of states as seen from the iron site to be the same as the "bulk" density of fracton states, $x = 3 + 2\bar{\bar{d}}$. Stapleton determines $\bar{d}$ from structural X-ray measurements. He seems to find a high correlation between $(x - 3)/2$ ($\bar{\bar{d}}$?) and $\bar{d}$:

	$(x - 3)/2$	$\bar{d}$
Cytochrome C551	1.43 ± 0.09	1.44 ± 0.04
metMb OH	1.61 ± 0.05	1.53 ± 0.03
Cytochrome C	1.66 ± 0.05	1.65 ± 0.05

For a (d = 3) self-avoiding random walk, $\bar{\bar{d}} = 1$ and $\bar{d} = 5/3$. For a (d = 3) percolating network, $\bar{\bar{d}} = 4/3$ and $\bar{d} = 2.5$. Clearly, some problems remain if these T_1 measurements are to be explained in terms of fracton excitations.

Finally, we point out the unusual thermal properties of amorphous systems. KELHAM and ROSENBERG [11] have examined the thermal conductivity and specific heat of epoxy-resin from 0.1 - 80 K. Their results are quite consistent with those of ZAITLIN and ANDERSON [12] who reported on "phonon thermal transport in noncrystalline materials". Briefly, the specific heat of epoxy-resin was analyzed in terms of a vibrational state density,

$$N(\omega) \sim \omega^2 \quad , \quad \hbar\omega/k < 8\ \mathrm{K} \tag{8a}$$

$$N(\omega) \sim \omega - \text{constant}, \quad 8\ \mathrm{K} < \hbar\omega/k < 50\ \mathrm{K} \tag{8b}$$

$$N(\omega) \sim \text{constant}' \quad , \quad 50\ \mathrm{K} < \hbar\omega/k < \hbar\omega_{max}/k \ . \tag{8c}$$

The "phonon" mean free path was extracted from the thermal conductivity. Using the standard kinetic theory formula, it was found to have the remarkable form:

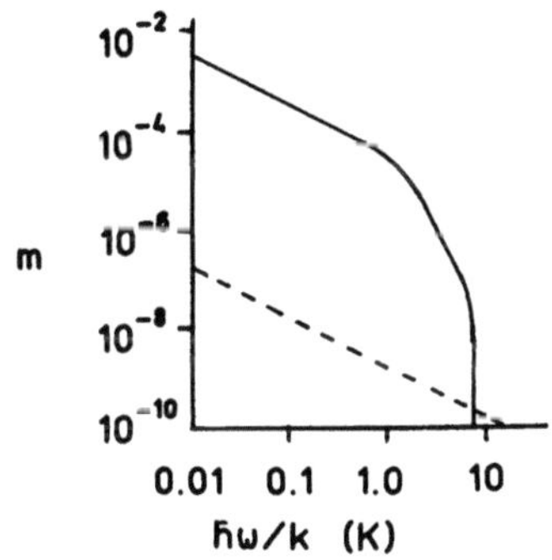

Fig. 2. The variation of the mean free path (solid line) with frequency at 100 K which was used to fit the thermal conductivity of epoxy-resin [11]. The corresponding phonon wavelengths are shown by the broken line

Note that the phonon mean free path falls precipitously to zero at 8 K, the same (crossover?) temperature at which $N(\omega)$ changes from ω^2 to ω. Note also that the "mean free path" is less than the corresponding phonon wavelength at $\hbar\omega/k > 8$ K, a condition which implies the phonon is no longer a satisfactory description of the vibrational properties of the system.

We therefore propose [4] that the fundamental thermal excitations in epoxy-resin, and perhaps in all non-crystalline materials, from the similar behaviors noted in [12] for energies greater than approximately 8 K may be fractons. From (2b), $N(\omega) \sim \omega^{\bar{\bar{d}}-1}$, so that $\bar{\bar{d}} = 2$ from (8b). Rammal and Toulouse have examined the conductance of fractals, finding a length scale dependent conductance $\sum(r) \sim r^{(\bar{d}/\bar{\bar{d}})(\bar{\bar{d}} - 2)}$. If we associate this exponent with the β of ABRAHAMS et al. [13], we see that $\bar{\bar{d}} = 2$ leads to $\beta = 0$, or marginally localized states. Hence, excitations in epoxy-resin with energies greater than 8 K may best be described as localized fractons. That transport that does occur of modes in the fracton range may best be described as a kind of Mott variable range-rate hopping rather than a wavelike propagation.

We emphasize that nearly all amorphous materials behave similarly [12]. Hence, fracton excitations may be a very general feature of these systems [14].

III. Thermodynamics of Fractal Structures

We derive some useful relations for the density of states on a fractal using the results of [4] and [6]. These will be applied to percolating networks at the end of this Section, and a speculation will be made as to the origin of the two level systems in amorphous materials.

The microscopic vibrational frequency scale is set by

$$\omega_0^2 = K/M \quad , \tag{9}$$

where K is a (range-dependent) force constant, and M the mass within that scale of range. Now we know that M scales with distance as $r^{\bar{d}}$ and RAMMAL and TOULOUSE [6] find the conductance to scale with $r^{(\bar{d}/\bar{\bar{d}})(\bar{\bar{d}}-2)}$. YU et al. [5] have shown that the force constant scales with the conductance, so that

$$\omega_0^2 \sim r^{-2\bar{d}/\bar{\bar{d}}}$$

or

$$\omega_0 \sim r^{-[1 + (\theta/2)]} \quad . \tag{10}$$

This is a kind of "Debye" law for fractons ($r^{-1} \sim k$).

This strange form for ω_0 is completely consistent with (2b) for $N(\omega)$ in the fracton regime, but it raises questions concerning the precise crossover in $N(\omega)$ between the phonon (long length scale) and fracton (short length scale) regimes. We quote from [4] to avoid tedious algebra. The phonon and fracton densities of states, suitably normalized, are:

$$N_{ph}(\omega) = d(L/a)^{d} \, [(\omega)^{d-1}/(\omega_D)^{d}] \quad , \quad \omega < \omega_{c.o.} \quad , \tag{11}$$

$$N_{fr}(\omega) = \bar{d}(L/a)^{\bar{d}} \, [(\omega)^{\bar{d}-1}/(\omega_{fD})^{\bar{d}}] \quad , \quad \omega_{c.o.} < \omega < \omega_{fD} \quad , \tag{12}$$

in an obvious notation. The crossover frequency is given by

$$\omega_{c.o.} = (a/L)\omega_D \quad , \tag{13}$$

so that, at crossover,

$$N_{fr}(\omega_{c.o.})/N_{ph}(\omega_{c.o.}) = \bar{\bar{d}}/d \quad . \tag{14}$$

For percolating networks, $\bar{\bar{d}} < d$, and one would find a sharp discontinuity in $N(\omega)$ at $\omega_{c.o.}$. Unfortunately, the direction of the discontinuity appears to be opposite to that found in amorphous materials. Thus, for epoxy-resin, (8) implies that $\bar{\bar{d}} = 2$, and (14) would require a drop in $N(\omega)$ by a third at $\omega_{c.o.}$. Instead, experimentally $N(\omega)$ jumps by about a factor of five at what we interpret to be $\omega_{c.o.}$ (i.e., at $\hbar\omega/k = 8$ K). We have no explanation for this discrepancy at present.

We now apply fracton theory to percolating networks. We shall arrive at results which are suggestive of amorphous system behavior. That is, we shall exhibit "missing" vibrational modes which are to be associated with the center of mass motion of the finite clusters. We speculate that these degrees of freedom may be related to the two level systems found in almost all amorphous structures.

For percolating networks away from p_c (i.e., ξ_p finite), (12) is to be associated with the density of states for vibrations of the infinite cluster. [4] shows the analogous quantitiy for the finite clusters to be

$$N_{fr}^{f.c.}(\omega) = L^{d}(\bar{d}/\omega)(\omega/\omega_{fD})^{\bar{d}} \, [(a/R(\omega))^{(\bar{d}-d)} - (a/L)^{(\bar{d}-d)}] \quad , \tag{15}$$

where $R(\omega)$ is the cluster radius such that only clusters of size $R > R(\omega)$ contain modes of frequency ω:

$$R(\omega)/a = (\omega/\omega_{fD})^{-2/(2+\theta)} \quad . \qquad (16)$$

Adding (15) to (12) one finds the total fracton density of states on a percolating network to equal

$$N_{fr}{}^{total}(\omega) = \bar{\bar{d}}(L/a)^{d}(1/\omega)(\omega/\omega_{fD})^{2d/(2+\theta)} \quad . \qquad (17)$$

Integrating (17) over frequencies from $\omega_{c.o.}$ to ω_{fD}, we find only $\bar{d}/d$ modes per atom. We show in [4] that the totality of these modes equal the number of finite clusters. Thus, we can associate these modes with bulk cluster motion, reminiscent of the macroscopic hindered rotation or tunneling sometimes associated with two level systems in amorphous systems. The mass distribution of the clusters is easily calculated:

$$\rho(M) \sim M^{-(d+\bar{d})/\bar{d}}$$

but without an explicit coupling term for finite cluster motion, one cannot calculate the frequency distribution. This makes it difficult to compare with the energy spread and distribution of two level systems. It remains, nonetheless, an interesting speculation.

IV. Summary and Conclusions

We have exhibited the three dimensionalities required to describe the dynamics of physical systems on a fractal structure. We have shown how the density of vibrational states depends on frequency on the "fracton" regime, i.e., for length scales less than that over which the system appears homogeneous. We have discussed the scaling properties of percolating networks. And we have attempted to make a connection between fracton thermodynamics and transport properties and the physical properties of amorphous solids.

There are many questions left to be resolved. The first and most important is whether any physical system does exhibit fractal structure. If one believes that amorphous materials do, up to a limited length scale (~60 Å), then how can the discrepancy in the direction of density of states discontinuity at $\omega_{c.o.}$ for amorphous systems be resolved? Is there a relationship between the center of mass motion of the finite clusters in percolating networks and the two level systems found in most amorphous systems? When attempting to fit the T_1 data of STAPLETON [3], one must ask if the local fracton density of states is the same as for the bulk. And how do fractons contribute to impurity-vibrational interactions and excitation transfer?

In summary, this paper may have raised more questions that it answered. Its purpose has been to examine whether random systems obey a "higher symmetry" which may enable one to predict physical properties where no obvious general relationships had been thought possible before.

This work has been supported by U.S. National Science Foundation and the U.S. Office of Naval Research.

References

1. B.B. Mandelbrot, "The Fractal Geometry of Nature" (Freemann, San Francisco, 1982), and "Fractals: Form, Chance and Dimension" (Freeman, San Francisco, 1977).

2. D. Ben-Avraham and S. Havlin, J. Phys. A: Math. Gen. 15, L-691 (1982); R.F. Voss, R.B. Laibowitz, and E.I. Allessandrini, Phys. Rev. Lett. 49, 1441 (1982); A. Kapultnik and G. Deutscher, Phys. Rev. Lett. 49, 1444 (1982).

3. H.J. Stapleton, J.P. Allen, C.P. Flynn, D.G. Stinson, and S.R. Kurtz, Phys. Rev. Lett. 45, 1456 (1980); J.P. Allen, J.T. Colvin, D.G. Stinson, C.P. Flynn, and H.J. Stapleton, Biophysical J. 38, 299 (1982); J.T. Colvin, G.C. Wagner, J.P. Allen, and J.J. Stapleton, Bull. of Biophysical Society, February, 1983, meeting.

4. S. Alexander, C. Laermans, R. Orbach, and H.M. Rosenberg, submitted for publication to the Physical Review, 1983.

5. P.G. de Gennes, J. Physique-Letters 37, L-1 (1976); "Scaling Concepts in Polymer Physics" (Cornell University Press, Ithaca, New York, 1979), p. 138; Kin Wah Yu, P. Chikin and R. Orbach, submitted to Rapid Communications, Physical Review, 1983.

6. R. Rammal and G. Toulouse, J. Physique-Letters 44, L-13 (1983).

7. S. Alexander and R. Orbach, J. Physique-Letters 43, L-625 (1982).

8. D. Stauffer, Phys. Rep. 54, 1 (1979).

9. M.P.M. den Nijs, J. Phys. A: Math. Gen. 12, 1857 (1979).

10. B. Derrida and J. Vannimenus, J. Phys. A: Math. Gen. 15, L-557 (1982).

11. S. Kelham and H.M. Rosenberg, J. Phys. C: Sol. St. Phys. 14, 1737 (1981).

12. M.P. Zaitlin and A.C. Anderson, Phys. Rev. B 12, 4475 (1975)

13. E. Abrahams, P.W. Anderson, D.C. Licciardello, and T.V. Ramakrishnan, Phys. Rev. Lett. 42, 673 (1979).

14. Very recent Mössbauer measurements of P. Boolchand on Sn sites in GeS_2 glass (private communication) exhibit a temperature dependence different from that expected for a simple Einstein plus Debye vibrational spectrum. It is possible that an $N(\omega)$ similar in form to Eq. 8 may provide a more satisfactory fit.

Hopping Transport in Disordered Systems

J. Klafter

Corporate Research Science Laboratories, Exxon Research and Engineering Company, Linden, NJ 07036, USA

Abstract

We present a study of transport in various disordered systems based on three different formalisms. We focus on the questions of diffusion of excitations and charge carriers in homogeneously disordered systems and on the conductivity in inhomogeneous (granular) systems. We derive first the generalized master equation as the basic equation for transport in a disordered system and then relate it to the continuous time random walk formalism and to the effective medium approximation approach. In the inhomogeneous case we use the critical path method to calculate the temperature dependence of the d.c. conductivity.

I. Introduction

Recently there has been considerable interest in various aspects of incoherent transport in condensed disordered media [1-9]. Both charge transport [4-6] and electronic energy transfer [1-3,8,9] have been extensively studied, resulting in interesting short time (a.c. limit) and long time (d.c. limit) behaviors. The disordered nature of the medium over which transport takes place has been shown to influence strongly experimentally accessible basic quantities like exciton lifetimes, the electron hole recombination rates and the a.c. and d.c. conductivities.

In this paper we will consider various random environments and review the methods used to analyze transport in these environments. We will concentrate on three cases: (a) Site disorder arising from the substitutional occupancy of sites on an ordered lattice by impurities,(b) Bond disorder, where the values of nearest-neighbor transfer rates on a lattice are assigned according to a given distribution function,(c) Granular disorder which results from inhomogeneous mixtures of metal and insulator. For each of these disorder classes one has to apply a different theoretical technique in order to calculate the relevant quantities.

In the case of site disorder systems we will derive the time-dependent diffusion coefficient in terms of the continuous time random walk (CTRW) which has been shown to be a powerful way for studying the general problems of transport in disordered systems [4,5]. For bond disorder systems we apply the effective medium approximation (EMA) [7,10-12]. The EMA is closely related to the coherent potential approximation for the electronic properties of alloys [13]. In the case of granular systems we will calculate the temperature-dependent d.c. conductivity using the critical path method [14,15]. The purpose of this paper is to provide a framework for the problems of transport in the various random systems. In Sec. II we consider the transport of excitations on a lattice randomly occupied by guest impurities. We derive the exact equation governing the transport averaged over all configurations. In Sec. III we show that this equation is equivalent to the CTRW equation and we calculate the diffusion coefficient D(t) in the CTRW framework. In Sec. IV we treat the bond disorder model and make connection to the previous approach. In Sec. V we study the temperature dependence of conductivity for high resistivity granular systems. It is shown that the low field conductivity $\exp(-A/T^{\alpha})$ with $\alpha=1/2$ is obeyed over large temperature ranges.

II. The Generalized Master Equation

In this section we derive the basic equation for transport on a lattice which is fractionally occupied by substituional impurities in a random manner. We assume that an electronic excitation migrates only among the impurities. We refer to this case as site disorder. The equation of motion of the probability $P_n(t)$ of an excitation (or an electron) being on site n at time t is the master equation

$$\dot{P}_n(t) = \sum_{m\neq n} [W_{mn}P_m(t) - W_{nm}P_n(t)] \tag{2.1}$$

where W_{mn} is the transition rate from site m to site n, and for convenience we will assume the high-temperature limit so that $W_{mn}=W_{nm}$. We may write Eq.(2.1) as a matrix equation

$$\dot{\underset{\sim}{P}}_n(t) = \sum_m (\underset{\approx}{V})_{mn}\underset{\sim}{P}_m(t) \tag{2.2}$$

where

$$(\underset{\approx}{V})_{mn} = (1-\delta_{nm})W_{mn} - \delta_{mn}\sum_{q\neq n} W_{nq} \,. \tag{2.3}$$

Eq. (1) holds for every configuration of the impurity molecules, but with transition rates which depend on the configuration. Each configuration gives rise to a set of interimpurity distances on the lattice, which in turn gives rise to a set of W_{mn}. For each configuration we may solve for $P_n(t)$, assuming that the excitation was on site n=o at t=o,

$$P_n(t) = \{\exp(t\underset{\approx}{V})\}_{no} \,. \tag{2.4}$$

Averaging over all configurations, we obtain

$$\langle P_n(t)\rangle = \langle \exp(t\underset{\approx}{V}\rangle_{no} \tag{2.5}$$

and by Laplace transform

$$\langle P_n(u)\rangle = \langle (u-\underset{\approx}{V})^{-1}\rangle_{no} \,. \tag{2.6}$$

An exact form for this can be derived using the ZWANZIG projection operator method (See Refs. [8,16] for details). Defining a projection operator as

$$Q... = \langle ....\rangle, \tag{2.7}$$

we have for the configurational average of $P_n(u)$,

$$\langle P_n(u)\rangle = [u-\langle \underset{\approx}{V}\rangle - \langle \delta\underset{\approx}{V}[u-(1-Q)\underset{\approx}{V}]^{-1}\delta\underset{\approx}{V}\rangle]^{-1}_{no}$$
$$= [u-\underset{\approx}{M}(u)]^{-1}_{no} \tag{2.8}$$

where $\delta\underset{\approx}{V} = \underset{\approx}{V}-\langle\underset{\approx}{V}\rangle$ and $\underset{\approx}{M}(u)$ is the selfenergy matrix. Rewriting (2.8) we now have

$$\langle P_n(u)\rangle = [u^{-1} + u^{-1}\ \underset{\approx}{M}(u)\ (u-\underset{\approx}{M}(u))^{-1}]_{no} \tag{2.9}$$

and by conservation of probability $\sum_{\ell} M_{n\ell}(u)=o$,

$$u\langle P_n(u)\rangle = \delta_{no} + \sum_{\ell} M_{n\ell}(u)\langle P_\ell(u)\rangle$$
$$= \delta_{no} + \sum_{\ell\neq n}\{M_{n\ell}(u) \langle P_\ell(u)\rangle - M_{\ell n}(u)\langle P_n(u)\rangle\}; \tag{2.10}$$

one then arrives at a generalized master equation (GME) for the $\langle P_n(t)\rangle$ with selfenergy matrix elements that contain the information about disorder:

$$\langle \dot{P}_n(t)\rangle = \sum_{\ell\neq n}\int_o^t d\tau\{M_{n\ell}(t-\tau)\ \langle P_\ell(t)\rangle - M_{\ell n}(t-\tau)\ \langle P_n(t)\rangle\} \tag{2.11}$$

with the initial condition $\langle P_n(o)\rangle=\delta_{no}$. For a discussion on different initial conditions see Ref. [17]. Thus, the basic equation for transport in a disordered system is a GME, Eq. (2.11). It is worth noticing that the GME in Laplace representation, Eq. (2.10), is identical to the Laplace transform of Eq. (2.1) with the random rates W_{mn} replaced by the <u>effective</u> (non random) rates $M_{mn}(u)$.

Upon averaging, the system is translationally invariant, representing an effective system, therefore

$$M(\underset{\sim}{k},t) = \sum_{n\neq\ell}\exp[-i\underset{\sim}{k}\cdot(\underset{\sim}{n}-\underset{\sim}{\ell})]M_{n\ell}(t) \tag{2.12a}$$
$$M(\underset{\sim}{k}=o,t) = \sum_{n\neq\ell} M_{n\ell}(t) = -M_{nn}(t). \tag{2.12b}$$

Eq. (2.10) in $\underset{\sim}{k}$ representation is

$$u\langle P_{\underset{\sim}{k}}(u)\rangle = 1 + [M(\underset{\sim}{k},u) - M(\underset{\sim}{o},u)]\ \langle P_k(u)\rangle \tag{2.13}$$

so that

$$\langle P_{\underset{\sim}{k}}(u)\rangle = [u + M\ (\underset{\sim}{o},u) - M(\underset{\sim}{k},u)]^{-1} \tag{2.14a}$$

and

$$\langle P_n(u)\rangle = N^{-1}\sum_{\underset{\sim}{k}}\frac{e^{i\underset{\sim}{k}\cdot\underset{\sim}{n}}}{u+M(\underset{\sim}{o},u)-M(\underset{\sim}{k},u)}$$
$$= \frac{N^{-1}}{u+M(\underset{\sim}{o},u)}\sum_{\underset{\sim}{k}}\frac{e^{i\underset{\sim}{k}\cdot\underset{\sim}{n}}}{1-[M(\underset{\sim}{k},u)/(u+M(\underset{\sim}{o},u)]}\ . \tag{2.14b}$$

III. Diffusion in the Framework of the CTRW

We now consider the relation between the GME of Sec. II and the CTRW and study diffusion over a disordered system in the CTRW framework [8]. Defining a structure function $\lambda(\underset{\sim}{k},u)$ by

$$\lambda(\underset{\sim}{k},u) \equiv \frac{M(\underset{\sim}{k},u)}{u+M(\underset{\sim}{0},u)} \tag{3.1}$$

we rewrite Eq. (2.14b) as

$$\begin{aligned}\langle P_n(u)\rangle &= \frac{N^{-1}}{u+M(\underset{\sim}{0},u)} \sum_{\underset{\sim}{k}} \frac{e^{i\underset{\sim}{k}\cdot\underset{\sim}{n}}}{1-\lambda(\underset{\sim}{k},u)} \\ &= \frac{1-\lambda(\underset{\sim}{0},u)}{N.u} \sum_{\underset{\sim}{k}} \frac{e^{i\underset{\sim}{k}\cdot\underset{\sim}{n}}}{1-\lambda(\underset{\sim}{k},u)} \quad .\end{aligned} \tag{3.2}$$

One can now calculate $\psi(\underset{\sim}{r},u)$ of the CTRW [4,16] from

$$\psi(\underset{\sim}{r},u) = N^{-1} \sum_{\underset{\sim}{k}} \lambda(\underset{\sim}{k},u)e^{i\underset{\sim}{k}\cdot\underset{\sim}{r}} \, . \tag{3.3}$$

$\psi(\underset{\sim}{r},u)$ is the Laplace transform of $\psi(\underset{\sim}{r},t)$ which is the probability density that the displacement between hops is $\underset{\sim}{r}$ at the time t. We will refer to $\psi(\underset{\sim}{r},t)$ as the stepping time distribution. The following relationships are now easily verified:

$$\begin{aligned}\psi(u) \equiv \sum_{\underset{\sim}{r}\neq 0} \psi(\underset{\sim}{r},u) &= M(\underset{\sim}{0},u)\ [u+M(\underset{\sim}{0},u]^{-1} \\ &= \lambda(\underset{\sim}{0},u)\end{aligned} \tag{3.4}$$

so that

$$M(\underset{\sim}{0},u) = u\ \psi(u)/[1-\psi(u)] \quad . \tag{3.5}$$

From Eqs. (3.1) and (3.3) we obtain

$$\psi(\underset{\sim}{r},u) = M_{\underset{\sim}{r}0}\ [u+M(\underset{\sim}{0},u)]^{-1} \quad . \tag{3.6}$$

Eqs. (3.4)-(3.6) constitute the exact relationship between the self-energies of Sec. II and the stepping time distribution function of the CTRW. This connection allows us to calculate the mean square displacement of an excitation, $\langle r^2(t)\rangle$ and therefore the diffusion coefficient, D(t), in a variety of ways. The GME can be used, wherein a microscopic calculation of the self-energy, $M(\underset{\sim}{k},u)$, must be made. This route has been taken by HAAN and ZWANZIG [1], by GOCHANOUR, ANDERSEN and FAYER [2] and by MOVAGHAR [18]. It is related to the effective medium approaches derived by ODAGAKI and LAX [19] and by WEBMAN [10]. Alternatively, the CTRW can be used by calculating $\psi(\underset{\sim}{r},t)$ as proposed by SCHER and LAX [4] for the problem of conduction in amorphous semiconductors and studied further by GODZIK and JORTNER [3] for the problem of electronic energy transfer in mixed crystals. In what follows we will consider diffusion in the framework of the CTRW.

The Laplace transform of the mean square displacement $\langle r^2(t)\rangle$ of a walker (an excitation or a charge carrier) is given by (L represents the Laplace transform)

$$\langle r^2(u)\rangle = L\ \langle r^2(t)\rangle = -\nabla^2_{\underset{\sim}{k}}\ P_{\underset{\sim}{k}}(u)|_{\underset{\sim}{k}=o} \quad . \tag{3.7}$$

From Eqs. (3.2) and (3.4) - (3.6) we obtain

$$P_{\underset{\sim}{k}}(u) = \frac{1-\psi(u)}{u}\ \frac{1}{1-\lambda(\underset{\sim}{k},u)} \tag{3.8}$$

so that

$$\langle r^2(u)\rangle = \frac{1}{u[1-\psi(u)]}\ \sum_{\underset{\sim}{r}} r^2\psi(\underset{\sim}{r},u) \quad . \tag{3.9}$$

The time-dependent diffusion coefficient D(t) of dimensionality Δ is defined as

$$D(t) \equiv \frac{1}{2\Delta}\ \frac{d}{dt}\ \langle r^2(t)\rangle \tag{3.10}$$

therefore

$$D(u) = \frac{1}{2\Delta[1-\psi(u)]}\ \sum_{\underset{\sim}{r}}\ r^2\psi(\underset{\sim}{r},u) \quad . \tag{3.11}$$

Thus, $\psi(\underset{\sim}{r},t)$ is sufficient to determine in the CTRW approach the diffusion coefficient D(t) uniquely. We will use the approximation suggested by SCHER and LAX [4] to calculate $\psi(t)$ for an excitation migrating among impurities as in the case discussed in Sec. II:

$$\psi(t) = -\frac{d}{dt}\,\Psi \tag{3.12}$$

where $\Psi(t)$ is the configurational average of the probability of an excitation on an initially excited impurity not to decay at time t to other impurities. The funtion $\Psi(t)$ has been extensively studied in many fields [20-22] where it has been shown that

$$\Psi(t) = \prod_{\underset{\sim}{r}}{}'\ [1-p + pe^{-tw(\underset{\sim}{r})}] \quad . \tag{3.13}$$

Here $w(\underset{\sim}{r})$ denotes the transfer rate from an impurity at the origin to an impurity at position $\underset{\sim}{r}$ and p is the probability of each site being occupied by an impurity. Eq. (3.13) is exact and valid for all times and all concentrations of the impurities among which migration takes place. It also holds regardless of the particular form of $w(\underset{\sim}{r})$. For derivations see Ref. [22]. In the continuum approximation one obtains for small p (p<<1)

$$\Psi(t) = \exp(-p\rho\int\{1-\exp[-tw(\underset{\sim}{r})]\}d\underset{\sim}{r}) \tag{3.14}$$

where ρ is the density of molecules in the medium. Thus, $p\rho$ is the concentration of the impurities participating in the transport. From Eqs. (3.12) and (3.13) follows [3,4,23]:

$$\psi(t) = \sum_{\underset{\sim}{r}} pw(\underset{\sim}{r})e^{-tw(\underset{\sim}{r})}\ \{\prod_{\substack{\underset{\sim}{r}'\neq\underset{\sim}{r}\\ \underset{\sim}{r}'\neq o}} [1-p+pe^{-tw(\underset{\sim}{r}')}]\} \equiv \sum_{\underset{\sim}{r}} \psi(\underset{\sim}{r},t) \quad . \tag{3.15}$$

In the continuum approximation and for low p,

$$\psi(\underset{\sim}{r},t) \simeq pw(\underset{\sim}{r}) \exp[-tw(\underset{\sim}{r})]\Psi(t) \tag{3.16}$$

where we replaced the product in Eq. (3.15) by its form (3.14). Returning to Eq. (3.11) and using Eq. (3.16) we have [23]

$$\sum_{\underset{\sim}{r}} r^2\psi(\underset{\sim}{r},u) = L[\Psi(t)\Omega(t)] \tag{3.17}$$

with

$$\Omega(t) = p\rho\int d\underset{\sim}{r}r^2w(\underset{\sim}{r})e^{-tw(\underset{\sim}{r})} \tag{3.18}$$

from which Eq. (3.11) can be evaluated as:

$$D(u) = L[\Psi(t)\Omega(t)]/[2\Delta u\Psi(u)] \tag{3.19}$$

where $\Psi(u)$ is the Laplace transform of $\Psi(t)$. In the SCHER-LAX approximation we can calculate D(t) by essentially knowing the configurational averaged probability $\Psi(t)$. We now consider for example multipolar interactions in three dimensions. In the multipolar case $w(\underset{\sim}{r})$ in Eq. (3.13) is

$$w(\underset{\sim}{r}) = \frac{1}{\tau}\left(\frac{d}{r}\right)^s \tag{3.20}$$

where we have chosen to scale $w(\underset{\sim}{r})$ in terms of the transfer rate τ^{-1} of an impurity pair at nearest distance d. The parameter s equals 6 for dipole-dipole, 8 for dipole-quadrupole,etc. Substituting $w(\underset{\sim}{r})$ of Eq. (3.20) into Eq. (3.14) one has [22]:

$$\Psi(t) = \exp[-C_2t^{3/s}] \tag{3.21}$$

with

$$C_2 = \Gamma(1-3/s)\,\frac{4\pi}{3}\,p\rho\left(\frac{d^s}{\tau}\right)^{3/s} \tag{3.22}$$

and with Eqs. (3.18) and (3.20)

$$\Omega(t) = C_1t^{5/s-1} \tag{3.23}$$

with

$$C_1 = \Gamma(1-5/s)\,\frac{4\pi p\rho}{s}\left(\frac{d^s}{\tau}\right)^{5/s} \quad . \tag{3.24}$$

In order to obtain the diffusion coefficient one generally has to evaluate the inverse Laplace transform of Eq. (3.19). In our case, for short and long times, the behavior of D(t) can be found directly from $\Omega(t)$ and $\Psi(t)$ using Tauberian theorems [24]. For short times $\Psi(t)$ is a slowly varying function of t, so that

$$\Psi(u) \sim \frac{1}{u} \tag{3.25}$$

and, using Eq. (3.23)

$$L[\Psi(t)\Omega(t)] \sim \frac{C_1\Gamma(5/s)}{u^{5/s}} \quad . \tag{3.26}$$

From Eq. (3.19) it now follows:

$$D(t) \sim L^{-1}\left[\frac{C_1\Gamma(5/s)}{6u^{5/s}}\right] \sim \frac{C_1}{6}\, t^{\,5/s-1}$$

$$= \Gamma(1-5/s)\,\frac{2\pi p\rho}{3s}\left(\frac{d^s}{\tau}\right)^{5/s} t^{\,5/s-1} . \tag{3.27}$$

This result concurs with the expression obtained by expanding the functions in powers of 1/u by GODZIK AND JORTNER [3].

For long times (small u) one has

$$L[\Psi(t)] = \sum_{j=o}^{\infty} \frac{(-u)^j}{j!}\,\tau_{j+1} \tag{3.28}$$

and

$$L[\Omega(t)\ \Psi(t)] = C_1 \sum_{j=o}^{\infty} \frac{(-u)^j}{j!}\,\tau_{j+5/s} \tag{3.29}$$

with

$$\tau_k \equiv \int dt t^{k-1}\Psi(t) = \frac{s}{3}\,\Gamma(ks/3)\,C_2^{-ks/3} . \tag{3.30}$$

Since the moments τ_j of Eqs. (3.28) and (3.29) exist, one has from the Tauberian theorems for longer times

$$D(t) \sim \frac{C_1\tau_{5/s}}{6\tau_1} + \Theta(1/t) \tag{3.31}$$

typical for time-independent diffusion. The crossover from the short time behavior to the diffusive limit has been analyzed in [2,3]. The diffusive limit occurs at later times if s gets larger and is also sensitive to the concentration: an increase of the impurity concentrate pushes the diffusive limit to shorter times.

IV. The Effective Medium Approximation (EMA)

In this section we discuss another method for solving Eq. (2.1) in random systems, the self-consistent EMA [10-12,19]. We study the problem of diffusion in a lattice characterized by random values of the transfer rates W_{nm} in Eq. (2.1) between pairs of nearest-neighbor sites. This is a bond problem where the values W assigned to the bonds are random variables distributed according to a probability density $\rho(W)$. The bond problem may provide a tractable model for the site disorder case discussed in Secs. II and III. Instead of solving the actual site disorder problem, one can map the system onto an ordered lattice where the nearest-neighbor transfer rates obey a distribution $\rho(W)$ which is constructed from the transfer law between two sites and the distribution of nearest-neighbor distances in the original problem [7,8]. The EMA for diffusion in random systems has been proposed for the one-dimensional case by ALEXANDER et al. [12] and for all dimensions by WEBMAN [10] and by ODAGAKI and LAX [19]. We follow here the approach proposed by WEBMAN [10] and further studied by WEBMAN and KLAFTER [11].

In terms of the GME of Sec. II, in the EMA we replace the effective rates $M_{n\ell}(u)$ of Eq. (2.10) by a nearest-neighbor effective rate $W_M(u)$ so that

$$u\langle P_n(u)\rangle = \delta_{no} + \sum_\ell W_M(u) \langle P_\ell(u)\rangle \tag{4.1}$$

and from Eqs. (2.12)

$$M(\underset{\sim}{k},u) = W_M(u) \sum_{n\neq\ell} \exp[-i\underset{\sim}{k}\cdot(\underset{\sim}{n}-\underset{\sim}{\ell})] \tag{4.2a}$$

$$M(\underset{\sim}{k}=o,u) = ZW_M(u) \tag{4.2b}$$

where Z is the lattice coordination number. From Eqs. (3.1) and (3.3) we now obtain that the corresponding stepping time distribution of Sec. III is decoupled in space and time. The Laplace transform of Eq. (2.1) can be rewritten in matrix representation as

$$\underset{\approx}{A}(u)\,\underset{\sim}{P}(u) = \underset{\sim}{1} \tag{4.3}$$

where using the "bracket" notation [25]

$$\underset{\sim}{P}(u) = \sum_n P_n(u)|n\rangle \tag{4.4}$$

and restricting summations to nearest neighbors

$$\underset{\approx}{A}(u) = \sum_{k,\ell} |k\rangle\, [(u + \sum W_{ik})\delta_{k\ell} - W_{k\ell}]\langle\ell| . \tag{4.5}$$

The effective equation (4.1) can be rewritten in the same way

$$\underset{\approx}{A}_M(u)\underset{\sim}{P}(u) = 1 \tag{4.6}$$

with

$$\underset{\approx}{A}_M(u) = \sum_{k,\ell} |k\rangle[(u+ZW_M(u))\delta_{k\ell} - W_M(u)\Delta_{k\ell}]\langle\ell| \tag{4.7}$$

where

$$\Delta_{k\ell} = \begin{cases} 1 & k,\ell \quad NN \\ 0 & \text{otherwise} \end{cases} .$$

In order to determine $W_M(u)$ we represent $\underset{\approx}{A}(u)$ in Eq. (4.5) as a sum of the effective part and a term which contains the random fluctuations and which we want to minimize on the average, an idea similar to the coherent potential approximation (CPA) for electronic states in mixed systems [7,13]

$$\underset{\approx}{A}(u) = \underset{\approx}{A}_M(u) - \delta\underset{\approx}{A}(u) \quad . \tag{4.8}$$

Here [25]

$$\delta\underset{\approx}{A}(u) = \sum_{k\ell} 2\,(W_{k\ell} - W_M(u))\underset{\approx}{Q}_{k\ell} \tag{4.9a}$$

$$\underset{\approx}{Q}_{k\ell} = 1/2\,[|k\rangle - |\ell\rangle]\,[\langle k| - \langle\ell|] \quad . \tag{4.9b}$$

$\underset{\approx}{A}^{-1}(u)$ can now be expressed as [7,25]

$$\underset{\approx}{A}^{-1}(u) = G_M + G_M\, T\, G_M \tag{4.10}$$

where $G_M = \underset{\approx}{A}_M^{-1}$ is the lattice Green's function for the effective medium

$$T = \sum_{k\neq\ell} t_{k\ell} + \sum_{k,\ell\neq m,n} t_{k\ell} G_M t_{mn} + \sum_{\substack{k,\ell\neq m,n\\ m,n\neq p,q}} t_{k\ell} G_M t_{mn} G_M t_{pq} + \dots \tag{4.11}$$

and the t matrix for the kl bond is

$$t_{k\ell} = \frac{[|k\rangle - |\ell\rangle](W_M - W_{k\ell})[\langle k| - \langle\ell|]}{1-[\langle k| - \langle\ell|]G_M[|k\rangle - |\ell\rangle](W_M - W_{k\ell})} \quad . \tag{4.12}$$

From Eq. (4.10) we obtain

$$\langle\underset{\approx}{A}^{-1}\rangle^{-1} = \underset{\approx}{A}_M - \frac{\langle T\rangle}{1+G_M\langle T\rangle} \quad . \tag{4.13}$$

The effective medium condition for $W_M(u)$ is [7,10,11]

$$\langle t\rangle = \int t(W,W_M)\rho(W)dW = 0 \quad . \tag{4.14}$$

Due to the summation restrictions in Eq. (4.11), and the random assignment of values of W to different bonds, this condition leads to the vanishing of the first three terms in ⟨T⟩ so that the first nonzero term is of fourth order in the t matrices.

Using Eq. (4.12), Eq. (4.14) can be written explicitly for any dimensionality

$$\int\rho(w)\,\frac{(W-W_M)dW}{W(1-u\overset{0}{G}_M(u))+W_M(Z/2 - 1 + u\overset{0}{G}_M(u))} = 0 \tag{4.15}$$

where

$$\overset{0}{G}_M(u) = \langle 0|G_M(u)|0\rangle \quad .$$

As an example we take a one-dimensional lattice for which Z=2. Eq. (4.15) then reduces to [11,12]

$$\int\rho(W)\,\frac{(W-W_M)dW}{W+(W_M-W)u\overset{0}{G}_M(u)} = 0 \tag{4.16}$$

and

$$\overset{0}{G}_M(u) = \frac{1}{[u(u+4W_M(u))]^{1/2}} \quad . \tag{4.17}$$

We assume randomness which is characterized by a non singular $\rho(W)$ [12] and we solve for small u (long times). From Eqs. (4.16) and (4.17) we obtain the following result for $W_M(u)$ up to order u [11]

$$W_M(u) = W_M(o) + W_{M1}\sqrt{u} + W_{M2}u + \dots \tag{4.18}$$

where

$$W_M(o) = \langle\frac{1}{W}\rangle^{-1}$$

$$W_{M1} = 1/2 \langle(\frac{1}{W} - \frac{1}{W_M(o)})^2\rangle W_M^{5/2}(o)$$

$$W_{M2} = 3/8 \langle (\frac{1}{W} - \frac{1}{W_M(o)})^2 \rangle^2 W_M^4(o) - 1/4 \langle(\frac{1}{W} - \frac{1}{W_M(o)})^3\rangle W_M^3(o) . \quad (4.19)$$

As discussed in Ref. [11] this small u behavior, Eq. (4.18), obtained by EMA is exact to order u and the lowest order correction to the EMA result is $\sim u^{3/2}$. Similar results have been recently obtained also by different methods [26,27]. The mean square displacement is given in the framework of the EMA by

$$\langle r^2(u)\rangle = ZW_M(u)/u^2. \quad (4.20)$$

This is the EMA equivalent of Eq. (3.9) when $\psi(\underset{\sim}{r},u)$ is decoupled, $\psi(\underset{\sim}{r},u)=f(\underset{\sim}{r})\psi(u)$ with $\psi(u)=ZW_M(u)[u+ZW_M(u)]^{-1}$.

We conclude that the EMA is a well-controlled approximation which can be systematically extended and the correction to which can be obtained.

V. Hopping Conduction in Granular Systems

While in the previous sections we were interested in transport among randomly distributed molecules or due to random bond values, here we present an analysis of conduction in granular systems which consist of inhomogeneous mixtures of metal and non metal [28,29]. We focus on the limit of small metal concentrations which results in small metallic islands (grains) embedded in an insulating matrix. In this case the electrical conductivity is small and highly activated [28-30] and it is believed to be due to the charge hopping among the metallic grains. An interesting feature of these granular systems in the activated regime is the temperature dependence of the d.c. conductivity which follows the $T^{-1/2}$ law: $\ln\sigma \propto -T^{-1/2}$, where σ is the conductivity, and T the temperature [28,29,31-33].

Although there has been strong experimental evidence that hopping conduction in disordered materials follows a temperature dependence of $\ln\sigma \propto -T^{-\alpha}$, where α is a constant ranging from 1/4 to 1 [14,15,34], only the α=1/4 and α=1/3 behaviors have been justified in terms of the variable range hopping proposed by MOTT [34]. However the α=1/2 temperature dependence, which has been observed in diverse granular systems with no apparent common feature, has no unique explanation.

Recently the same $T^{-1/2}$ law has been observed in doped polyacetylenes [35,36] suggesting that the same mechanism of hopping among conducting islands, as in granular systems, may apply also in this case. It has also been demonstrated that conducting patterns can be generated in thin films of various aromatic compounds using electron beam irradiation

[37]. The dependence of conductivity on temperature again follows the $T^{-1/2}$ law over the entire range of temperatures investigated, and the conductivity increases by many orders of magnitudes compared to the nonirradiated films.

In what follows we discuss the conductivity in the activated regime of granular systems, when the systems consist of isolated metallic grains. We will concentrate on the consequences of the distribution of grain sizes and the distribution of the separation of the grains on the temperature behavior.

The methods described in Secs III and IV cannot be directly applied to the present case because of the additional distribution of site energies that we have to include and will be discussed later. We will then follow the critical path method (CPM) of AMBEGAOKAR, HALPERIN and LANGER [14] and SHANTE [15] to analyze the transport problem in random media characterized by both random hopping distances and random grain energies. In the CPM, one denotes by G_{ij} the conductance between any two localized sites (or two conducting grains) i and j. The conductance of the macroscopic medium is then determined by the following argument: Pick a value of conductance G and consider any two sites as connected if $G_{ij}>G$. If G is sufficiently large, the resulting connected sites will only form disjoint clusters. As G is lowered, however, the connected clusters are expected to increase in size until at $G=G_c$, the percolation conductance, an infinite network of connected sites is formed. Since G_c is the largest value of the conductance at which conduction over a continuous path becomes possible, it is identified as the macroscopic conductance of the sample and is assumed to dominate its temperature dependence. We assume [14,15] the conductivity to be $\sigma=G_c/L$, where L is some characteristic length which may be temperature dependent. For possible corrections to this percolation condition, which do not change the dominant temperature dependence, see ref. [15]. The hopping conduction is generally a problem of correlated bond-site percolation. In the CPM, however, correspondence is made with the bond percolation problem. G_c is then determined by the condition for bond percolation [15]: in a Δ-dimensional system, there should be on the average $b_c=\Delta/(\Delta-1)$ allowed bonds per grain (site) [15,38]. For a three-dimensional lattice $b_c = 3/2$ and in the case of an amorphous material b_c approaches the limiting value of 2.8 [38]. The electrical conduction in granular systems results from tunneling of electrons and holes from charged grains to neutral grains [28-30]. To generate the charge carriers, electrons have to be removed from one neutral grain to another. This process requires at least an energy E_c, which is the <u>charging energy</u> [28-30] $E_c=e^2/KD$, where e is the electronic charge, D is the grain size, and K is the dielectric constant of the inhomogeneous system. In the low field limit, which we are interested in here, the charge carriers are thermally activated; the conductance G_{ij} between two grains i and j is then [14,15]

$$G_{ij}=G_o \exp[-2\chi S_{ij} - E_{ij}/kT] \qquad (5.1)$$

where χ is the tunneling parameter, $E_{ij}=1/2(|E_i-E_j|+|E_i|+E_j|)$, and S_{ij} denotes the distance between the grain surfaces along the line joining their centers. The condition $G_{ij}>G_c$ mentioned above is expressed as follows [14,15]:

$$\frac{S_{ij}}{S_m} + \frac{1}{2}\ \frac{|E_i| + |E_j| + |E_i-E_j|}{E_m} \leq 1 \qquad (5.2)$$

where $S_m = (1/2\chi)(\ln)(G_o/G_c)$ and $E_m = kT{\cdot}\ln(G_o/G_c)$. S_m and E_m are the maximum tunneling distance and the maximum energy, respectively, that any initial or final state can have, so that $G_{ij} \geqslant G_c$.

From the above discussion, it is clear that the condition for a critical path requires the following equation to be satisfied:

$$b_c = \frac{\int_{-E_m}^{E_m} dE_i \rho_i(E_i) \int_{-E_m}^{E_m} dE_j \rho(E_j) \int_0^{S_a(E_i,E_j)} dS_{ij} \int_0^{\infty} dr_{ij}\, h(r_{ij},S_{ij},\frac{D_i+D_j}{2})}{\int_{-E_m}^{E_m} dE_i \rho(E_i)} \tag{5.3}$$

where $h(r_{ij},S_{ij},\frac{D_i+D_j}{2}) = 4\pi r_{ij}^{\ 2}\, g(r_{ij})\delta(r_{ij} - \frac{D_i+D_j}{2} - S_{ij})$.

Here $\rho(E)$ is the density of states, $S_a(E_i,E_j)$ is the maximum allowed non-negative value of S_{ij} satisfying the inequality (5.2) for a given set of E_i and E_j. r_{ij} is the center-to-center separation between particles i and j, and

$$g(r_{ij}) = \begin{cases} N & r_{ij} > (D_i+D_j)/2 \\ 0 & r_{ij} \leq (D_i+D_j)/2 \end{cases} \tag{5.4}$$

is the hard sphere correlation function, where N denotes the number of particles per unit volume. Carrying out the r_{ij} and the S_{ij} integrations in Eq. (5.3) and approximating $(D_i+D_j)/2$ by the average diameter D_o, we get

$$b_c = N \frac{\int_{-E_m}^{E_m} dE_i \rho(E_i) \int_{-E_m}^{E_m} dE_j \rho(E_j)\, \frac{4\pi}{3}\, [S_a^{\ 3} + 3S_a^{\ 2} D_o + 3S_a D_o^{\ 2}]}{\int_{-E_m}^{E_m} dE_i \rho(E_i)}\ . \tag{5.5}$$

For the case of point impurities (D_o=0) and constant $\rho(E)$, Eq. (5.5) can be written in the following form as obtained by SHANTE [15]:

$$b_c = E_m S_m^{\ 3} f, \tag{5.6}$$

where f is a constant. From Eq. (5.6) and from the definitions of E_m and S_m the $\ln G_c \sim -T^{-1/4}$ relation of variable range hopping in three dimensions is directly recovered [34].

In order to carry out the calculation for random systems with conducting grains, we have to consider the role of the charging energy and its relation to $\rho(E)$. Since the charging energy E_c is just the capacitance energy required to put/remove a charge on/from a neutral conducting grain, it defines the minimum energy a charge carrier can have on any given grain. For a typical grain, the existence of the charging energy therefore implies that the density of state for the grain has the form $\rho_s(E) = \theta(E-E_c)/\bar{\Delta}$, where $\bar{\Delta}$ is the average electronic level separation inside conducting grains, and $\theta(E)$ is the step function. Generally $\bar{\Delta}$ might be energy dependent, which will modify $\rho(E)$ but will not effect our conclusions. To obtain $\rho(E)$ for an assembly of grains, we have to integrate $\rho_s(E)$ with the distribution of E_c, $P(E_c)$. If we assume a log-normal distribution for the size D of the grains [39], then $P(E_c)$ will also be a log-normal distribution

$$P(E_c) = \frac{1}{\sqrt{2\pi}} \frac{1}{\ln\mu} \exp\left[-\frac{(\ln(E_c/E_0))^2}{2(\ln\mu)^2}\right] \frac{1}{E_c} , \tag{5.7}$$

where μ is the parameter controlling the width of the distribution, and $E_0 \equiv e^2/KD_0$ is the most probably value of E_c. Given $P(E_c)$, we get for the density of states of the grains $\rho_G(E)$:

$$\rho_G(E) = \frac{1}{\bar{\Delta}} \int_0^E P(E_c)dE_c \quad . \tag{5.8}$$

It should be noted that for E=0, $\rho_G(E)=0$. However, in a composite material there could be states other than those in the conducting grains. For example, there may be impurities which would contribute a finite density of states at E=0 so that at the Fermi level these states, instead of those inside the grains, would give the main contribution to conduction. Assuming that there is a constant density of state ρ_0 due to these states, we get $\rho(E)$ as

$$\rho(E) = \rho_G(E) + \rho_0 \quad . \tag{5.9}$$

The value of ρ_0 may of course depend on the concentration of the metallic component. Qualitatively, $\rho(E)$ has the behavior of increasing from ρ_0 at E=0 to about $\rho_0 + 1/2\bar{\Delta}$ at $|E| = E_0$. For $|E| > E_0$, $\rho(E)$ slowly approaches the asymptotic value of $\rho_0+(1/\bar{\Delta})$. Using this form of $\rho(E)$ in Eq. (5.5) we find that $\ln G_c$ behaves as $-T^{-\alpha}$ over a significant temperature range. In the case of granular aluminum we could fit the $\alpha=1/2$ dependence in the temperature range $T \simeq$ 2K-400K [40] which reproduces the experimental results of CHUI et al. [31]. However, Eq. (5.5) can also result in $\alpha=1/4$ behavior. By varying the value E_0 in Eq. (5.7) one realizes that decreasing the grain size D_0 (increasing E_0) extends the $\alpha=1/4$ behavior to higher temperatures, so that $\alpha=1/4$ is a better fit for the temperature dependence. On the other hand increasing D_0 (decreasing E_0) makes the $\alpha=1/2$ a better fit [40]. Such a crossover with changing of the grain sizes has been recently reported for conductivites in doped polyacetylenes [36].

The $\alpha=1/2$ behavior and the crossover to $\alpha=1/4$ which result from Eq. (5.5) can be understood in the following way. For T→0K hopping occurs over large distances and between states close to the Fermi level so that $\rho(E)\simeq\rho_0$; in this case we obtain the $\alpha=1/4$ dependence. At high temperatures hopping is between nearest neighbors and $\rho(E)$ is in the monotonically increasing region. If we assume $\rho(E) \sim E^\beta$, $\beta > 0$, then from Eq. (5.5) $b_c \propto E_m^{\beta+1} S_m$ which corresponds to $\alpha > 1/2$. The $\alpha=1/2$ behavior which is observed experimentally so often is then a result of the interpolation between these two limiting cases over an intermediate temperature range [40].

Acknowledgements

The collaborations with R. Silbey, A. Blumen, G. Zumofen, I. Webman, and P. Sheng in establishing the results presented here are gladly acknowledged.

References

1. S.W. Haan and R. Zwanzig, J. Chem Phys. 68, 1879 (1978)
2. a) C.R. Gochanour, H.C. Andersen and M.D. Fayer, J. Chem. Phys. 70, 4254(1979)
   b) A. Blumen, J. Klafter and R. Silbey, J. Chem. Phys. 72, 5320 (1980)
3. K. Godzik and J. Jortner, J. Chem. Phys. 72, 4471 (1980)
4. H. Scher and M. Lax, Phys. Rev. B. 7, 4491, 4502 (1973)
5. H. Scher and E. Montroll, Phys. Rev. B 12, 2455 (1975)
6. G. Pfister and H. Scher, Adv. Phys. 27, 747 (1978)
7. M. Lax and T. Odagaki, in Macroscopic Properties of Disordered Media, edited by R. Burridge, S. Childress and G. Papanicolaou (Springer Verlag, Berlin 1982)
8. J. Klafter and R. Silbey, J. Chem. Phys. 72, 843 (1980)
9. A. Blumen and G. Zumofen, J. Chem. Phys. 77, 5127 (1982)
10. I. Webman, Phys. Rev. Lett. 47, 1496 (1981)
11. I. Webman and J. Klafter, Phys. Rev. B26, 5950 (1982)
12. S. Alexander, J. Bernasconi, W.R. Schneider and R. Orbach, Rev. Mod.Phys. 53, 175 (1981)
13. R.J. Elliot, J.A. Krumhansl and P.L. Leath, Rev. Mod. Phys. 46, 465 (1974)
14. V. Ambegaokar, B.I. Halperin and J.S. Langer, Phys. Rev. B4, 55 2612 (1971)
15. V.K. Shante, Phys. Rev. B16, 2597 (1977)
16. J. Klafter and R. Silbey, Phys. Rev. Lett. 44, 55 (1980)
17. J. Klafter and R. Silbey, Phil Mag B47, 337 (1983)
18. B. Movaghar, J. Phys. C13, 4915 (1980)
19. T. Odagaki and M. Lax, Phys. Rev. B24, 5284 (1981)
20. M. Inokuti and F. Hirayama, J. Chem. Phys. 43, 1978 (1965)
21. D.G. Thomas, J.J. Hopfield and W. M. Augustyniak, Phys. Rev. 140, A202 (1965)
22. A. Blumen, Nuovo Cimento 63B, 50 (1981)
23. J. Klafter and A. Blumen, in Random Walks in the Phys. and Bio. Sciences,edited by M. Shlesinger and B. West (AIP Conference Proceedings 1983)
24. W. Feller, An Introduction to Probability Theory and Its Applications (Wiley, New York, 1968), Vol II, p. 445
25. S. Kirkpatrick, Rev. Mod. Phys. 45, 574 (1973)
26. S. Alexander and R. Orbach, Physica (Utrecht) 107B, 675 (1981)
27. J. Machta, Phys. Rev. B24, 5260 (1981)
28. P. Sheng, B. Abeles and Y. Arie, Phys. Rev. Lett. 31, 44 (1973)
29. B. Abeles, P. Sheng, M. D. Coutts and Y. Arie, Adv. in Phys. 24, 407 (1975)
30. C.A. Neugebauer and M.B. Webb, J. Appl. Phys. 33 74 (1962)
31. T. Chui, G. Deutscher, P. Lindelfeld and W.L. McLean, Phys. Rev. B23, 6172 (1981)
32. N. Savvides, S.P. McAlister and C.M. Hurd, Can. J. Phys. 60, 1484 (1982)

33. S. Barzilai, Y. Goldstein, I. Balberg and J.S. Halman, Phys. Rev. B23, 1809 (1981)
34. N.F. Mott, Metal-Insulator Transition (Taylor and Frances, London (1974)
35. Kell Mortensen, M.L.W. Thewalt, Y. Tomkiewicz, T.C. Clarke and G.B. Street, Phys. Rev. Lett. 45, 490 (1980) B25 5574 (1982)
36. E.K. Sichel, M. Knowles, M. Rubner and J. Georges, Jr., Phys. Rev. B25, 5574 (1982)
37. S.R. Forrest, M.L. Kaplan, P.H. Schmidt, T. Venkatesan and A.J. Lovinger, App. Phys. Lett. 41, 708 (1982)
38. G.E. Pike and C.H. Seeger, Phys. Rev. B10, 1421 (1979)
39. C.G. Granqvist and R.A. Buhrman, J. Appl. Phys, 47, 2200 (1976)
40. P. Sheng and J. Klafter, Phys. Rev. B27, 2583 (1983)

Energy Transfer and Relaxation Processes as Studied by Picosecond Fluorescence Spectroscopy

D. Schmid

Universität Düsseldorf, Physikalisches Institut IV
D-4000 Düsseldorf, Fed. Rep. of Germany

1. Introduction

The excited state of molecular crystals is usually described by the concept of delocalized states or excitons. The degree of delocalization manifests itself most prominently in the energy transfer phenomena which are observed in these crystals, such as sensitized fluorescence or exciton-exciton annihilation processes. Consequently experimental studies of these phenomena have been performed for many years in order to get information about the nature of the exciton.

In experiments on sensitized fluorescence, the crystal under investigation, called the host, is doped with small amounts of another substance, the guest, and after selective excitation of the host or the guest the luminescence intensities or their time dependencies are observed. For the interpretation of these observations one usually deduces a so-called energy transfer rate k_{HG} from the experiments and then derives from it the diffusion constant. In performing this interpretation of various experimental results the so-called hopping model has been quite successful [1]. However, initiated by the early activities on time-resolved spectroscopy of the sensitized fluorescence in doped crystals [2] the question was posed whether the hopping model is adequate to explain the experimental observations or whether more sophisticated theoretical models have to be invoked [3-6]. It is a common feature of the different theoretical concepts that they predict a rate of energy transfer which depends on the time elapsed since pulse excitation into the host exciton band has taken place. The actual time dependencies, however, are predicted differently in different theoretical approaches. The challenge to the experimentalist therefore consists in the question, whether such a time dependence can be detected experimentally and whether it is possible to distinguish between the different models on the basis of such experiments. There have been numerous attempts to answer these questions by investigating the time dependence of the sensitized fluorescence [1, 7-11]. The results of these various efforts were quite contradictory. Whereas the majority of the experiments did not reveal any information on the time dependence of k_{HG}, Powell and coworkers [1] reported observations which indicated a significant time dependence. They interpreted their results in terms of a "long-range resonance transfer" model.

On the other hand in [8] a complete set of experimental data was reported covering a wide range of temperature (4.2 to 300 K) and of concentration ($9 \cdot 10^{-7}$ to $4 \cdot 10^{-5}$ mole/mole) for naphthalene crystals doped with anthracene. Whereas in the high and medium concentration limit the experiments could be interpreted in terms of a time-independent energy transfer rate, the low-concentration, low-temperature results apparently required the assumption of a time-dependent rate of energy transfer, formally equivalent to that used by Powell and coworkers. However, when attempting to interpret the results in terms of the concept of Powell and coworkers, it was necessary to assume that both the trapping radius of the guest molecules and the diffusion constant were strongly dependent on the guest concentration, a hypothesis which is obviously unreasonable.

The primary aim of the work reported in this contribution was to establish whether the time dependence of the sensitized fluorescence in tetracene-doped anthracene following a picosecond pulse excitation necessitated the assumption of a time-dependent rate of energy-transfer anywhere in the accessible range of temperature (1.6 to 300 K) and of concentration ($1 \cdot 10^{-6}$ to $2.3 \cdot 10^{-3}$ mole/mole). Since the anthracene excited-state lifetime is only about 5 ns, a good time resolution was required. In the present experiments an optimum time resolution of about 10 ps was established. The most important result of this systematic study is that at no concentration and at no temperature within the above-mentioned limits did the experimental results force us to invoke a time-dependent rate of energy transfer [12].

2. Experimental Procedure

Picosecond transients were studied by exciting the crystals with single pulses of a mode-locked Nd-glass laser (system 2000, JK) or with harmonics thereof. In some experiments we used single pulses from a passively mode-locked flash-lamp pumped dye laser (SUA 33, Electro Photonics) for excitation. The time dependence of the resulting fluorescence was recorded using a picosecond streak-camera (Imacon 675/II, Hadland) the output of which was monitored with an optical multichannel analyzer (1205 A with SIT-vidicon, PAR) and processed using an on-line computing system (2117 F, Hewlett Packard). The overall time resolution of this setup was 5 ps (Nd-glass) and 10 ps (dye-laser), respectively. The dynamical range at these resolutions was 24. Additional specifications of the setup are given in [13].

Nanosecond events were excited with a nitrogen-laser/dye-laser combination (M 100 and FL 100 A, Lambda Physics) and recorded using a fast photomultiplier (C 31024, RCA) and a transient digitizer (R 7912, Tektronix).

Fig. 2 presents three typical fluorescence spectra of tetracene-doped anthracene. We define the quantum flux ratio

$$Q_G/Q_H = {}_G\int I_G(\bar{\nu})\,d\bar{\nu} / {}_H\int I_H(\bar{\nu})\,d\bar{\nu}, \qquad (1)$$

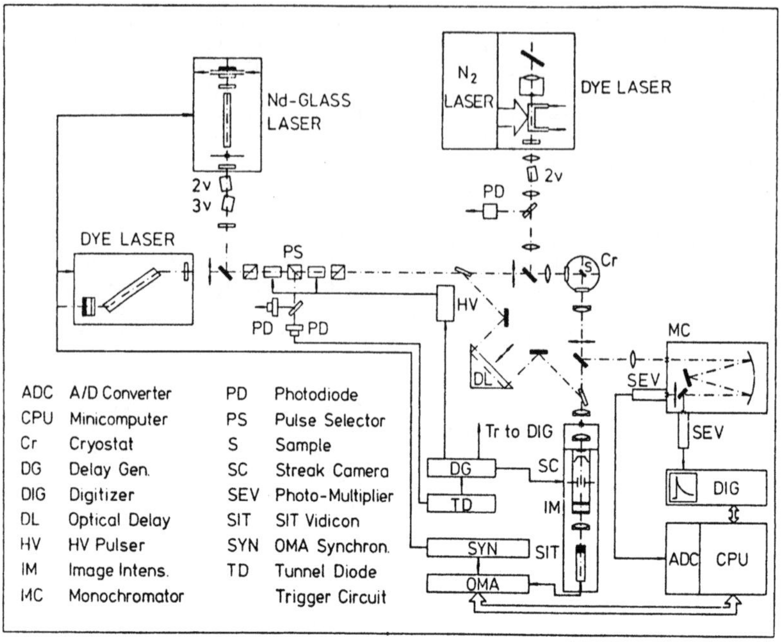

Fig. 1 Schematic diagram of the experimental setup

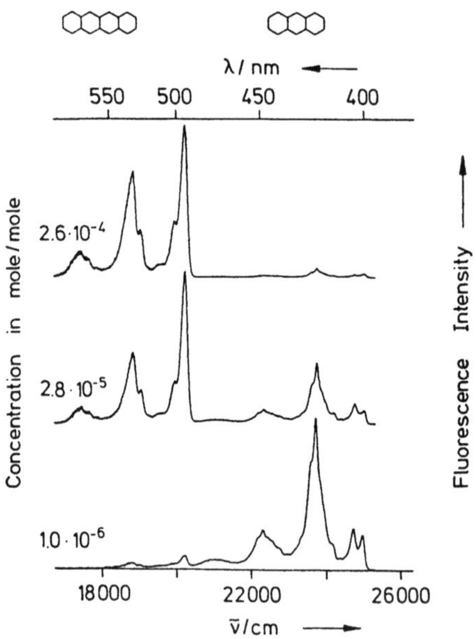

Fig. 2. Fluorescence spectra of tetracene-doped anthracene for various dopant concentrations at T = 60 K

where $_G\int$ or $_H\int$ indicate an integration over the spectral ranges relevant for the respective fluorescence intensities (16.800 to 20.400 cm^{-1} for the guest fluorescence and 20.400 to 25.000 cm^{-1} for the host fluorescence). From this experimentally accessible quantity we can determine the guest concentration in the sample using the relation [11,12]

$$Q_G/Q_H = 9.8 \cdot 10^4 \; c_G \pm 30\ \%, \qquad (2)$$

where c_G is the guest concentration in mole guest/mole host.

3. Experimental Results

It is well known that the fluorescence spectra of anthracene depend critically on the crystal quality and purity [14,15] It was shown in [11] that the contribution of chemical and structural defects is strongly enhanced in the region near to the surface and that their influence can be reduced significantly by using two-photon excitation. Therefore in most of the experiments reported in this contribution we used a two-photon excitation with single pulses from a mode-locked dye laser at a wavelength of 614.5 nm. Furthermore, although the influence of reabsorption is more pronounced than in experiments using conventional one-photon excitation, it is more readily corrected for in experiments using two-photon excitation: independent of the detection wavelength the time constant of the purely exponential fluorescence decay of pure anthracene crystals decreases gradually from 15.5 ns at 300 K to 9.7 ns at 150 K, 6.2 ns at 60 K and approaches a limiting value of 5 ns at 20 K. This temperature dependence can be attributed to a change in the spectral overlap of absorption and fluorescence spectra, whereas the intrinsic lifetime of the anthracene fluorescence is $\tau_0 = (5\pm0.5)$ ns.

The time dependence of the sensitized fluorescence of tetracene-doped anthracene was studied in an analogous way at temperatures between 1.6 and 300 K and for dopant concentrations between $1\cdot10^{-6}$ and $2.3\cdot10^{-3}$ mole/mole.

Fig. 3 presents the results obtained for the sample, with the highest dopant concentration at 1.6 K. The host fluorescence (Fig. 3a) rises within less than 10 ps, the overall response

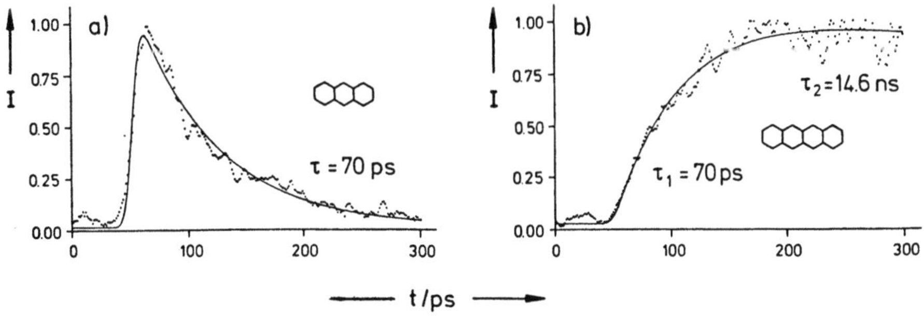

Fig. 3 Time dependence of the sensitized fluorescence of tetracene-doped anthracene ($c_G = 2.3\cdot10^{-3}$ mole/mole) at 1.6 K following two-photon pulse excitation (λ_{exc} = 614.5 nm, pulse duration 10 ps, pulse energy 10 µJ, initial exciton density $10^{15}\,cm^{-3}$). a) Host fluorescence b) Guest fluorescence

time of the setup. Following this the host fluorescence decays exponentially with a time constant of 70 ps. The full line in Fig. 3a is a convolution of this exponential with the temporal response function of the entire setup. The guest fluorescence (Fig. 3b) rises exponentially with the host fluorescence decay time (70 ps) and then decays exponentially with a time constant of 14.6 ns. This decay process is too slow to show up on the time scale of this figure.

The time dependencies obtained at lower dopant concentrations are qualitatively similar to Fig. 3. The rise time of the guest fluorescence is always identical to the host-fluorescence decay time, which at low concentrations approaches the decay time of the pure anthracene fluorescence. The decay time of the guest fluorescence is identical to the lifetime of the excited guest molecules. (The latter statement is based on measurements of the guest-fluorescence decay following a selective excitation of the guest molecules with a laser pulse at 482 nm. The lifetime of tetracene molecules in their excited state turns out to depend only weakly on the temperature: (17.3±2) ns at 300 K and (14.6±2) ns at 1.6 K.)

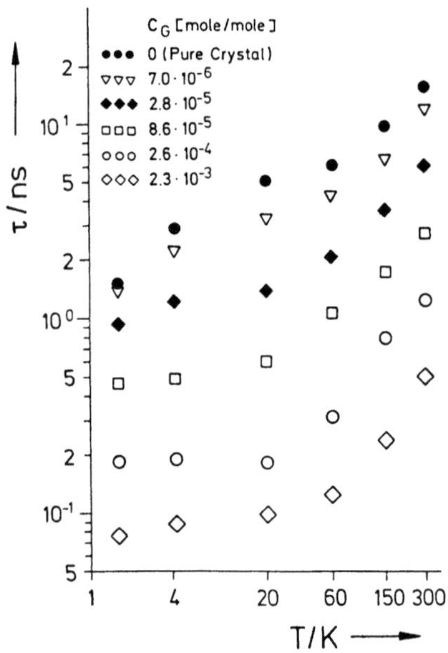

Fig. 4 Survey of the observed host-fluorescence decay times

Fig. 4 presents a survey of the observed host-fluorescence decay times (which are identical to the guest-fluorescence rise times). In all cases we observed an exponential decay of the host fluorescence, and an exponential rise of the guest fluorescence followed by an exponential decay, provided the two-photon excitation intensity was low enough to avoid annihilation and guest saturation.

4. Discussion

The primary aim of the work reported in this contribution was to answer the question to what extent time-resolved studies of the sensitized fluorescence can contribute to the understanding of the energy transfer process, in particular whether a significant time dependence in the rate of energy transfer can be observed. Disappointingly the answer to the latter question is negative. Within the limits of the time resolution of our experiments (10 ps) all the experimental results could be explained on the basis of a time-independent energy transfer rate, since in that concept one expects an exponential decay of the host fluorescence and a superposition of an exponential rise and an exponential decay for the guest fluorescence. From our experiments we can only extract the time-independent part of the energy transfer rate. This quantity, extrapolated to a fictitous guest concentration 1 mole tetracene/mole anthracene is plotted in Fig. 5 as a function of the temperature.

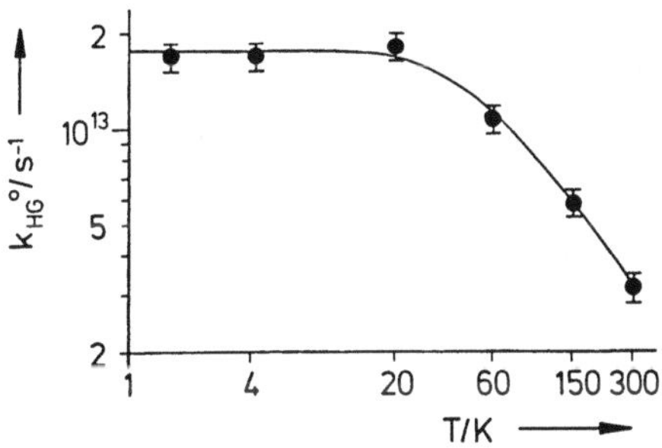

Fig. 5 Temperature dependence of the normalized energy-transfer rate, $k_{HG}^{o} = k_{HG}/c_G$ for tetracene-doped anthracene. The full line is obtained from fitting the solutions of the rate equations based on the model presented in Fig. 6 to the experimental time dependencies. (Fitting parameters: $k_{HT} = 1/t_h = 1.7 \cdot 10^{13}\ s^{-1}$, $\varepsilon = 43\ cm^{-1}$)

The observed temperature dependence can be accounted for using the tentative model illustrated in Fig. 6. We assume that the exciton trapping at the guest molecules takes place via a trapping energy level, which is slightly below the exciton level by an amount ε. As soon as the exciton falls into such a localized pre-trapping state, it is trapped in an "energy funnel", in which it can only decay into states with lower energy levels and a greater degree of localization, except if thermal reactivation via exciton-phonon interaction takes place and promotes the exciton into energetically higher mobile states. For the experiments reported in this work it is reasonable to make the following assumptions: $k_H \cong k_T$, $k_{TG} \cong k_{HT}$ and $c_G \cong c_T \leq 10^{-4}$ mole/mole. Using these assumptions it can be shown that the respective rate constants for trapping and detrapping are related to each other by

$$k_{TH} = k_{HT} \exp(-\varepsilon/kT) . \qquad (3)$$

The resulting rate equations for n_H, n_T and n_G can be solved easily and the results can be compared to the experimentally observed time dependencies. In order to account for the observed temperature dependence, ε is used as a fitting parameter. The full line in Fig. 5 has been obtained in this fashion by assuming that $k_{HT} \cong k_{HG}^{o} = 1/t_h = 1.7 \cdot 10^{13} s^{-1}$ and $\varepsilon \cong 43\ cm^{-1}$. (For the three-species rate equations a third exponential is

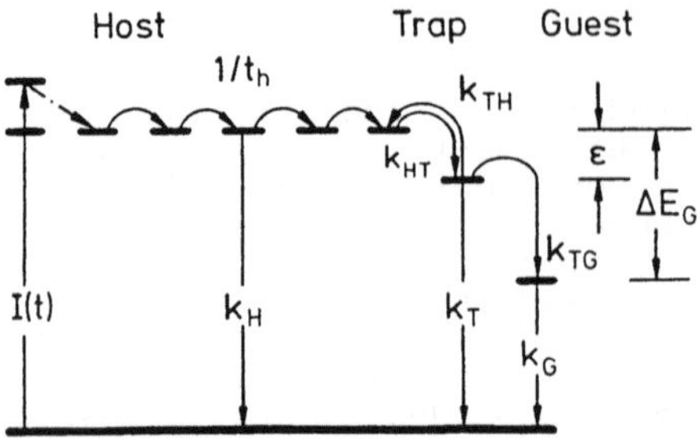

Fig. 6 Schematic energy level diagram for the exciton-trap-guest model defining the relevant rate constants. k_H, k_T and k_G are the inverse lifetimes of the excited host, trap and guest molecules, respectively. t_h is an "effective exciton hopping time". k_{HT} is the energy-transfer rate from the host exciton band to the trap molecule, k_{TG} that for transfer from the trap to the guest

expected in the solutions of the rate equations. However, it can be shown that under the present experimental conditions its contribution is too small to be detectable.)

The nature of the pre-trapping state is not known. For instance, it could be either a highly excited vibronic level of the guest molecule or an X trap in the neighbourhood of the guest molecule.

The observed temperature dependence of k_{HG} leads to another problem which has been pointed out in [16]: The energy-tranfer rate k_{HG} is closely related to the exciton-exciton annihilation constant γ, which takes account of the fact that excitons at a sufficiently high density can interact with each other destructively. The two processes are usually described in terms of the following phenomenological equations, which describe the evolution of the exciton density n_H:

$$dn_H/dt = -n_H/\tau_H - k_{HG}n_H \qquad (4)$$

$$dn_H/dt = -n_H/\tau_H - \gamma n_H^2. \qquad (5)$$

The first term on the right-hand side of (4) and (5) describes the exciton decay, τ_H being the lifetime. The last term represents energy transfer in (4) and annihilation in (5).

It has been pointed out by Kenkre [6, 17, 18] that k_{HG} and γ are generally given by the following expressions:

$$k_{HG} = c_G\ (1/C + 1/M)^{-1} \equiv c_G k' \qquad (6)$$

$$\gamma/V = (1/B + 1/M)^{-1} \equiv \gamma' . \qquad (7)$$

V is the volume per molecule in the host crystal, C is the rate at which the elementary capture process occurs at the guest molecule, 2B is the rate at which the elementary process of annihilation takes place, and M is the motion rate defined as

$$M = \{\int_{-\infty}^{+\infty} \exp(-t/\tau)\ \psi_0(t)dt\}^{-1} \qquad (8)$$

where $\psi_0(t)$ is the exciton self-propagator, i.e. the probability that an exciton initially placed at a site is at that site at a later time t, in the pure captureless and annihilationless host crystal. In a crude way one may illustrate (6) and (7) by stating that 1/k' represents the sum of the motion time 1/M, i.e. the time to get the exciton within the trapping influence of a guest molecule and the elementary capture time 1/C. Similarly 1/γ' may be said to represent the sum of the motion time and the time for the actual destruction process 1/B. It is generally believed that both k_{HG} and γ are motion controlled whereas the characteristics of the elementary capture of annihilation processes drop out because they are too rapid. However, the observed temperature dependence of k_{HG} contradicts this assumption strongly, since no comparable temperature dependence of γ is observed. In Fig. 7 the experimentally observed quantities are compiled for naphthalene and anthracene: Whereas k' depends strongly on the temperature, both in naphthalene and in anthracene, for both crystals γ' is independent of the temperature within the limits of error [8, 11, 12].

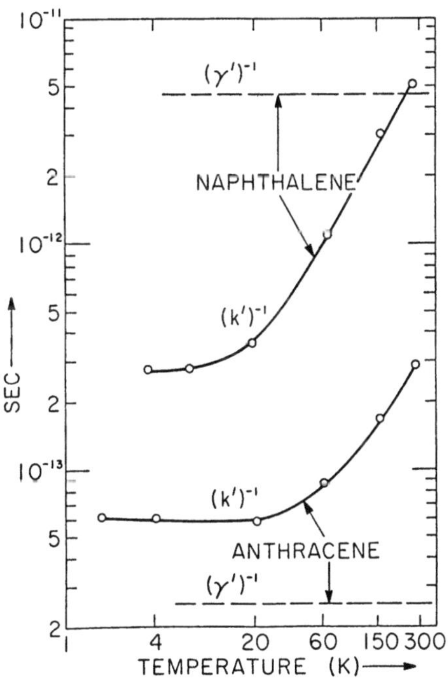

Fig. 7. The reciprocal of the rates k' and γ' (see text for definition) plotted logarithmically as functions of the temperature for anthracene and for naphthalene

From (6) and (7) one sees that 1/M is smaller than both 1/γ' and 1/k'. Therefore, for anthracene we can conclude that $1/M < 2.3\cdot10^{-14}$ s or $M > 4.2\cdot10^{13}\,s^{-1}$. For naphthalene we have $1/M < 3\cdot10^{-13}$ s ($M > 3\cdot10^{12}\,s^{-1}$) at 4 K and $1/M < 5\cdot10^{-12}$ s ($M > 2\cdot10^{11}\,s^{-1}$) at 300 K. Furthermore in the case of anthracene, at least at $T \geq 60$ K the energy transfer rate is essentially capture limited, since $k' \ll \gamma'$. In naphthalene an analogous statement holds for γ' in the low-temperature range: Since $\gamma' \ll k'$ in this range it is evidently destruction limited.

Thus, at least in the case of anthracene it is certainly not justified to assume that the energy transfer is motion limited. In fact in this case it is not even possible to extract information about the motion from energy transfer measurements, at least not in the high-temperature region ($T > 60$ K) since here the energy transfer is capture limited. For the interpretation of the observed rate constants both processes should always be taken into account. A similar statement holds for the naphthalene annihilation results.

5. Acknowledgements

This work was performed in cooperation with members of the Universität Stuttgart (Physikalisches Institut, Teil 3), the University of Rochester (Department of Physics and Astronomy) and the Universität Düsseldorf (Physikalisches Institut, Lehrstuhl IV). In particular important contributions are due to Dr. A. Braun, Professor V. M. Kenkre and Professor H. C. Wolf. Financial support by the Deutsche Forschungsgemeinschaft via Sonderforschungsbereich 67 (Defektstrukturen in festen Stoffen) and via research project Schm 270/4-1 is gratefully acknowledged.

References

1. H. C. Wolf, Adv. Atomic Mol. Physics 3, 119 (1967)
2. R. C. Powell and Z. G. Soos, J. Luminescence 11, 1 (1975)
3. R. Silbey, Ann. Rev. Phys. Chem. 27, 203 (1976)
4. R. Kopelman in "Radiationless Processes in Molecules and Condensed Phases", ed. F. K. Fong, Springer Verlag, Berlin 1976, page 33
5. A. Blumen and R. Silbey, J. Chem. Phys. 70, 3707 (1979)
6. V. M. Kenkre, Springer Tracts in Modern Physics, Vol. 94, Springer Verlag, Berlin 1982, p.1
7. A. J. Campillo, S. L. Shapiro and C. E. Swenberg, Chem. Phys. Letters 52, 11 (1977)
8. H. Auweter, A. Braun, U. Mayer and D. Schmid, Z. Naturforsch. 34a, 761 (1979)
9. N. J. Bridge and D. P. Solomons, J.C.S. Faraday II, 76, 472 (1980)
10. J. O. Williams and A. C. Jones, J. Luminescence 24/25, 723 (1980)
11. A. Braun, U. Mayer, H. Auweter, H. C. Wolf and D. Schmid, Z. Naturforsch. 37a, 1013 (1982)
12. A. Braun, Dissertation, Universität Stuttgart, 1981
13. U. Mayer, H. Auweter, A. Braun, H. C. Wolf and D. Schmid, Chem. Physics 59, 449 (1981)
14. E. Glockner and H. C. Wolf, Z. Naturforsch. 24a, 943 (1969)
15. E. Glockner and H. C. Wolf, Chem. Physics 10, 479 (1975)
16. V. M. Kenkre and D. Schmid, Chem. Phys. Letters 94, 603 (1983)
17. V. M. Kenkre, Phys. Rev. B 22, 2089 (1980)
18. V. M. Kenkre, Z. Physik B43, 221 (1981)

Theoretical Methods for the Analysis of Exciton Capture and Annihilation

V.M. Kenkre

Department of Physics and Astronomy, University of Rochester
Rochester, NY 14627, USA

1. Introduction

The theoretical activity in the field of exciton dynamics in molecular crystals may be considered as being divided into three parts: (i) the search for an appropriate evolution equation or transport instrument to serve as a point of departure for the analysis of exciton motion, (ii) the construction of a practical formalism or scheme to take one from the point of departure to observable quantities, and (iii) the application of the consequences of the scheme to particular experimental results. Such a division is not only analogous to similar logical classifications in entire branches of physics, e.g., in statistical mechanics itself, but, in the case of Frenkel exciton dynamics, it is also, in large part, chronological. In 1932 PERRIN found definite disagreement with experiments - on fluorescence depolarization - on choosing the Schrödinger equation among sharp molecular site states as the transport instrument [1]. In 1948 FÖRSTER showed that the disagreement stemmed not from the particular transport mechanism assumed by PERRIN - dipole-dipole interactions - but from the transport instrument itself [2]. On replacing the Schrödinger equation by the Master equation, i.e., recognizing that the molecular site states of the exciton were "broadened" into groups of states by both interactions and using the Fermi Golden Rule among these "smeared-out" states, FÖRSTER found excellent agreement with experiment. These two transport instruments, the Schrödinger equation and the Master equation, were known to be valid in the two extreme limits known as <u>coherent</u> and <u>incoherent</u> respectively, and a unified description was sought [3]. Treatments of the coexistence or unification of these limits were given by HAKEN, STROBL and REINEKER [4,5], by GROVER and SILBEY [6,7] and by KENKRE and KNOX [8,9] from three different viewpoints. The relations among these three viewpoints have been investigated in great detail [10,11] and it is by now amply clear that for the analysis of certain observational situations a particularly convenient point of departure is the generalized master equation (GME) obeyed by the probabilities $P_m(t)$ that the molecular site m is occupied at time t:

$$\frac{dP_m(t)}{dt} + \frac{P_m(t)}{\tau} = \int_0^t dt' \sum_n [W_{mn}(t-t')P_n(t') - W_{nm}(t-t')P_m(t')]. \quad (1.1)$$

Here τ is the lifetime of the exciton and the memory functions W_{mn} contain the dynamics. In this context the GME (1.1) is by no means a replacement for the treatments of refs. [4-7] but, on the contrary, provides the most convenient framework to express and utilize the physics contained in those (or alternative) treatments. Indeed, the memory functions $W_{mn}(t)$ corresponding to the analysis of refs. [4-7] have been calculated explicitly [10-12]. The results of such work leading to the microscopic derivation of the W's have been presented in the author's recent reviews [11,12] and will not be repeated here. Of the three activities in exciton dynamics mentioned above, the present article deals exclusively with (ii). The point of departure will be the GME (1.1) and the construction of the practical formalism to arrive at measurable quantities will be demonstrated for two specific classes of exciton observations: capture by traps and mutual annihilation. These correspond to well-defined methods for the experimental study of exciton motion.

2. Capture by Traps: Coherence Effects

If the (host) crystal is doped with guest molecules which can trap the excitons when they wander within the influence of the traps during their motion through the host, one may study the exciton motion by monitoring the luminescence of the traps and/or the host. This is the well-known method of sensitized luminescence [13]. The simplest way to describe this situation is to append a term $-c\sum_r{}'\delta_{m,r}P_m(t)$ to the right-hand side of (1.1). The primed summation covers those host sites which have trap molecules near them and feed them at the rate c. More complex models of the trapping process may be found elsewhere [11]. Equation (1.1) with the above trapping term may be rewritten as

$$\tilde{P}_m(\varepsilon) = \sum_n \tilde{\psi}_{m-n}(\varepsilon')P_n(0) - c\sum_r{}'\tilde{\psi}_{m-r}(\varepsilon')\tilde{P}_r(\varepsilon) \qquad (2.1)$$

where ε is the Laplace variable, tildes denote Laplace transforms, ψ_m is the probability propagator for the decayless trapless host, and $\varepsilon' = \varepsilon+1/\tau$.

The solution of (2.1) requires the evaluation of a determinant whose size is determined by the number of trap-influenced host sites r. We shall denote the ratio of the latter to the total number N of host sites by ρ. There exists a class of systems wherein ρ is perforce small ($<10^{-4}$) as a result of solubility restrictions. The study of the dilute trap case is therefore of experimental relevance. In this section we restrict the analysis to the dilute case with the focus on manifestations of coherence and return to the high concentration case in section 3.

Consider therefore the case of a single trap [14] with the understanding that factors 1/N are to be replaced by the concentration ρ. Exact solution of (2.1) for all P_m's is straightforward. The primary experimental observables are the time-dependent host luminescence intensity which, apart from unimportant constants, equals the total host excitation probability

$n_H(t) \equiv \sum_m P_m(t)$, the host quantum yield ϕ_H which is the ratio of total number of excitons emerging radiatively from the host to the number initially put into the host through illumination, and the corresponding guest (i.e., trap) quantities. They are respectively given by

$$\tilde{n}_H(\varepsilon) = \frac{1}{\varepsilon'}\{1 - \frac{\rho}{\varepsilon'[(1/c)+\tilde{\psi}_0(\varepsilon')]}\} \tag{2.2}$$

$$\phi_H = 1 - \frac{\rho\tau}{(1/c)+\tilde{\psi}_0(1/\tau)} \tag{2.3}$$

where, for simplicity, we assume that the radiative lifetime equals the total lifetime.

Numerical Laplace inversions of (2.2) give [15] the luminescence intensities in the time domain explicitly. No inversions are required for the quantum yield (2.3). The effect of transport characteristics on the luminescence observables appears simply and solely in the propagator ψ_0 in (2.2) and (2.3). Detailed studies in this direction, particularly of the effect of coherence, have been carried out [11,14,15]. The assumed memory functions $W_{mn}(t)$, or equivalently the quantities $A_{mn}(t)$ which equal $-W_{mn}(t)$ for $m \neq n$, with $A_{mm}(t) = \sum_n W_{nm}(t)$, give the propagators $\tilde{\psi}_m(\varepsilon)$ through

$$\tilde{\psi}_m(\varepsilon) = (1/2\pi)^d \int dk e^{-ik \cdot m} [\varepsilon + \tilde{A}^k(\varepsilon)]^{-1} \tag{2.4}$$

where d is the number of dimensions of the crystal, A^k is the (discrete) Fourier transform of the A_{mm}'s, and $k \cdot m$ is a dot product. The case m=0 of (2.4), substituted in (2.2) and (2.3), finally produces the explicit luminescence observables in terms of the memory functions.

We stress that in the light of this formalism, it is not necessary to shift from a band analysis in one limit to a hopping analysis in the other, changing the evolution equation in the process. Memory functions are available and propagators (and thence immediately the observables through (2.2), (2.3)) can be calculated for arbitrary degree of coherence. A detailed general discussion of how coherence enters into luminescence observables has been given elsewhere [11]. Explicit expressions and plots versus the degree of coherence for particular models are also available [14,15]. The overall conclusions of those investigations are that, while the degree of coherence does influence sensitized luminescence observables in the above kind of experiment quantitatively to a significant extent, the results are not dramatic. As expected, purely coherent motion leads to _much_ more efficient capture by traps than motion with the same intersite interaction but with scattering. This is reflected in the considerable decrease of guest observables as the exciton mean free path changes from being much larger than the lattice constant to smaller than it. However, there are no _qualitative_ changes in the shapes of the observables as a function of time as the degree of coherence is changed. Such qualitative changes are expected in a different kind of capture experiment wherein excitons are created at one end of the crystal and captured at the other end rather than in the bulk of the crystal. Work on such an analysis is under way.

3. Capture by Traps: Multitrap Effects

For systems in which the trap concentration is not low, the single-trap analysis is not adequate. The problem of the solution of (2.1) may be approached [11,16] by realizing that the quantity of interest in the evaluation of the experimental observables $n_H(t)$ and ϕ_H is the sum of the left side of (2.1) over the trap-influenced sites. If we make the assumption that the function $\nu_s(t)$ defined as

$$\nu_s(t) \equiv \sum_r{}' \psi_{r-s}(t) \qquad (3.1)$$

is independent of s, i.e., $\nu_s \equiv \nu$, it is possible to obtain n_H and ϕ_H explicitly from (2.1). The results are identical to (2.2) and 2.3) with the single replacement of the self-propagator ψ_0 by the ν function.

$$\tilde{n}_H(\varepsilon) = \frac{1}{\varepsilon'}\{1-\frac{\rho}{\varepsilon'[(1/c)+\tilde{\nu}(\varepsilon')]}\} \qquad (3.2)$$

$$\phi_H = 1-\frac{\rho\tau}{(1/c)+\tilde{\nu}(1/\tau)} \; . \qquad (3.3)$$

The prescription offered by this multitrap formalism [15,16] is thus as follows. The memory functions lead to the propagators ψ_m through (2.4) as explained above, the ν function is constructed from (3.1), and the experimental observables are then given by (3.2) and (3.3). The ν function, even before the assumption of the s independence, equals $\psi_0(t)$ for short times but tends to the concentration ρ at long times in contrast to $\psi_0(t)$ which tends to (1/N) and therefore to 0 in a large (infinite) system. The ν function can also be shown to be simply related to $n_H(t)$ asymptotically [15] and it can be reexpressed as

$$\nu(t) = \sum_m \psi_m(t) p_m \; . \qquad (3.4)$$

The sum in (3.4) is unrestricted and one sees here explicitly how $\nu(t)$ contains information about the dynamics of exciton motion in the pure host (through the propagators ψ_m) combined in a straightforward way with information about the statics of the trap placement (through p_m). The quantity p_m is the probability that the mth host-site is trap influenced given that the 0th site is trap influenced.

In addition to an explicit description of the dependence of luminescence variables on the trap concentration ρ [15], the present multi-trap formalism has also resulted [17] in an analysis of the effect of cooperative interactions among traps.

Absence of such interactions leads to random placement [15] and to

$$\nu(t) = \rho+(1-\rho)\psi_0(t). \tag{3.5}$$

For extreme repulsive trap-trap interactions capable of producing a periodic placement, $\nu(t)$ equals $\psi_0(t)$ corresponding to a smaller lattice of sites $1/\rho$ [15] and, in the particular case of a 1-d crystal in which the excitons move with nearest-neighbour transfer rates F, one gets [11,16]

$$\varepsilon\tilde{\nu}(\varepsilon) = [\tanh(\xi/2)][\tanh(\xi/2\rho)]^{-1} \tag{3.6}$$

where $\cosh\xi = 1+\varepsilon/2F$. Further analytical results on the effect of cooperative trap interactions can be obtained with the help of a lattice gas (or Ising) model to describe the interaction [17]. In a 1-d crystal, if one characterizes the interaction by an energy which equals infinity if the two trap-influenced host sites coincide, $-\varepsilon_0$ if they are nearest-neighbour, and zero in all other cases, one can prove [17]

$$p_m = 1-(1-\rho)(1-x^{|m|}) \tag{3.7}$$

$$x = (y-1)/(y+1) \tag{3.8}$$

where $y^2 = 1-4\rho(1-\rho)[1-\exp(\varepsilon_0/k_BT)]$, where T is the temperature of relevance to the trap placement and k_B is the Boltzmann constant. Equation (3.4) gives

$$\nu(t) = \rho+(1-\rho)\sum_{m=-\infty}^{\infty} x^{|m|}\psi_m(t) \tag{3.9}$$

and (3.9) can be evaluated in closed form for several kinds of exciton motion [17].

This formalism can be used to understand the deviation from 1 of the exponent of ρ in the energy transfer rate reported in many observations [13,18,19]. However, the experimental data is not certain as the reported deviations could vanish within experimental errors.

4. Mutual Annihilation: the Process Itself

Two separate investigations concerning the mutual annihilation of excitons are described in this article. The analysis of observations directed at the annihilation process itself in a way analogous to the capture observations described in sections 2 and 3 occupy us in the present section, whereas section 5 describes a separate class of experiments in which annihilation serves only as a probe.

Two moving excitons may undergo mutual annihilation when within each other's influence. In the case of singlets it merely leads to the depletion of their population. For triplets the destruction of the excitons can produce singlets and therefore the relatively dramatic phenomenon of delayed fluorescence.

The experimental observables are the yield and the luminescence intensities as in the capture case. For medium exciton densities, the construction of the formalism to take one to the observables proceeds [11,20] via the consideration of a two-exciton system and the analysis of the system-point in the corresponding 6-dimensional space. By thus converting the annihilation problem into a capture problem in the space of twice the number of dimensions as the actual system, the methods of section 3 can be used and the observables calculated. Details are not shown here for space reasons. The general result for the well-known "annihilation constant" γ is

$$\gamma = v[(1/b)+\tilde{\psi}_0(1/\tau)]^{-1} \tag{4.1}$$

where the actual destruction process is assumed to be short-ranged as in the capture case (for simplicity) and to proceed at rate 2b, and v is the volume of the unit cell. The general prescription follows the pattern of the capture case. Time-dependent luminescence intensities may also be obtained by Laplace inversion of appropriate expressions [11,20].

It is of interest here to exhibit the simplest transport evolution equation [21] capable of a description of an arbitrary degree of coherence:

$$\frac{\partial \rho_{mn}}{\partial t} = -iV(\rho_{m+1\ n}+\rho_{m-1\ n}-\rho_{m\ n+1}-\rho_{m\ n-1})-\alpha(1-\delta_{m,n})\rho_{mn}\ . \tag{4.2}$$

Here ρ_{mn} is the density matrix element in the representation of localized (Wannier) states m,n in a 1-d crystal, V is the interaction matrix element which is proportional to the bandwidth or to the average velocity in the absence of scattering, and α is the scattering rate. Recently this equation has been used extensively for the description of coherence effects [11,22]. The degree of coherence is measured by V/α, which is proportional to the ratio of the mean free path to the lattice constant. The propagator ψ_0 to be substituted in (4.1) or in the corresponding capture expressions in section 3 is, in the respective coherent ($\alpha=0$) and the incoherent ($\alpha\to\infty$, $V\to\infty$, $2V^2/\alpha=F$) limits,

$$\tilde{\psi}_0(1/\tau) = (2\tau/\pi\sqrt{1+16V^2\tau^2})\ \mathbb{K}\ (4V\tau/\sqrt{1+16V^2\tau^2}) \tag{4.3}$$

$$\tilde{\psi}_0(1/\tau) = \tau/\sqrt{1+4F\tau} \tag{4.4}$$

where $\mathbb{K}$ is the elliptic integral of the first kind. The intermediate case is complicated but explicitly available [11].

A recent application [23] of the combined formalism for capture and annihilation has led to serious doubts being cast on whether existing sensitized luminescence and annihilation observations are really measuring exciton motion. Observed differences in the behaviour of γ and of the energy transfer rate in the same crystals and the same temperature range motivated that investigation. Thus it appears in the light of existing data [24] that in anthracene the energy transfer rate is capture-limited at least above 60 K and that values of quantities such as the diffusion constant obtained by applying usual recipes to these observations are probably without content.

5. Mutual Annihilation: as a Probe for the Motion

An ingenuous method [25] of studying triplet exciton motion consists of creating a spatial inhomogeneity in the exciton population by illuminating the crystal through Ronchi rulings (arrays of alternating opaque and transparent strips) and detecting the time evolution of the population inhomogeneity through delayed fluorescence (DF) signals arising from the mutual annihilation of the excitons. Recent analysis [22] has shown that this kind of experiment is capable of manifesting qualitative effects of the degree of coherence.

The experiment employs illumination intensities which are large enough to make the DF signal observable but small enough so that the effect of annihilation on the motion can be neglected. The evolution equation is therefore (1.1) with a source term (representing the illumination) added to its right side. The signal on the other hand is proportional to $\sum_m P_m^2(t)$ since DF arises from the mutual annihilation. Fourier-transforming (discretely) (1.1) with the source term one gets

$$(1/G^k)P_b^k(t) = (1/\tau)\int_0^t dt' e^{-t'/\tau}\psi^k(t') \tag{5.1}$$

$$(1/G^k)P_d^k(t) = [(1/\tau)\tilde{\psi}^k(1/\tau)]\psi^k(t)e^{-t/\tau} \tag{5.2}$$

$$(1/G^k)P_s^k(t) = (1/\tau)\ \tilde{\psi}^k(1/\tau)\ . \tag{5.3}$$

The 3 subscripts b,s,d, refer to the build-up stage when an illumination source constant in time is applied, the steady-state value reached as a result, and the decay stage when the source is subsequently cut off. All these stages of the experiment can be separately used for transport investigations. The quantity G^k is the product of the illumination intensity, the absorption coefficient and a factor g^k which is the Fourier transform of the spatial variation of the source, i.e. of the Ronchi ruling. The factor g^k is representative of the square wave [25].

The striking result of the analysis is that the DF signal, which is proportional to $\sum_k |P^k|^2$, measures directly the Fourier transform of the propagator, i.e. $\psi^k(t)$. This is particularly true in the decay stage, the integral to time t appearing in the build-up stage and to ∞ in the steady state. Since the spatial nature of the source (i.e. g^k) selects the k's appearing in (5.1)-(5.3), the experimentalist can span the ψ^k's by varying the ruling period. Coherence effects appear in the DF signals if the ruling period is comparable to the mean free path of the exciton. These effects manifest themselves as oscillations and in clear changes in the shape of the signal as a function of time. The analysis based on (4.2) leads immediately to these effects since ψ^k is given by

$$\psi^k(t) = J_0(c_k t)e^{-\alpha t}+\int_0^t du\ \alpha e^{-\alpha(t-u)} J_0(c_k\sqrt{t^2-u^2}) \qquad (5.4)$$

in the general case, (c_k = 4Vsin(k/2)), with the coherent limit $\psi^k(t) = J_0(c_k t)$ and the incoherent limit $\exp[-4Ft\sin^2(k/2)]$. Coherence is represented by the oscillations of the Bessel function and an initial curvature in the shape of the signal. Detailed analysis has shown [22] that the second effect is particularly discernible for realistic values of the exciton bandwidth and the ruling periods.

In contrast to sensitized luminescence which measures effects of the self-propagator ψ_0 or of a sum of propagators ψ_m, the Ronchi ruling experiment measures the Fourier transform ψ^k. The latter quantity has strong qualitative differences in the coherent and incoherent limits and therefore the ruling experiment, or the related transient grating experiment [26] which we have analyzed earlier [11], is particularly appropriate for coherence investigations. It is hoped that such experiments will be carried out in the near future.

6. Concluding Remarks

Theoretical work of the kind (ii) referred to in section 1 is, as we have seen, proceeding actively in various fields of exciton transport. New light on old issues is also appearing as in the case of coherence and the role of the capture process in sensitized luminescence.

Acknowledgments

It is a pleasure indeed to thank Prof. H.C. Wolf for his invitation to and his excellent hospitality at the Universität Stuttgart. Discussions with Prof. D. Schmid were invaluable in this research. The work was supported by NSF grant no. DMR-8111434 and by a grant from the NSF for USA-Germany cooperative research.

References

1. F. Perrin, Ann.Physique 17, 283 (1932)
2. Th. Förster, Ann.Phys. (Leipzig) (b) 2, 55 (1948)
3. Th. Förster, in Comparative Effects of Radiation, J.S. Kirby-Smith, J.L. Magee ed. (Wiley, New York 1960) Chap. 13
4. H. Haken and G. Strobl, Z.Phys. 262, 135 (1973); H. Haken and P. Reineker, Z.Phys. 249, 253 (1972)
5. P. Reineker, in Exciton Dynamics in Molecular Crystals and Aggregates, G. Höhler ed. (Springer-Verlag, Berlin, 1982)
6. M. Grover and R. Silbey, J.Chem.Phys. 54, 4843 (1971)

7. R. Silbey, Ann.Rev.Phys.Chem. 27, 203 (1976)
8. V.M. Kenkre and R.S. Knox, Phys.Rev. B9, 5279 (1974); Phys.Rev.Lett. 33, 803 (1974)
9. V.M. Kenkre, in Statistical Mechanics and Statistical Methods in Theory and Application, U. Landman ed. (Plenum, New York, 1977)
10. V.M. Kenkre, Phys.Rev. B11, 1741 (1975); 12, 2150 (1975); Phys.Lett. 63A, 367 (1977); 65A, 391 (1978)
11. V.M. Kenkre, in Exciton Dynamics in Molecular Crystals and Aggregates, G. Höhler ed. (Springer-Verlag, Berlin, 1982)
12. V.M. Kenkre, J.Stat.Phys. 30, 293 (1983); Proceedings of the Fourth International Seminar on Energy Transfer, J. Pantoflicek ed. (Prague, 1981) p. 54
13. H.C. Wolf, in Adv. in Atomic and Molecular Physics vol. 3, D.R. Bates and I. Esterman ed. (Acad. New York 1967)
14. V.M. Kenkre and Y.M. Wong, Phys.Rev. 23, 3748 (1981)
15. V.M. Kenkre and P.E. Parris, Phys.Rev. B6, 3221 (1983)
16. V.M. Kenkre, Chem.Phys.Lett. 93, 260 (1982)
17. V.M. Kenkre, P.E. Parris and S.M. Phatak, University of Rochester preprint
18. W. Klöpfer, H. Bauser, F. Dolezalek and Gr. Naundorf, Mol.Cryst.Liq.Cryst. 16, 229 (1972)
19. H. Port, H. Schneckenburger and H.C. Wolf, Z.Naturforsch. 36a, 697 (1981)
20. V.M. Kenkre, Phys.Rev. B22, 2089 (1980); Z.Phys. B43, 221 (1981)
21. P. Avakian, V. Ern, R.E. Merrifield and A. Suna, Phys.Rev. 165, 974 (1968)
22. V.M. Kenkre, A. Fort and V. Ern, Chem.Phys.Lett. 96, 658 (1983); Phys.Rev. B, scheduled for publication
23. V.M. Kenkre and D. Schmid, Chem.Phys.Lett. 94, 603 (1983)
24. A. Braun, U. Mayer, H. Auweter, H.C. Wolf and D. Schmid, Z.Naturforsch. 37a, 1013 (1982)
25. V. Ern, P. Avakian and R.E. Merrifield, Phys.Rev. 148, 862 (1966)
26. J. Salcedo, A.E. Siegman, D.D. Dlott and M.D. Fayer, Phys.Rev.Lett. 41, 131 (1978)

Excitation Transport in Naphthalene Aggregates: Mixed Crystals, Amorphous Thin Films and Polymeric Glasses

R. Kopelman

The University of Michigan, Ann Arbor, MI 48109, USA

There is much interest in the phenomena and mechanisms of energy transport (ET) in molecular aggregates - from basic solid-state considerations (diffusion, percolation, Anderson transition)[1] to photosynthesis and luminescent solar concentrators[2]. ET can also be used for the characterization of molecular aggregates - in mixed crystals, polymeric glasses, thin films and biomembranes.

Our program spans various forms of aggregates and disorder: neat crystals, isotopically mixed crystals, chemically mixed crystals, disordered semi-crystalline films, doped polymeric glasses and polymeric matrices that mimic photosynthetic aggregates and luminescent solar concentrators. We use a number of ET techniques (including, for instance, time and spectrally resolved delayed fluorescence due to heterofusion) to characterize the transport; spectroscopic techniques to characterize the electron and phonon spectra of the aggregates and preparation techniques to define and characterize the materials. For instance, a perfect binary or ternary crystal is first investigated at 4[K] and then sublimed and re-crystallized both as a sublimation flake and as a rapidly deposited film. Both the flake and film are investigated again at 4[K]. The film then undergoes *in situ* annealing cycles and is re-investigated at 4[K].

The semi-crystalline films are prepared by sublimation from an effusion micro-oven onto a cooled substrate located within the sample zone of a liquid helium cryostat. The sample is deposited in vacuum at about 50 [K] and then kept under nitrogen or helium atmosphere, depending on the temperature. For the preparation of the doped polymeric glassy films a solvent casting technique is used. A typical glassy film consists of extensively zone refined naphthalene doped into polymethylmethacrylate (PMMA, medium molecular weight), using a spectrograde dichloromethane solvent. The mixed crystals are of the quality and purity used before in our laboratory[3,4]. The excitation methods and measurements are also very similar to those used in our laboratory before[3,4].

Our semi-crystalline and glassy films show previously observed features such as naphthalene and anthracene excimers (dimers) and inhomogeneously broadened "monomer" lines. These films have been produced in a manner that is fairly reproducible in its properties both from sample to sample and from one part of the sample to another. Optically heterogeneous films have been rejected.

The essential ET features of heavily doped isotopic mixed naphthalene crys--tals, i.e.,the concentration and temperature dependence, remain qualitatively intact but show significant quantitative changes in the disordered crystalline films and polymeric glasses. Also, the ET properties of freshly deposited pure or doped naphthalene films may show less resemblance to the parent pure or doped crystal than to films of naphthalene-doped "plexiglass" (PMMA).

Binary mixed crystals of naphthalene in perdeuteronaphthalene with a trace of betamethylnaphthalene have been excited with a xenon lamp. They show a significant and sharp reduction in phosphorescence lifetime and intensity (at 1.7 [K])at about 10% naphthalene concentration. Higher naphthalene concentrations and/or temperatures are required to show a similar effect for the deposited naphthalene films and naphthalene-doped glasses (PMMA). The phosphorescence decay times of the freshly deposited films (monitored on supertrap emission lines) decrease upon annealing as they do upon increasing the naphthalene guest concentration. The phosphorescence decay times of polymeric glasses with a high concentration of naphthalene in PMMA show a decrease by a factor of about 3 upon increasing the naphthalene concentration from 1 to 20% at liquid nitrogen temperature (Table 1). Similar polymeric glasses, containing in addition to naphthalene from 0.01% to 16% anthracene, show an increase of anthracene fluorescence with anthracene concentration. At the same time both the naphthalene phosphorescence and fluorescence intensities are reduced. The same samples show anthracene fluorescence decay times of the order of a second and the naphthalene phosphorescence decays are also measured in seconds (exact decay times cannot be given due to extreme deviations from exponentiality). Our interpretation of the above is based on a "two-population" model: An excited naphthalene molecule is either connected or not connected to an anthracene supertrap via an ET path. The excitons in any connected naphthalene quasi-lattice are quickly quenched by the anthracene. However, in any "unconnected" quasi-lattice the naphthalene excitons continue to enjoy a long life. A similar situation has been observed for perfect isotopic mixed crystals where the naphthalene fluorescence and the betamethylnaphthalene (supertrap) fluorescence each exhibited its natural decay time[6].

Table 1

Phosphorescence (5000 A^{o}) decay times: naphthalene in poly(methylmethacrylate)

Naphthalene conc.(%)	77K time(sec)	4K time(sec)
1	2.0	2.3
5	1.7	2.1
10	1.0	2.2
20	0.6	1.9

All our samples show, under the appropriate conditions, "spectral diffusion", due to triplet exciton trapping, as well as "spectral fusion" due to triplet exciton annihilation, i.e.,typical ET phenomena. When one uses supertrap species (e.g.,betamethylnaphthalene or anthracene in naphthalene), the spectral diffusion is transformed, in part, to "supertrapping" and the quantum yield (supertrap/total emission) characterizes the ET for both triplet and singlet excitons[1]. Similar information is obtained from the rise and fall times of the supertap and guest (donor) spectrally resolved emission[1]. The fusion process includes homofusion (free-free exciton annihilation) and heterofusion (free-trapped exciton annihilation). Both trapping and fusion require significant ET of the free excitons. The shortening of the supertrap emission times results from the heterofusion process. As both the heterofusion and the homofusion processes cannot occur without ET of the free excitons the shortening of the supertrap phosphorescence times is due to fusion and thus becomes a reliable ET indicator.

Possibly the most dramatic consequence of the confining of excitons to molecular clusters is exhibited by the unusual orders (exponents X) of reaction for the triplet fusion process: T + T = S, where $[S] \propto [T]^X$. Isotopic mixed crystals of naphthalene showed[3] values of X from 2 to 30. Preliminary experiments have indicated a similar behaviour for isotopic mixed naphthalene films and for the chemically mixed polymeric films thus indicating that the naphthalene excitons are confined within naphthalene molecular aggregates.

Because of the short-ranged nature of the exchange type exciton interactions, triplet ET is more sensitive to the local structure and order (disorder) compared to singlet ET (usually due to dipole-dipole interactions). Most previous studies on organic films involved singlet ET[5]. On the other hand, room temperature studies[7] indicate a triplet ET in polymeric glasses containing anthracene that is much faster than in the crystal. Our lowest temperature studies do not support an increase in ET from crystal to glass. In fact, our studies show that the ET is hampered due to new additional disorder, compared to the crystalline samples. Obviously, we do expect semi-crystalline and glassy samples to show positional, environmental and energetic disorders in addition to the substitutional disorder found in isotopic mixed crystals.

Acknowledgments

The experiments were carried out by Paul W. Klymko (mixed crystals), Laurel A. Harmon (semi-crystalline films) and E. Irene Newhouse (polymeric films). We thank for support the U. S. Public Health Service (NIH Grant No. R01 NS 08116-15).

References

1. R. Kopelman: In Modern Problems in Solid State Physics, Vol.4, eds. V.M. Agranovich and R.M. Hochstrasser (North-Holland, Amsterdam 1983) pp. 139 - 184
2. P.S. Friedman: Optic. Eng. 20, 887 (1981)
3. P.W. Klymko and R. Kopelman: J. Phys. Chem. 86, 3686 (1982)
4. S.T. Gentry and R. Kopelman: J. Chem. Phys. 78, 373 (1983)
5. H. Bässler: Phys. Stat. Sol. (b) 107, 9 (1981)
6. R.P. Parson and R. Kopelman: Chem. Phys. Lett. 87, 528 (1982)
7. R.D. Burkhart and E.R. Lonson: Chem. Phys. Lett. 54, 85 (1978)

Part VII

Electronic Excitation of Impurities in Glasses and Polymers

Dynamical Linewidth Effects of Hole Burning of Free Base Phthalocyanine in Polymers: Spectral Diffusion and Exchange Narrowing

A. Gutierrez, G. Castro, G. Schulte *, and D. Haarer*

IBM Research Laboratory, San Jose, CA 95193, USA

* Physikalisches Institut der Universität Bayreuth, P.O.B. 3008
D-8580 Bayreuth, Fed. Rep. of Germany

Abstract: The temperature dependence of the hole burning linewidth of free base phthalocyanine is investigated in polystyrene (PS) and polyethylene (PE) matrices. The linewidth in both polymers differs by more than a factor of 10. The temperature dependence in PS can be explained with theories which are applicable to other amorphous systems. The temperature dependence in PE shows, in a limited temperature range, an increase of linewidth with falling temperature, which we attribute to motional effects.

INTRODUCTION

The technique of photochemical hole burning (PHB) (1,2) provides a convenient experimental method for extending the optical spectral resolution well below the inhomogeneous bandwidth. In many instances the gain in optical resolution is better than 10^3 and therefore the technique allows one to measure spectral features which are too small to be measured by straightforward optical spectroscopy (3) or to investigate macromolecular broadband systems under high resolution conditions (4). The use of PHB to measure homogeneous linewidths and thus various relaxation and dephasing processes of excited states has been the subject of extensive study (5). Of particular interest has been the observation that the homogeneous linewidths of absorbers imbedded in amorphous hosts are broader than in crystalline environments. This has been reported for inorganic glasses (6), organic glasses (7), and polymer hosts (8,9). Various theoretical models have been proposed to explain these fast dephasing processes and their temperature dependence. All of the theories involve the interaction of the local absorber with a distribution of two level systems (TLS), or tunnelling states which are characteristic of the amorphous host. These theories predict temperature dependencies of the homogeneous linewidth ranging from linear (7) to quadratic (10) to a combination of both (11), but as yet have not provided a suitable framework to interpret the diverse experimental results. In the following, free base phthalocyanine (H_2Pc) (2) dissolved in polyethylene (PE) and polystyrene (PS) are the PHB systems used to study temperature and matrix effects on the lineshape of photochemical holes. These polymer hosts were chosen because they are prototype polymers whose structure and dynamics, although as yet poorly understood, have been more fully investigated. The photochemical mechanism which leads to hole burning in the H_2Pc system is believed to be similar to the one observed for free base porphyrin (12), namely a phototautomerization of the central protons of the molecule (9). This is symbolically depicted in Fig.1.

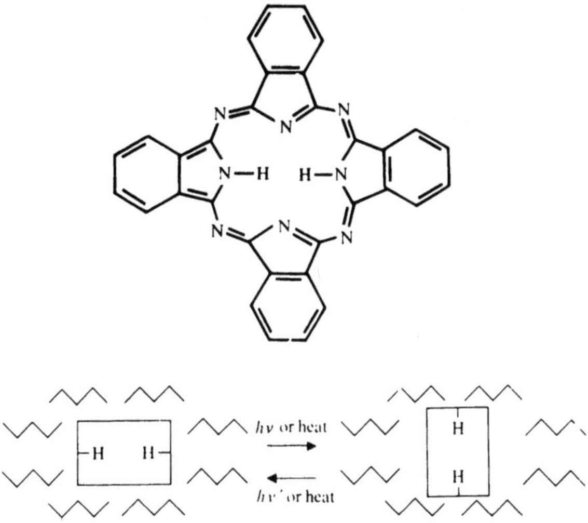

Fig. 1

Proposed phototautomerization of H_2Pc

RESULTS AND DISCUSSION

In the following we present some experimental results for PHB linewidths of H_2Pc in PS and PE matrices covering the temperature range between 2K and 20K. The experimental data on PS can be well understood with the theories on optical linewidths which have been successfully applied to other amorphous PHB systems. The experimental results for PE, however, whose macroscopic structure is characterized by both amorphous and crystalline domains, differ considerably from other comparable systems.

Polystyrene

The optical lineshape of a hole $L^{\tau}(\omega)$ can be regarded as a convolution of a true molecular lineshape function $g(\omega-\omega')$ and a site population function $N^{\tau}(\omega')$ (13).

$$L^{\tau}(\omega) = \int_{-\infty}^{\infty} N^{\tau}(\omega')g(\omega-\omega')d\omega'. \qquad (1)$$

The index τ in the above equation characterizes the burning time and, hence, the depth of the photochemical hole. In order to avoid deviations from the 'ideal' lineshape through groundstate depletion phenomena, all data are extrapolated to the short burning time limit (13), in which one can expect to measure the 'true' linewidth.

Figure 2 shows the measured hole width at three temperatures as a function of the total number of photons irradiated. The figure shows that the short burning time limit can be experimentally well achieved.

For the investigation of the temperature dependence of photochemical holes, it should be noted that different experimental approaches will, in general, yield different PHB linewidths. This is a consequence of

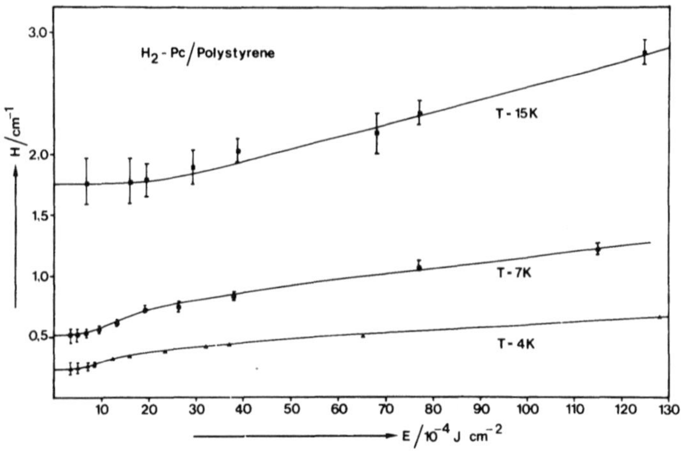

Fig. 2 Hole width in PS as a function of irradiation at three different temperatures

the relation given in equation (1), in which the site population function $N^{\tau}(\omega')$ plays the role of a memory. $N^{\tau}(\omega')$ acts as a memory for the temperature at which the photochemical hole was produced. It also acts as a memory for irreversible spectral diffusion processes (14) during any temperature cycle of the experiment. Both facts have been accounted for theoretically (15), yet, up to now, experimental data, allowing the verification of Equ.(1) are quite scarce. Based on Equ. (1), one can conclude that there are two rather different ways to measure the temperature dependence of photochemical holes. In the first kind of experiment the hole is burnt at the lowest temperature T_o of a temperature cycle and then the linewidth is measured as the temperature is raised. In the second kind of experiment a virgin hole is burnt at each of the various temperatures T_i of a complete temperature cycle. In the first experiment $N^{\tau}(\omega')$, the 'site memory function' is characteristic of the lowest temperature T_o of the cycle.

$$N^{\tau}(\omega')_o = N^{\tau}(\omega')/_{T = T_o} \quad . \tag{2}$$

In the second type of experiment, the site memory function is produced at each temperature T_i .

$$N^{\tau}(\omega')_i = N^{\tau}(\omega')/_{T = T_i} \quad . \tag{3}$$

It is quite obvious that Equs.(2) and (3) are identical for the lowest temperature point of each cycle. This behaviour is shown in Fig. 3. Here the lines drawn with large slopes are the measured hole widths as a function of temperature for four experimental cycles starting at 4K, 7K,10K and 15K. A detailed numerical evaluation of the various cycles shows that Equ.(1) accounts for the data displayed in Fig. 3 in a semiquantitative fashion, correlating the different datasets within 20% of the measured linewidth.

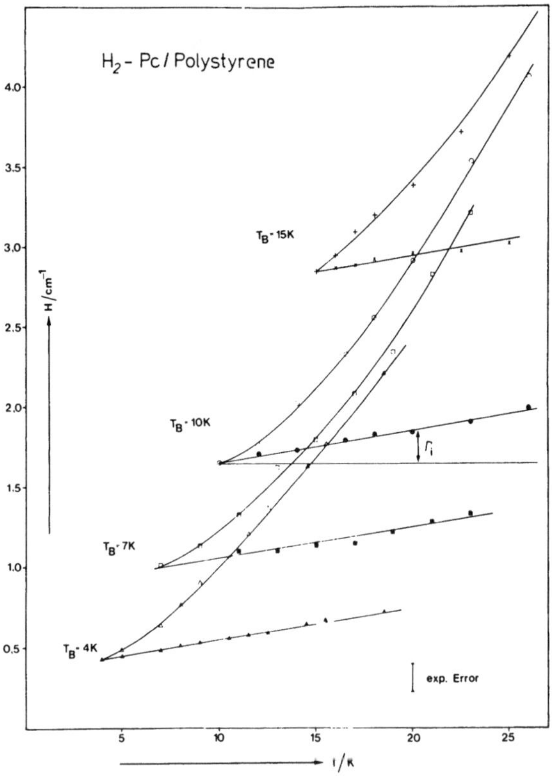

Fig. 3 Hole width in PS measured with temperature cycles (see text)

So far the site memory function $N^{\tau}(\omega')$ was discussed only in terms of an empirical parameter. One can show, however, that $N^{\tau}(\omega')$ contains two contributions (15). A reversible contribution which is present in both, crystalline and amorphous solids and which is dominated by the optical relaxation rates at the various temperatures T_o at which the site memory function is produced via photochemistry. This reversible contribution causes the factor of two relating the measured hole width and the homogeneous linewidth Γ_o at the lowest temperature T_o of a cycle. At this temperature both $N^{\tau}(\omega')$ and $g(\omega-\omega')$ are characterized by the homogeneous linewidth Γ_o and, hence, the convolution integral of Equ.(1) yields a hole width for $L(\omega)$ of $2\Gamma_o$. A second contribution to $N^{\tau}(\omega')$ was found to be present in amorphous materials reflecting their non-equilibrium state. It is irreversible and was interpreted in terms of a redistribution process among the tunnelling levels (TLS) describing the amorphous state (14,15). This part of $N^{\tau}(\omega')$ can be measured by recycling the sample at each temperature T_i of the experiment (Fig.3) to the lowest temperature T_o of the cycle. The measured irreversible contributions Γ_i to the linewidth in polystyrene are given in Fig.3 by the solid lines with small slope originating at each of the various starting temperatures. They give an irreversible contribution Γ_i to the linewidth which increases in a linear fashion with

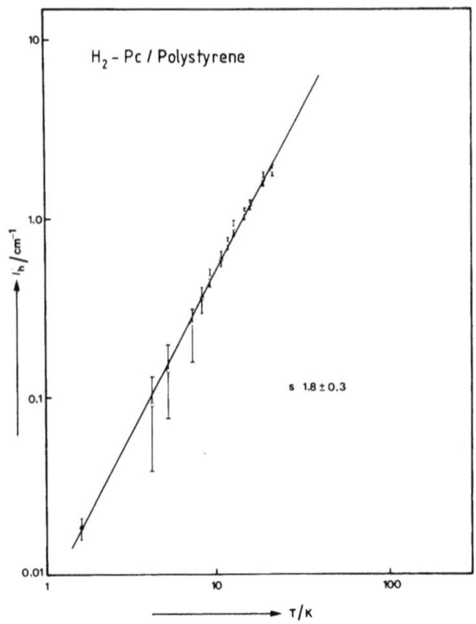

Fig. 4

Hole width in PS versus temperature (double log plot)

temperature in the range between $4K<T<26K$ and which amounts to $\Gamma_i = 21 \pm 2 \cdot 10^{-3} cm^{-1}/K$. The origin of this irreversible linewidth contribution has been attributed to the energy asymmetry parameters of tunnelling centers in amorphous hosts (14). For a more exhaustive microscopic interpretation of the irreversible linewidth contributions more experimental data on various materials are required. It would be especially important to find out whether or not the linear temperature dependence which we have found for PS levels off at elevated temperatures. Experiments along these lines are difficult, because at temperatures which are well above 25K the PHB photochemistry shows signs of thermal reversibility. At these elevated temperatures the integrated hole area decreases making a quantitative interpretation of the data rather difficult.

Based on thermal cycling experiments, as shown in Fig.3, one can separate the two linewidth contributions $N^{\tau}(\omega')$ and $g(\omega-\omega')$ experimentally. This allows an investigation of the reversible part of the linewidth $g(\omega-\omega')$. The linewidth Γ_r of $g(\omega-\omega')$ is only due to relaxation processes and, hence, can be used to check the applicability of the various linewidth theories (10,11). Figure 4 shows a double logarithmic plot of the measured $g(\omega-\omega')$ versus temperature. The experimental data, which cover one temperature decade, can be described by a power law

$$\Gamma_r(T) = \text{const}\ T^S . \qquad (4)$$

The measured power law (Equ.(4)) for H_2Pc in PS yields $S=1.8\pm0.3$. A power law with $S=2$ has been postulated by ORBACH's theory (10). A more detailed treatment of REINEKER et al. (11,16), which assumes strong coupling between the PHB center and the TLS of the glass, can account for exponents between $S=1$ and $S=4$. Various experimental studies of amorphous systems have yielded rather different temperature exponents (7,14,17,18) between 1 and 2. Yet it is too early to establish systematic trends which link properties of the glass matrix to the measured temperature exponents.

Polyethylene

The linewidth of photochemical holes in polystyrene was analyzed with the aid of Equ.(1) in order to separate phenomena which are due to site distribution or redistribution as described by the function $N^{\tau}(\omega')$ and dephasing parameters as described by the function $g(\omega-\omega')$. Other PHB systems (13,14) like quinizarine in hydrogen-bonded hosts or in boric acid (14,15) seem to be well described by this rather straightforward model.
Among the PHB systems which are known to us, H_2Pc in PE is an exception, because it deviates from the above model. The deviation from the model calculation is twofold. First, it does not reproduce the $N^{\tau}(\omega')$ of Equ.(1) and the system behaves like having a δ-function like site distribution. Second, the system shows an anomalous temperature dependence of the measured hole width in a certain temperature range. In this range the linewidth increases with both rising and falling temperature, having a minimum at or close to the burning temperature. The increase in linewidth with rising temperature corresponds to the temperature behavior of other known PHB systems. The increase in linewidth with falling temperature, however, seems to be a strong evidence for motional narrowing phenomena at very low temperatures.
In the following we want to assume an exchange model and discuss the measured linewidth features. The exchange model was originally de-

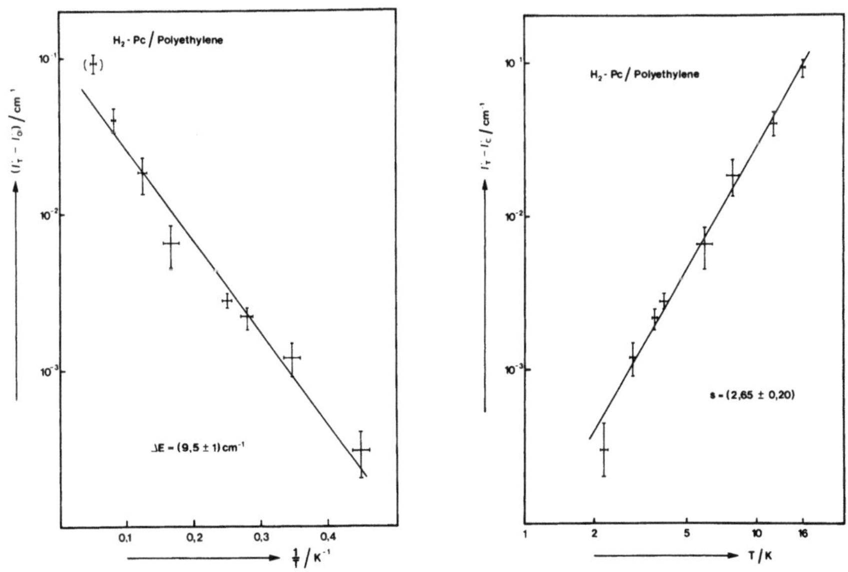

Fig. 5 Hole width in PE versus temperature
5a log Γ versus 1/T
5b log Γ versus log T

rived for nuclear magnetic resonance experiments by KUBO et al. (19). It was lateron adopted to spin-echo systems (20) and to optical experiments (21). In one case of a PHB system, namely porphin in crystalline octane, a case of slow exchange was observed experimentally (22,23). The basic equation governing the exchange model is (19,21)

$$\Gamma_{eff} - \Gamma_{o} = \frac{(\delta\omega)^2 \cdot \tau}{1+(\delta\omega)^2\tau^2} \exp(-\Delta E/kT). \qquad (5)$$

$(\Gamma_{eff} - \Gamma_o)$ is the effective linewidth corrected for the lifetime contribution Γ_o. $\delta\omega$ is the difference in optical 'Larmor frequencies' between the two exchanging levels. τ is the lifetime of the exchanging level which is detuned by the amount $\delta\omega$ and ΔE is the energy splitting of this level (21).

Assuming constant $\delta\omega$ and τ values one would expect that the effective linewidth can be described by an activated process as shown in Fig.5a. The experimental data fit such a plot in the temperature range of $2K<T<16K$ yielding a value of $\Delta E=9.5\pm1cm^{-1}$. Figure 5b shows that a power law with the slope of $S=2.6\pm0.2$ cannot be ruled out due to the limited temperature range of the experiment.

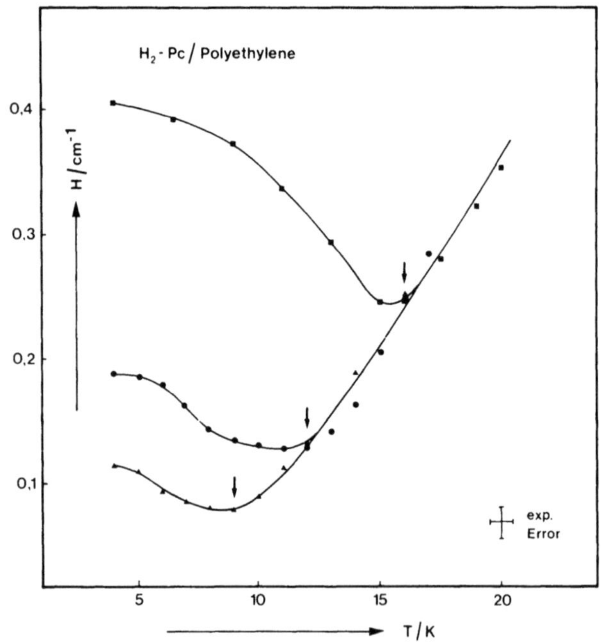

Fig. 6 Hole width in PE measured with temperature cycles. The arrows mark the burning temperatures

Figure 6 shows PHB experiments with thermal cycling starting at three different burning temperatures (see Fig.3 for comparison). Quite in contrast to the same experiments in the PS matrix, there is no measurable 'memory effect' of $N^{\tau}(\omega')$. This means that both high and low burning temperatures yield the same result as long as the temperature at which the hole is being measured is the same. We interpret this result as being caused by rapid exchange processes between the various levels contributing to the $N^{\tau}(\omega')$ distribution. If one takes, for instance, the data point at 16K and a typical $\delta\omega$ width on the order of $1cm^{-1}$ one would arrive at a $\delta\omega\tau$ value of 1/5. This would correspond to a τ value of 1 psec. The most difficult aspect of the PE experiment is the measured increase in optical linewidth upon cooling the sample after PHB chemistry at 'high temperatures' (e.g. 16K). Obviously Equ.(5) would not describe such a linewidth behavior without making the additional assumption that

$$\tau = \tau(T) \neq \text{const.} \tag{6}$$

Under this assumption an increase of τ would explain an increase in linewidth because, in the fast exchange regime ($\delta\omega\tau > 1$), an increase in τ leads to an increase of the measured linewidth.
Since a temperature-dependent τ makes Equ.(5) almost intractable, more data have to be collected for a complete understanding of the various line-broadening phenomena. One conclusion can, however, be drawn from the above experiments; the conclusion that the site distribution function $N^{\tau}(\omega')$ which had been thus far treated as 'frozen in' can show dynamical effects. These dynamical effects lead to an effective reduction in linewidth in the fast exchange regime. If the temperature is lowered, one can observe an increase in linewidth due to an increase of the correlation time τ. This is a situation which is quite unusual in the optical regime; it is analogous to the one encountered in the field of ESR and NMR.

Acknowledgements: We would like to thank Dr. Friedrich for helpful discussions and the Stiftung Volkswagenwerk for financial support.

Note added in proof:

The measured temperature exponent of the homogeneous linewidth of H_2Pc in P.S. of 1.8 comes close to the exponent of 1.75 which has been calculated for a dipole-quadropole model by S.K. Lyo (24).

LITERATURE

1. B.M. Kharlamov, R.I. Personov and L.A. Bykovskaya: Opt. Comm. 12, 191 (1974)

2. A.A. Gorokhovskii, R.K. Kaarli and L.A. Rebane: J.E.T.P. Lett. 20, 216 (1974)

3. S. Voelker, R.M. Macfarlane and J.H. van der Waals: Chem. Phys. Lett. 53, 8 (1978)

4. J. Friedrich, H. Scheer, B. Zickendraht-Wendelstadt and D. Haarer: J. Am. Chem. Soc. 103, 1030 (1981)

5. L.A. Rebane, A.A. Gorokhovskii and J.V. Kikas: Appl. Phys. B29, 235 (1982)

6. J. Hegarty and W.M. Yen: Phys. Rev. Lett. 43, 1126 (1979)

7. G. Small in "Spectroscopy and Excitation Dynamics of Condensed Molecular Systems" ed. by V.M. Agranovich and R.M. Hochstrasser, North Holland 1983

8. R.J.A.P. Marchetti, M. Scozzafava and R.H. Young: Chem. Phys. Lett. 51, 424 (1977)

9. A.R. Gutierrez, J. Friedrich, D. Haarer and H. Wolfrum: IBM J. Res. Develop. 26, 198 (1982)

10. S.K. Lyo and R. Orbach: Phys. Rev. B22, 4223 (1980)

11. P. Reineker and H. Morawitz: Chem. Phys. Lett. 86, 359 (1982)
Solid State Comm. 42, 609 (1982)

12. S. Voelker and J.H. van der Waals: Mol. Phys. 32, 1703 (1976)

13. J. Friedrich and D. Haarer: J. Chem. Phys. 76, 61 (1982)

14. J. Friedrich, H. Wolfrum and D. Haarer: J. Chem. Phys. 77, 2309 (1982)

15. J. Friedrich, D. Haarer and R. Silbey: Chem. Phys. Lett. 95, 119 (1983)

16. P. Reineker, private communication

17. H.P.H. Thijssen, A.I.M. Dicker and S. Voelker: Chem. Phys. Lett. 92, 7 (1982)

18. R. Jankoviak and H. Bässler: Chem. Phys. Lett. 95, 310 (1983)

19. R. Kubo and K. Tomita: J. Phys. Soc. Jap. 9, 888 (1954)

20. C.A. van't Hof and J. Schmid: Chem. Phys. Lett 36, 457 (1975)

21. C.B. Harris, R.M. Shelby and P.A. Cornelius: Phys. Rev. Lett. 38, 1415 (1977)

22. S. Voelker, R.M. Macfarlane and J.H. van der Waals: Chem. Phys. Lett. 53, 8 (1978)

23. A.I.M. Dicker, J. Dobrowski and S. Voelker: Chem. Phys. Lett. 84, 415 (1981)

24. S.K. Lyo: Phys.Rev.Lett. 48, 688 (1982)

Theory of Dephasing of Impurities in Glasses *

S.K. Lyo

Sandia National Laboratories, Albuquerque, NM 87185, USA

1 Introduction

Twelve years ago ZELLER and POHL [1] showed that below 1 [K] the thermal conductivity and the heat capacity of amorphous solids behave in a strikingly different way from those of crystalline solids. Since then a number of studies found that other low-temperature properties such as acoustic, dielectric and spin-relaxation properties are also unusual [2]. These results indicate that there is a continuum of low-energy excitations in the amorphous systems which is absent in the crystalline solids. It is generally recognized that the low-energy modes arise from tunneling of an atom or a group of atoms between two energy minima of a double-well through a potential barrier. A random distribution of the energy asymmetry of the well depths and the barrier heights gives a broad distribution of energies and a nearly constant density of states for the so-called two-level systems (TLS) [3,4]. This model is essentially phenomenological and its justification lies ultimately in the fact that it explains most of the data successfully. In this paper we discuss another unusual effect, namely, the anomalous dephasing of optically active impurities which seems to arise from interactions with TLS at low temperatures.

Recent homogeneous optical linewidth ($\Delta\omega$) measurements in amorphous organic and inorganic glasses and amorphous polymers at low temperatures showed that the linewidths are orders of magnitude larger than those in crystalline solids and follow unusual power-law temperature dependences

$$\Delta\omega \propto T^m \tag{1.1}$$

with the exponent $1<m<2.2$. The power-law dependence is obeyed in a wide range of temperatures, suggesting that a single line-broadening mechanism is likely to be responsible for the anomalous linewidth. The homogeneous linewidths have been measured in inorganic glasses by fluorescence line narrowing. The exponents in (1.1) are found to be $m = 1.8 \pm 0.2$ in the range 7 - 80 [K] for Eu^{3+}-doped silicate glass [5] and $m = 1.85 \pm 0.2$ and $m = 1.88 \pm 0.2$, respectively, for Pr^{+}-doped BeF_2 and GeO_2 glasses in the range 6 - 300 [K] [6]. The linewidths are about two orders of magnitude larger at lower ends of the above temperature ranges than those in crystalline solids. The nearly quadratic temperature dependence is also observed at higher temperatures (200 - 300 [K]) for Eu^{3+}-doped silicate glasses [7] and for Eu^{3+} in different inorganic glasses (295 - 800 [K]) [8]. Recent hole burning measurements in silicate glasses show that the $m = 1.8$ behavior for Eu^{3+} ions in [5] continues down to very low temperatures (1.5 - 22 [K]) [9]. On the other hand, for Pr^{3+} ions the exponent is found to be $m = 1.0$ in the same temperature range for the same system [9].

* This work performed at Sandia National Laboratories is supported by the U.S. Department of Energy under contract number DE-AC04-76DP00789.

In organic glasses measurements have been made by photochemical [10, 11] and non-photochemical [12,13] hole burning. Holewidths of 0.5 to a few [cm^{-1}] were measured at T = 1.8 [K] which are two to three orders of magnitude larger than those in inorganic glasses. The holewidths of dihydroxyanthraquinone in ethanol/methanol and in boric acid [11] follows a T^2 dependence up to 20 [K], while those of tetracene in ethanol/methanol and in glycerol/DMSO/DMF glasses show a linear T behavior between 4 - 20 [K] [13]. On the other hand a recent photochemical hole burning study of free-base porphin in several organic glasses [14] shows a $T^{1.3 \pm 0.1}$ law between 1.2 - 20 [K]. In this case much smaller holewidths (~0.01 [cm^{-1}] at 1 [K]) were found.

The anomalous linewidth data in organic and inorganic glasses discussed above cannot be explained in terms of the standard two-phonon Raman process much below the Debye temperature (θ_D), although it may account for the T^2 behavior at high temperatures. A scarce density of acoustic phonon modes and small strain yield a very small linewidth and a faster temperature dependence (i.e., m = 7). A direct resonant one-phonon process gives a small exponential contribution, for example, yielding $\Delta\omega \propto \exp(-216/T)$ with T in kelvins for Eu^{3+} in silicate glasses where the lowest level from the ground state is at a distance of 150 [cm^{-1}] [5]. The basic problem for any theoretical attempt to explain this phenomenon in glasses lies in the fact that not much is known microscopically about all the possible low-energy modes and the nature of their coupling to ions. Our view in the present paper is that probably a single mechanism is at work dominantly at low temperatures for these data [5-14] with other mechanisms possibly playing additional roles. Recently contributions from low-energy optical modes in silicate glasses [15] and libronic modes in organic glasses and amorphous polymers [16] have been suggested as possible explanations for the linewidth. However, while successful in some cases, they do not explain the data [5-14] consistently especially at low temperatures. On the other hand a theory based on the ion - TLS interaction to be presented below seems to explain the data [5-14]. The two-level modes have a large enough density of states (deduced from the specific heat data of [1] for inorganic glasses) even at low energies and couple sufficiently strongly to the ions through electric multipolar interactions. The line-broadening then arises from modulation of the ionic levels involved in the resonant optical transition by a rapid phonon-assisted tunneling motion between the two local potential wells in TLS. The tunneling motion may not necessarily be in the translational degree of freedom. It can occur in the rotational motion of a molecule.

The organization of the present paper is as follows. In the next section we introduce the model Hamiltonian by defining optical levels of an ion, TLS Hamiltonian, ion - TLS interaction and TLS - strain coupling. A typical ion - TLS interaction strength at a closest distance (~ 10[Å]) is assumed to be about a few hundred [cm^{-1}] or less which is on the order of the inhomogeneous linewidth or less in glasses [17] and to decay rapidly as $1/r^s$ (s = 3, 4, and 5) with distance (r). Also, it is useful to bear in mind that for a typical density of states of TLS (~ 10^{20}/[$eVcm^3$]) the average separation between TLS within the thermal energy, say, 10 [K] is about 100[Å]. In Section 3.1 perturbative theories which treat the ion - TLS coupling as a perturbation are discussed. It is pointed out that a perturbation theory breaks down for the most important range of ion - TLS coupling strengths for diagonal modulation which is mainly responsible for the linewidth. A rigorous non-perturbative theory for the linewidth is given in Section 3.2 in terms of a Green's function method and the same result is rederived by using a simpler and more intuitive analysis. The prediction of the theory is examined in Section 4 and compared with data. Finally a brief conclusion is given in Section 5.

2 Model Hamiltonian

The resonant optical transition occurs between the ground (n = 0) and excited (n = 1) levels of an impurity ion shown in Fig. 1. The transition energy, namely, the resonant photon energy $\hbar\omega_0 = \varepsilon_1 - \varepsilon_0$ is much larger than the Debye acoustic phonon energy ($\hbar\omega_D$) and the energy of TLS. The energy $\hbar\omega_0$ has a broad distribution of a few hundred [cm^{-1}] due to a static (i.e., inhomogeneous) broadening [17].

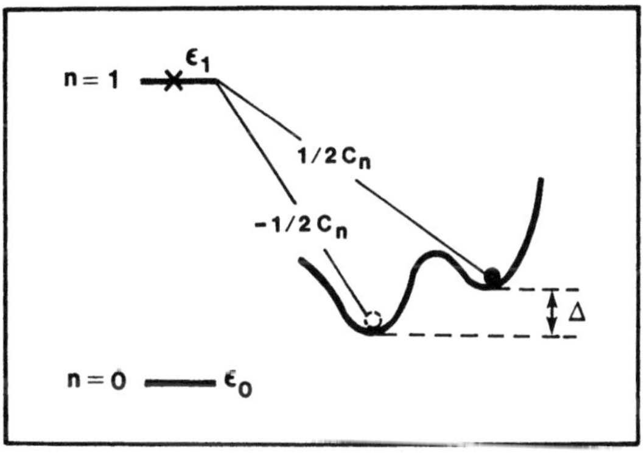

Fig. 1 Interaction of the ground (n = 0) and excited (n = 1) optical levels on the left side with TLS shown by an asymmetric double-well on the right side

The Hamiltonian for TLS shown in Fig. 1 is given, in the site (i.e., well) representation (denoted by the subscript "site"), by

$$\frac{1}{2}\begin{bmatrix} \Delta & t \\ t & -\Delta \end{bmatrix}_{\text{site}} . \tag{2.1}$$

In (2.1) Δ and $t/2$ are the energy asymmetry of the wells and the overlap energy, respectively.

The interaction between the n-th optical level and TLS illustrated in Fig. 1 is described, apart from a constant term, by

$$\frac{1}{2} C_n \begin{bmatrix} 1 & 0 \\ 0 & -1 \end{bmatrix}_{\text{site}} \tag{2.2}$$

where C_n is the difference in the coupling strengths between the n-th optical level and the two unperturbed wells. In (2.2) the off-diagonal term is neglected for simplicity. This approximation does not affect the final result appreciably.

The energy conservation for a tunneling motion between the wells is provided by the deformation potential [3]

$$\frac{B\,\varepsilon}{2}\begin{bmatrix} 1 & 0 \\ 0 & -1 \end{bmatrix}_{\text{site}} \quad . \tag{2.3}$$

Here ε is the strain and B the difference in the deformation potential constants for the two unperturbed wells.

The TLS Hamiltonian in (2.1) is easily diagonalized and is given in the new diagonal representation (denoted by the subscript "diag") by

$$\frac{1}{2}E\begin{bmatrix} 1 & 0 \\ 0 & -1 \end{bmatrix}_{\text{diag}} \equiv E\sigma^z/2 \tag{2.4}$$

where $E = (\Delta^2 + t^2)^{1/2}$ is the two-level energy separation and σ^z the z component of the Pauli matrix ($\vec{\sigma}$). Thus the total Hamiltonian reads in the new diagonal representation:

$$H = \sum_{n=0}^{1} \varepsilon_n \psi_n^{\dagger}\psi_n + 1/2E\sigma^z + 1/2 \sum_{n=0}^{1} \sum_{\alpha} V_n^{\alpha}\psi_n^{\dagger}\psi_n\sigma^{\alpha}$$

$$+ \sum_q \hbar\omega_q(n_q + 1/2) + 1/2 \sum_{\alpha} f^{\alpha}\varepsilon\sigma^{\alpha} \quad . \tag{2.5}$$

The first two terms and the fourth term in (2.5) describe the noninteracting optical ion under investigation, a single TLS and the phonons, respectively. The third and last terms therein represent the ion-TLS and TLS-phonon interactions introduced in (2.2) and (2.3). The fermion creation and annihilation operators $\psi_n^{\dagger}$ and ψ_n in (2.5) denote the electronic states of the ion and n_q the boson number operator for the phonon mode of momentum q. The superscript is summed over α = z, +, and -, with $\sigma^{\pm} = \sigma^x \pm i\sigma^y$. The coupling parameters in (2.5) are given by

$$V_n^z = C_n\Delta/E, \quad V_n^{\pm} = C_n\, t/2E \tag{2.6}$$

and

$$f^z = B\Delta/E, \quad f^{\pm} = Bt/2E \quad . \tag{2.7}$$

In the following we study the model Hamiltonian defined in (2.5).

3 Dephasing of Impurities by Two-Level Systems

The ion-TLS coupling term in (2.5) contains diagonal (proportional to $V_n^z\sigma^z$) as well as off-diagonal (proportional to $V_n^{\pm}\sigma^{\pm}$) terms. Dephasing is then caused by diagonal modulation of the optical levels arising from "spin density" (i.e., σ^z) fluctuation and also by off-diagonal modulation due to spin-flip (i.e., $\sigma^{\pm}$) transitions in TLS. The "spin" transitions are induced by a phonon-assisted tunneling motion.

3.1 Theoretical Background: Perturbation Theories

A simple natural approach to the problem is to treat the ion-TLS coupling term as a perturbation. For the purpose of a later discussion we start with the interesting work of Selzer et al. [5], who calculated the optical linewidth due to diagonal modulation exploiting the analogy between the present problem and the nuclear spin-relaxation via coupling to a spin-1/2 system. They obtained, using a relaxation time approximation [5]

$$\begin{aligned}\Delta\omega' &= \hbar^{-1}(V^z/2)^2 \int_0^\infty \langle \sigma^z(t)\sigma^z(0) - \langle\sigma^z\rangle^2\rangle dt \\ &= \frac{\tau}{4\hbar} (V^z)^2 \operatorname{sech}^2 (\beta E/2) \left[1 + (\omega\tau)^2\right]^{-1} (\omega = 0) \; .\end{aligned} \tag{3.1}$$

Here $\Delta\omega'$ is the contribution from a single "spin" to be summed over all TLS, τ the "spin" relaxation time to be given later (cf. (3.4)), $V^z = V_1^z - V_0^z$, $\beta = (k_BT)^{-1}$ and k_B is Boltzmann's constant. The angular brackets in (3.1) denote the thermodynamic average and $\sigma^z(t)$ is in the Heisenberg representation. The expression in (3.1), however, does not lead to the quadratic temperature dependence observed in silicate glasses [5]. Furthermore, a surprising aspect of the expression is that it is linear in τ. Namely, the linewidth ($\Delta\omega'$) becomes narrower, the faster the TLS-modulation is (i.e., for shorter τ). The answer to this puzzling question, as will be found later, lies in the fact that the perturbation result in (3.1) is valid only for extremely small $|V^z|$ (i.e., $\ll \hbar\tau^{-1}$) or for an extremely rapid motion in TLS. In this extreme weak coupling limit of perturbation, however, we find that $\Delta\omega'$ in (3.1) is so small that it is negligible. A correct form of the diagonal contribution in the opposite and important non-perturbative regime will be obtained later.

The contribution to the linewidth from off-diagonal modulation was calculated recently by using a second-order perturbation method pictured in Fig. 2, for example, for TLS in the upper level [18]. The transition from the upper to the lower level of TLS is induced by the off-diagonal elements $V_n^\pm$ after or before a resonant phonon emission. These processes together with those with TLS starting from the ground level yield [18]

$$\Delta\omega'' = \frac{\hbar}{2\tau} \sum_n (V_n^z/E)^2 \operatorname{sech}^2 (\beta E/2) \tag{3.2}$$

again to be summed over all TLS. The expression in (3.2) has two nice features. First, $\Delta\omega'' \propto \tau^{-1}$ and the line broadens as a result of TLS - modulation. Second, using $\tau^{-1} \propto E^3$ for a phonon-assisted tunneling [3] and a constant density of states, we find a quadratic temperature dependence (i.e., $\Delta\omega \propto T^2$) much below the Debye temperature [18,19] in agreement with the data of [5,6]. Note that it is crucial to have two powers of the intermediate energy E in the denominator of the expression in (3.2) to obtain such a temperature dependence. A drawback of (3.2) is that the quantity $(V_n^z)^2$ is expected to decay rapidly with distance and only the nearest-neighbor TLS (for which $|V_n| \sim 100[cm^{-1}]$ is necessary for a numerical fitting) plays an important role. For strongly coupled TLS (i.e., $|V_n| \gg E$), however, the TLS energy should be renormalized, as will be shown later, and the desired quadratic temperature dependence is lost. In this connection, a rough estimate of V_n for a nearest-neighbor TLS gives a few hundred $[cm^{-1}]$ which is consistent with typical inhomogeneous widths

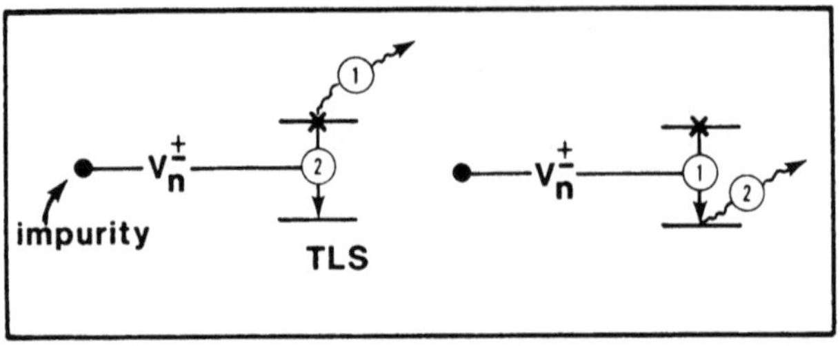

Fig. 2 Two-step processes for off-diagonal modulation. Directed wiggly lines and circled numbers indicate phonon emission and the sequence in the perturbation chain, respectively

in glasses [17]. Comparing (3.1) and (3.2) we find for a given TLS that $\Delta\omega'$ is much larger than $\Delta\omega''$ because $\Delta\omega'/\Delta\omega'' \sim (E/\hbar\tau^{-1})^2 \gg 1$. This relationship, however, applies only to an extremely weakly coupled TLS for which (3.1) is valid. Nevertheless we will show later that diagonal modulation plays a more important role than off-diagonal modulation at low temperatures where the condition $|V^z| > \hbar\tau^{-1}$ is satisfied even for TLS very far away from the ion (e.g., a few hundred [Å] at 5[K]).

3.2 Non-Perturbative Theory

The ion-TLS coupling strength can be quite large as compared to the energy scale $E \sim k_BT$ of the present problem. So it is best to avoid a perturbation approach. On the other hand the TLS-phonon coupling, namely, the last term in (2.5) (to be designated as H_{TLS-ph}) is small and will be treated as a perturbation. In fact it will be shown later that for diagonal modulation the perturbation approach [cf. (3.1)] breaks down even for such a small value of $|V^z|$ as $|V^z| > \hbar\tau^{-1}$ ($\sim 10^{-3}$ [cm^{-1}] at 5[K]) while it is still allowed for off-diagonal modulation. The diagonal contribution, however, is found to dominate the off-diagonal contribution at low temperatures.

The linewidth will be calculated first by using a rigorous Green's function method. Then the same result will be rederived in a simpler and more heuristic way. The spectral line shape is given by the imaginary part of the Fourier transform [$F(\omega)$] of the retarded Green's function [20]

$$F(t) = -i\theta(t)\langle[(\phi_1 \phi_0^\dagger)(t), \phi_0 \phi_1^\dagger]\rangle \tag{3.3}$$

where θ and the square brackets represent the unit step function and the commutator, respectively. The spectral function is evaluated by using the strain as a smallness parameter but for arbitrary V_n^α. The result will first be discussed for diagonal modulation in the absence of off-diagonal modulation (i.e., $V_n^{\pm} = 0$). We find

$$F(\omega) = \frac{\Omega + 1/2V^z\,(p_+ - p_-) - i\hbar\tau^{-1}}{\Omega^2 - 1/4(V^z)^2 - i\Omega\hbar\tau^{-1} - (1/4)\,iV^z\,(\Gamma_+ - \Gamma_-)} \tag{3.4}$$

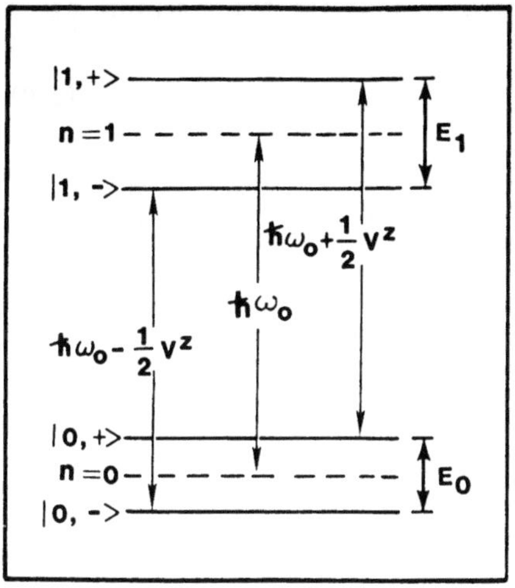

Fig. 3 Renormalized ion-TLS levels and possible optical transitions for diagonal modulation

where $\Gamma_{\pm} = \Gamma_0^{\pm} + \Gamma_1^{\pm}$, $\Omega = \hbar\omega - \hbar\omega_0$ and $\hbar\tau^{-1} = (\Gamma_+ + \Gamma_-)/2$. Here τ is the relaxation time introduced in (3.1) and $\Gamma_n^{\pm}$ the phonon-assisted width of the coupled ion-TLS doublets for the nth optical level (Fig. 3):

$$\Gamma_n^{\pm} = 2\pi \sum_q (f^+)^2 |\langle n_q \pm 1|\varepsilon|\, n_q\rangle|^2 \,\delta\,(\hbar\omega_q - E_n)\,. \tag{3.5}$$

The upper (lower) sign corresponds to the phonon emission (absorption) process for the upper (lower) level of the renormalized TLS with energy $E_n = |E + V_n^z|$. In (3.4) $p_{\pm}$ is the thermal occupation probability for the ground-state doublet:

$$p_{\pm} = \frac{1}{2} \exp\,(\mp\beta E_0/2)/\cosh(\beta E_0/2). \tag{3.6}$$

Although the expression in (3.4) looks quite complicated, its true meaning is revealed in the following two limiting interesting situations. In the extreme weak-coupling limit $|V^z| \ll \hbar\tau^{-1}$ corresponding to those TLS far away from the impurity center, (3.4) reduces to ($E_n \simeq E$)

$$F(\omega) = [\Omega - 1/2\; V^z\, (p_+ - p_-) - 1/2\; i\Delta\omega']^{-1} \tag{3.7}$$

yielding for the linewidth the motional-narrowing result $\Delta\omega'$, which was obtained earlier in (3.1) by a pertubation method. Note also that the shift [i.e., the real part of the denominator in (3.7)] equals the thermal average of the shifts for the two optical transitions shown in Fig. 3. These extremely distant and weakly coupled TLS give little contribution to the linewidth and will be neglected hereafter. Therefore, in the following discussions, even the phrase "weakly coupled TLS" will always indicate TLS with $|V^z| > \hbar\tau^{-1}$.

On the other hand in the opposite limit $|V^z| \gg \hbar\tau^{-1}$, we find from (3.4)

$$F(\omega) = \sum_{\pm} \frac{p_{\pm}}{\Omega \mp 1/2\; V^z - 1/2\; i\Gamma_{\pm}}\,. \tag{3.8}$$

Recalling that $\hbar\omega_0$ has a broad distribution, we find for the linewidth at any spectral position

$$\Delta\omega = \sum_{\pm} p_{\pm}\Gamma_{\pm} \quad . \tag{3.9}$$

It will be argued later that a major contribution to the total linewidth arises from distant TLS with a weak coupling to the ion (but with $|V^z| > \hbar\tau^{-1}$) and with small spectral shifts. The meaning of the spectral shifts $\pm 1/2V^z$ in (3.8) and the linewidth in (3.9) is illustrated in Fig. 3. The quantity $\Delta\omega$ in (3.9) is the thermal average of the sum (i.e., $\Gamma_{\pm}$) of the widths of the initial and final state levels for the optical transitions represented by the spectral shifts $\pm 1/2\ V^z$ in Fig. 3. This result is consistent with the fact that the line shape is Lorentzian. Eq. (3.9) also follows simply from (3.1) by using $\omega = V^z/2\hbar$ and the detailed balance.

The effect of off-diagonal modulation can also be studied in terms of the Green's function in a similar way. However, we will omit writing down the lengthy expression for the spectral function because its physical meaning is not transparent. Instead, we will derive the identical expression for the linewidth in the presence of both diagonal ($|V^z| > \hbar\tau^{-1}$) and off-diagonal modulations, using a simpler and more intuitive approach, already illustrated partly in Fig. 3. For this purpose the first three terms in (2.5) are diagonalized for the occupation of the nth level yielding a doublet $\varepsilon_n \pm 1/2\tilde{E}_n$ with

$$\tilde{E}_n = [(E + V_n^z)^2 + 4(V_n^+)^2]^{1/2} \ . \tag{3.10}$$

These four levels are drawn in Fig. 3 for the case $V_n^+ = V_n^- = 0$. The transition probability for the four possible optical transitions between these new diagonal bases (to be denoted as $|n, \pm\rangle$) is given by ($\alpha,\alpha' = \pm$)

$$S_{\alpha'\alpha} = |\langle 1,\alpha'|0,\alpha\rangle|^2 = 1/2\{\tilde{E}_0\tilde{E}_1 + \alpha'\alpha[(V_0^z + E)(V_1^z + E)$$

$$+ 4V_0^+V_1^+]\}/\tilde{E}_0\tilde{E}_1 \ . \tag{3.11}$$

In the absence of off-diagonal modulation the transition vanishes between different "spin" states (i.e., $S_{\alpha\alpha'} = \delta_{\alpha,\alpha'}$ for $V_n^{\pm} = 0$) as shown in Fig. 3.

The off-diagonal matrix element of the TLS-phonon interaction [i.e., the last term in (2.5)] within an (nth) optical level is then written as

$$\langle n,\pm|H_{TLS-ph}|n,\mp\rangle = \frac{\varepsilon}{\tilde{E}_n}\{f^zV_n^+ - f^+(E + V_n^z)\} \ . \tag{3.12}$$

The widths of the sublevels $|n,\alpha=\pm\rangle$ due to one-phonon-assisted processes are given by

$$\tilde{\Gamma}_n^{\alpha} = 2\pi \sum_q \left\{\frac{f^zV_n^+ - f^+(E + V_n^z)}{\tilde{E}_n}\right\}^2 |\langle n_q + \alpha|\varepsilon|n_q\rangle|^2\delta(\hbar\omega_q - \tilde{E}_n) \tag{3.13}$$

where $\alpha = \pm 1$ will be assigned to $\alpha = \pm$ for the strain matrices in (3.13) for emission and absorption processes, respectively. The total optical homogeneous linewidth is then a sum of these individual widths properly averaged over the ground-state doublet occupation probabilities ($\tilde{p}_\alpha$) and equals

$$\Delta\omega = \sum_E \sum_{\alpha=\pm} \tilde{p}_\alpha \sum_{\alpha'=\pm} S_{\alpha'\alpha} (\tilde{\Gamma}_0^\alpha + \tilde{\Gamma}_1^{\alpha'}) \tag{3.14}$$

where the first sum denotes summing over all TLS and

$$\tilde{p}_\alpha = 1/2 \exp(-\alpha\beta\tilde{E}_0/2)/\cosh(\beta\tilde{E}_0/2) . \tag{3.15}$$

The first and second terms in the curly brackets of (3.13) represent for the linewidth in (3.14) contributions from off-diagonal and diagonal modulations, respectively. The former leads for small V_n to $\Delta\omega''$ in (3.2) while the latter reduces to (3.9) for $V_n^\pm = 0$. The off-diagonal term in (3.13) cancels part of the diagonal term in view of (2.6) and (2.7), yielding

$$\tilde{\Gamma}_n^\alpha = \left(\frac{E}{\tilde{E}_n}\right)^2 2\pi \sum_q (f^+)^2 |\langle n_q+\alpha|\varepsilon|n_q\rangle|^2 \, \delta(\hbar\omega_q - \tilde{E}_n). \tag{3.16}$$

The expressions in (3.14) - (3.16) then constitute the central formal result for the linewidth.

The linewidth in (3.14) is to be summed over all TLS, which are assumed to have a uniform spatial distribution. In the following we calculate the temperature-dependent part of the linewidth [21]. For a subgroup of TLS with a given energy $E_n(<k_BT)$ a dominant contribution comes from the weakly coupled TLS with $\tilde{E}_n \simeq E >> V_n^\alpha$. This is not only because of the prefactor in (3.16) but also because majority of TLS are far away from the ion. For these weakly coupled TLS we can approximate $\tilde{E}_n \simeq E$, $\tilde{\Gamma}_n^\alpha \simeq \Gamma_n^\alpha \equiv \Gamma^\alpha$ (independent of n), $S_{\alpha'\alpha} \simeq \delta_{\alpha,\alpha'}$, $\tilde{p}_\alpha \simeq p_\alpha$, obtaining [21]

$$\Delta\omega = 2 \sum_E \sum_\alpha p_\alpha \Gamma^\alpha = 2 \int dE \, \rho(E) \left\langle \int_{r<r_c} d^3r \sum_\alpha p_\alpha(E) \, \Gamma^\alpha(E) \right\rangle_{av} . \tag{3.17}$$

According to the first equality, the optical linewidth due to a single TLS equals twice the thermal average of the width of the two levels. In (3.17) $\rho(E)$ is the density of states per volume of TLS and $\langle \; \rangle_{av}$ denotes the average over the TLS parameters for a given E. The E dependence is specifically shown for p_α and Γ^α. The width of TLS is given in the Debye approximation by [3]

$$\Gamma^\alpha = 1/4\pi(E/\hbar\omega_D)^3 \mathfrak{D}[n_E + (\alpha + 1)/2] \tag{3.18}$$

where n_E is the boson function and

$$\mathfrak{D} = \frac{3B^2}{Mc^2} \left(\frac{t}{E}\right)^2 .$$

Here M is the mass of a unit cell and c the sound velocity.

For an electric multipolar coupling of the form $C = |C_1 - C_0| = b/r^s$, the cutoff range r_c in (3.17) is a decreasing function of E and is determined by the condition

$$|V^z| = b\Delta/Er_c^s = \hbar\tau^{-1} = \Gamma^-(E) + \Gamma^+(E). \tag{3.19}$$

Assuming a density of states of the form $\rho(E) = \rho_0 E^\mu$ (ρ_0, μ are constants) for TLS, we find at temperatures much below θ_D and the cutoff energy of TLS

$$\Delta\omega = \frac{4}{3}\pi^2\tilde{D}\hbar\omega_D\delta_s I_{s,\mu}\,(4b/\pi\tilde{D})^{3/s}\,\rho_0(k_BT)^\mu\,(T/\theta_D)^{4-9/s} \tag{3.20}$$

where the integral

$$I_{s,\mu} = \int_0^\infty \frac{x^{3+\mu-9/s}\,e^x\,dx}{(e^x+1)^{1+3/s}\,(e^x-1)^{1-3/s}}$$

is of order unity (e.g., $I_{3,0} = 0.5$, $I_{4,0} = 0.684$) and

$$\langle(\Delta/E)^{3/s}\,\mathfrak{D}^{1-3/s}\rangle_{av} \equiv \delta_s\tilde{D}^{1-3/s}\,.$$

Here the quantity $\delta_3 = \langle\Delta/E\rangle_{av}$ is of order unity, $\delta_s = 1 (s \neq 3)$, and $\tilde{D}$ is of the order of $D \equiv \langle\mathfrak{D}\rangle_{av}$ which is estimated to be about 0.02 eV for silicate glasses [21]. The linewidth in (3.20) is an increasing function of the coupling strength b. This is due to the fact that, for a stronger coupling, a greater number of TLS farther away contribute to the linewidth.

For the special case of a dipolar coupling (i.e., s = 3) the quantities preceding the square brackets in (3.18) cancel out from the final result in (3.20) which is then independent of $\tilde{D}$ and ω_D. Namely, the latter is independent of the Debye approximation. This property follows from the fact that the spatial integral in (3.17) yields $r_c^3\Gamma^\alpha(E)$, which is independent of the quantities preceding the square brackets in (3.18) in view of the condition in (3.19). Extending this argument further, we conclude that the dipolar linewidth is independent of the nature of the boson modes responsible for the TLS relaxation and equals

$$\Delta\omega_{dip} = \frac{16}{3}\pi b\delta_3 I_{3,\mu}\rho_0(k_BT)^{1+\mu}\,. \tag{3.21}$$

4 Applications

The exponent "s" for the multipolar interaction is determined by the nature of the electrical moments of the ion and TLS. As we found in (3.21), the expression in (3.20) reduces to a quasi-linear temperature dependence for a dipolar coupling and small μ:

$$\Delta\omega \propto T^{1+\mu}\,. \tag{4.1}$$

If, on the other hand, there is a local inversion symmetry for the ion, the lowest possible form of the interaction is dipole quadrupolar (i.e., s = 4), yielding a quasi-quadratic temperature dependence for (3.20):

$$\Delta\omega \propto T^{1.75+\mu} . \tag{4.2}$$

Note that this result is also relatively insensitive to $\tilde{D}$ ($\Delta\omega \propto \tilde{D}^{1/4}$). It is interesting to note that the above two types of temperature dependences seem to explain the data [5-14] discussed in the introduction. We also mention here that for a quadrupole-quadrupole coupling (s = 5) the linewidth in (3.20) is proportional to $T^{2.2+\mu}$.

In the following an order of magnitude estimate is made for $\Delta\omega$ in (3.20) for the ${}^3P_0-{}^3H_4$ resonant transition in Pr^{+3}-doped GeO_2 glass [6]. Using θ_D = 300 [K], μ = 0, $\rho_0 = 2.0 \times 10^{20}$/[eVcm^3], $\delta_3 \simeq 1$, and choosing s = 4, b = 5.1×10^2 [eV(A°)][1,19], we fit the data at 20 [K] ($\Delta\omega$ = 0.15 [cm^{-1}]):

$$\Delta\omega = 7.9 \times 10^{-4}\, T^{1.75}\ [\mathrm{cm}^{-1}] \tag{4.3}$$

for T in kelvins. Defining "a" to be the approximate molecular dimension of TLS and the ion and "e" the unit electronic charge, and setting b = e^2a^3, we find a $\simeq$ 3.3 [A°]. Other sets of data with quasi-quadratic temperature dependence can be explained in a similar way. It is quite possible that a small deviation from a constant density of states for TLS (i.e., small μ) as well as other background line-broadening mechanisms may raise the exponent in (4.3) somewhat.

The recent data of linewidth in Pr^{3+}-doped silicate glasses show a linear temperature dependence [9] corresponding to s = 3 situation with μ = 0 in (3.2). A weak dipolar coupling may arise from an imperfect local symmetry (such as broken bonds) of the ion. We fit the data [9] using μ = 0, ρ = 1.3×10^{20}/[eVcm^3][1,3] and choosing b = 5.3 [eV(A°)^3] in (3.21):

$$\Delta\omega = 4.0 \times 10^{-3} T\ [\mathrm{cm}^{-1}] \tag{4.4}$$

for T in kelvins. Defining b = e^2a^2 as before, we find a = 0.61 [A°]. A similar interpretation can be applied to the linear temperature behavior of the holewidth of tetracene in organic glasses [13], which was explained in [13] by using a formula similar to (3.2) and assuming that the TLS energy cutoff is very small ($\sim$ 2[cm^{-1}]).

The holewidth data of free-base porphin in some organic glasses show a $T^{1.3\pm0.1}$ dependence [14] showing a significant deviation from the linear behavior and requiring $\mu \approx 0.3$ in the above analysis. At present the density of states of TLS is not known for these glasses. Recently, however, the above $T^{1.3}$ dependence was explained in the range 4- 30 [K] by combining the linear T behavior with an exponential-like activated contribution arising from low-lying libronic modes present in organic glasses [22].

5 Conclusion

The homogeneous optical linewidth arising from impurity-TLS interaction is examined for resonant optical transitions in glasses. A rigorous Green's function approach and a simpler and more intuitive method are found to yield the same result given in (3.14). The width is due to modulation of the optical levels caused by coupling to the diagonal as well as off-diagonal elements of TLS. The main contribution arises from diagonal modulation at low temperatures. In this case a perturbation theory is shown to be inadequate. For diagonal modulation the contribution to the linewidth from a typical weakly coupled (but with $|V^z| > \hbar\tau^{-1}$) TLS is larger than that of a strongly coupled TLS and is independent of the coupling strength to the ion: it equals twice the thermal average of the two level-widths of the TLS. Therefore the total linewidth arises dominantly from these abundant but weakly coupled TLS. For a flat density of states of TLS the low-temperature linewidth is proportional to T, $T^{1.75}$ and $T^{2.2}$, respectively, for a dipole-dipole, dipole-quadrupole and quadrupole-quadrupole coupling. The result is compared with recent data [5-14]. Other mechanisms such as local vibrational and librational modes are also expected to contribute to the linewidth at higher temperatures. More quantitative theoretical studies as well as experimental works are necessary to determine exact proportions of each contribution.

References

1. R. C. Zeller and R. O. Pohl, Phys. Rev. B4, 2029 (1971)
2. For review articles see Amorphous Solids: Low-Temperature Properties Ed. W. A. Phillips, Springer-Verlag (1981)
3. P. W. Anderson, B. I. Halperin and C. M. Varma, Phil. Mag. 25, 1 (1972)
4. W. A. Phillips, J. Low Temp. Phys. 7, 351 (1972)
5. P. M. Selzer, D. L. Huber, D. S. Hamilton, W. M. Yen and M. J. Weber, Phys. Rev. Lett. 36 813 (1976)
6. J. Hegarty and W. M. Yen, Phys. Rev. Lett. 43, 1126 (1979)
7. P. Avouris, A. Campion and M. A. El-Sayed, J. Chem. Phys. 67, 3397 (1977)
8. J. R. Morgan and M. A. El-Sayed, Chem. Phys. Lett. 84, 213 (1981)
9. R. M. Macfarlane and R. M. Shelby, Opt. Commun. 45, 46 (1983)
10. J. Friedrich and D. Haarer, Chem. Phys. Lett. 74, 503 (1980), J. Friedrich, H. Scheerer, B. Zickendraht-Wendelstadt and D. Haarer, J. Chem. Phys. 74, 2260 (1981) and J. Am. Chem. Soc. 103, 1030 (1981)
11. J. Friedrich, H. Wolfrum and D. Haarer, to be published
12. B. M. Kharlamov, R. I. Personov and L. A. Bykovskaya, Opt. Commun. 12, 191 (1974), E. Cueller and G. Castro, Chem. Phys. 54, 217 (1981)
13. J. M. Hayes, R. P. Stout and G. J. Small, J. Chem. Phys. 74, 4266 (1981)
14. H. P. H. Thijssen, A. I. M. Dicker and S. Völker, Chem. Phys. Lett. 92, 7 (1982); H. P. H. Thijssen, S. Völker, M. Schmidt and H. Port, Chem. Phys. Lett. 94, 537 (1983)
15. D. L. Huber, preprint
16. H. Morawitz and P. Reineker, Sol. St. Commun. 42, 609 (1982)
17. L. A. Riseberg, Phys. Rev. Lett. 28, 789 (1972)
18. S. K. Lyo and R. Orbach, Phys. Rev. B22, 4223 (1980)
19. P. Reineker and H. Morawitz, Chem. Phys. Lett. 86, 359 (1982)
20. D. E. McCumber and M. D. Sturge, J. Appl. Phys. 34, 1682 (1963)
21. S. K. Lyo, Phys. Rev. Lett. 48, 688 (1982)
22. B. Jackson and R. Silbey to be published

Part VIII

Conductivity and Superconductivity in Organic Materials

Spin Resonance and Conductivity of Fluoranthenyl Radical Cation Salts

W. Stöcklein, B. Bail, and M. Schwoerer
Physikalisches Institut, Universität Bayreuth, P.O.B. 3008
D-8580 Bayreuth, Fed Rep. of Germany
D. Singel and J. Schmidt
Huygens Laboratory, University of Leiden, NL-2300 Leiden, The Netherlands

1. Introduction

Fluoranthenyl radical cation salts show a high electrical conductivity. They belong to the new class of 2:1 charge transfer crystals, the donors of which are dimers of pure aromatic hydrocarbons Ar_2 without heteroatoms[1,2]. The acceptors are inorganic molecules X; typically $X=AsF_6$, PF_6, SbF_6 and others. Thus the general formula is $(Ar_2)^{\dot{+}}X^-$. The increasing interest in these crystalline organic conductors partially stems also from the idea that they might serve as model compounds for highly conductive polymers, such as polyacetylene $(CH)_x$[3]. In these materials, with their conjugated carbon backbone, doping with electron acceptors produces the conductivity; generation of radical cations $(CH)_x^{\dot{+}}$ presumably is the origin of this conductivity.

In this paper we report on new experimental results on $(FA_2)^{\dot{+}}PF_6^-$ where FA = Fluoranthene, $C_{16}H_{10}$. The peculiarities of this crystal, and of the other fluoranthene radical cation salts, are their comparatively good stoichiometry, their minute content of solvent, their stability, their simple crystal structure[4,5], and their exceptionally narrow ESR spectrum[6,7].

Their crystal structure has been analyzed by Enkelmann[4,5]. The aromatic planes are oriented perpendicular to the linear stack axis a. The nearest-neighbour distances of the donors along a at room temperature are 3.28 Å and 3.33 Å, respectively. Thus, the donors are slightly dimerized already at room temperature. Due to both the small intermolecular distance and the rotation of the nearest-neighbour molecules by 180° around the a axis the intermolecular overlap within the cation stack is very high. This is one of the reasons for the high conductivity perpendicular to the π-electron planes.

At about 190 K a structural phase transition occurs. It consists primarily in the disappearance of a mirror plane containing the stack axis. The FA molecules are rotated by about 8° from their original, room temperature orientation. Cooling down below this phase transition, according to Enkelmann, does not produce a further doubling of the unit cell along the stack axis.

The anions, PF_6^-, are also oriented in linear stacks situated in between the cation stacks. They are diamagnetic, closed shell systems. We believe that the conductivity is electronic and not ionic.

The simplest model of the electronic structure is suggested by the one-dimensional crystal structure, the stoichiometry and the assumed complete single ionization of PF_6 to PF_6^- anions leaving one half defect electron per FA molecule: if the repeat unit along the one-dimensional FA-stack were the nearest-neighbour distance d, the first Brillouin zone would terminate at π/d. The highest band would be a quarter filled defect electron band or, equivalently a 3/4 filled electron band. But as the FA molecules are dimerized already at room temperature the repeat unit is $a = 2d = 6.61$ Å, resulting in the Brillouin zone border at $\pi/a = 1/2(\pi/d)$, and in a band gap at $k = \pi/a$. This gap may be very small but it results in a half-filled conduction band with a Fermi wave vector $k_F = \pi/2a$. In the one-dimensional free electron model the Fermi energy is

$$\varepsilon_F = \frac{\hbar^2}{2m}\left(\frac{N\pi}{2L}\right)^2 = 0.22 \text{ eV} \mathrel{\hat{=}} 2500 \text{ K} \qquad (1)$$

($N/L = 1/6.61$ Å = line density of the charge carriers).

2. Crystal Growth

$(FA_2)PF_6$ crystals were grown electrochemically by the method described originally by Chiang et al.[8] and later by Fritz et al.[9] As solvent we used CH_2Cl_2 or CH_3COOCH_3. Large single crystals of different habitus but with identical crystal structure were grown by variation of both temperature and solvent. Fig.1 shows the four different habitus of $(FA_2)PF_6$ crystals which we have grown so far. The crystals do not always grow along the a axis which is the axis of maximum conductivity. Fig. 1, for comparison, also shows the habitus of an $(FA_2)SbF_6$ single crystal.

HABITUS	SOL-VENT	ANION	TYP. DIM./mm a×b×c*
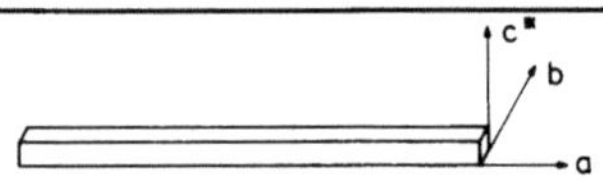	CH_3COOCH_3	PF_6	10×0.1×0.1
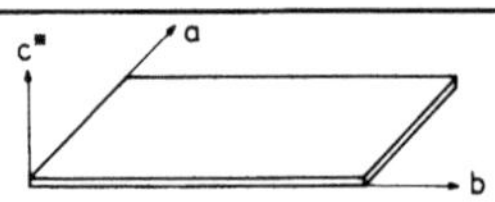	CH_3COOCH_3	PF_6	2×4×0.01
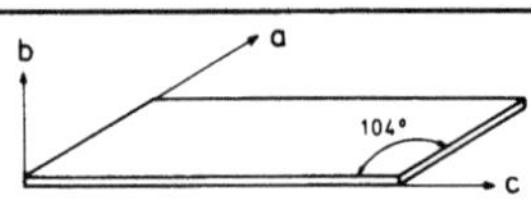	CH_3COOCH_3	PF_6	0.3×0.05×1
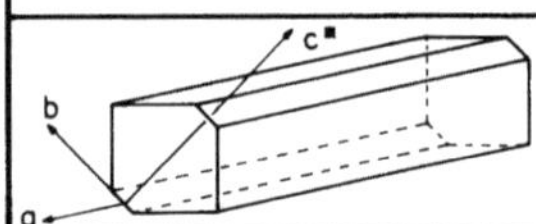	CH_2Cl_2	PF_6	2×1×1
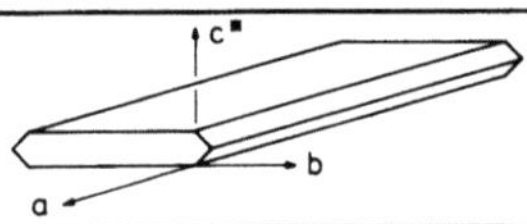	CH_2Cl_2	SbF_6	10×4×0.2

Fig. 1 Habitus and typical sizes of $(FA_2)PF_6$ and of $(FA_2)SbF_6$ single crystals grown electrochemically in two different solvents

3. DC Conductivity

Using standard four-probe technique we were able to measure the dc conductivity along the stack axis $\sigma_\parallel$ within the temperature range between room temperature and 100 K (Fig. 2). Above the phase transition temperature $\sigma_{\parallel}$ increases slightly with increasing temperature and reaches about $600 (\Omega cm)^{-1}$ at room temperature; $\sigma_{\parallel}$ decreases by at least two orders of magnitude and shows semiconducting behaviour below about 170 K:

$$\sigma_{\parallel} \propto \exp(-\Delta E/2kT). \tag{2}$$

For $(FA_2)PF_6$ the half activation energy is $\Delta E/2 = 0.1$ eV. Ohm's law was shown to be valid for current densities between 10^{-4} A/mm^2 and 10^{-2} A/mm^2 throughout the entire temperature range. Fig. 2 shows for comparison also the experimental values of the $(FA_2)AsF_6$ and of the $(FA_2)SbF_6$ crystals respectively. Both activation energy and room temperature conductivity are dependent on the crystal quality; we therefore are not able to identify a dependence on the variation of the anions.

Assuming a conductor with one free electron per FA dimer, we calculate a mobility $\mu = 2$ cm^2/Vs and a collision time of about 10^{-16}s at room temperature for $\sigma_{\parallel} = 600 (\Omega cm)^{-1}$.

The conductivity $\sigma_{\parallel}$ at room temperature is comparable to the value of TTF-TCNQ; it is less than the value measured by Schweitzer[2] for the corresponding perylene radical cation salt and

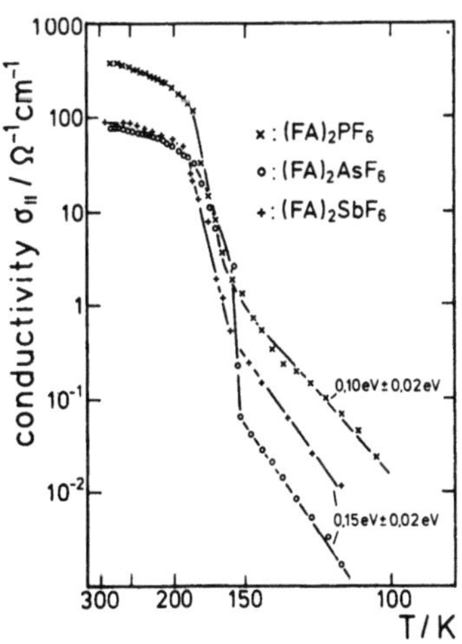

Fig. 2 Dc conductivity $\sigma_{\parallel}$ of $(FA_2)PF_6$, $(FA_2)AsF_6$, and $(FA_2)SbF_6$ single crystals as measured by standard four probe technique

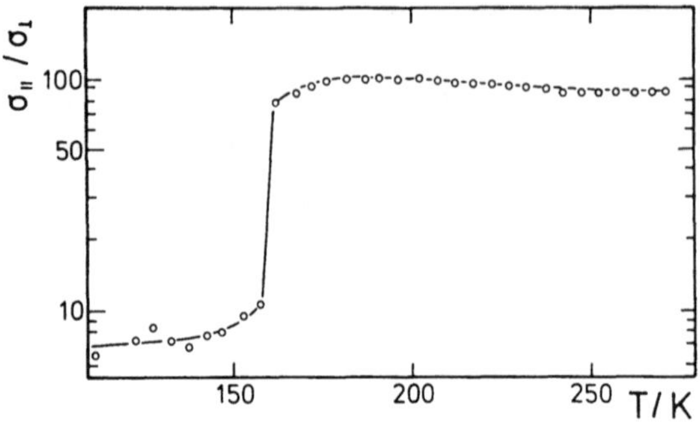

Fig. 3 Temperature dependence of the ratio of the dc conductivities parallel and perpendicular to the stack axis a, $\sigma_{\parallel}/\sigma_{\perp}$, as measured by the Montgomery technique

it is about three orders of magnitude less than the value of copper. In comparing these values one should emphasize however that the crystal growing is not optimized at all so far.

Using the Montgomery technique we were also able to measure the conductivity perpendicular to the stacks, $\sigma_{\perp}$ (Fig. 3). Although these results are not very well reproducible so far, the ratio $\sigma_{\parallel}/\sigma_{\perp}$ shows clearly that at 170 K the anisotropy drops from about 100 to a value of less than 10, the one-dimensional "quasi-metal" becomes an almost isotropic semiconductor.

4. Electron Spin Resonance

4.1 Spin Susceptibility

The cw-ESR experiments were performed with a 9.5 GHz spectrometer (Bruker, 200D SRC). For the variation of the temperature we used a variable temperature cryostat (Oxford, ESR 9, DT(2)). Because of the extremely small width of the ESR line of less than 15 mG special care was taken for the stability of both field and frequency; we used a field frequency lock and kept the modulations as low as possible (no AFC; field modulation frequency: 12,5 kHz; field modulation amplitude: 2 mG). For all experiments to be discussed below we used $(FA_2)PF_6$ single crystal samples.

By comparison with a ruby ESR standard we evaluated the ESR susceptibility at room temperature (T = 297 K):

$$\chi_{ESR} = (9,1 \pm 2)\ 10^{-5}\ \text{emu/mol.} \tag{3}$$

This value is in complete agreement with the values given by Höptner et al.[10], and by Dormann[11], who both used the Schumacher-Slichter method[12].

Fig. 4 shows the temperature dependence of the integrated ESR signal intensity, normalized to the room temperature value. There exist three different temperature ranges, two of which qualitatively correspond with the temperature ranges of the dc conductivity (Fig. 2):

- a quasi metallic range for $T>180$ K,
- a thermally activated range for $70\ K<T<180$ K
- a quasi Curie range for $T<70$ K.

The increase of the ESR susceptibility with increasing temperature for $70\ K<T<180$ K is described by

$$\chi_{ESR} \propto \exp(-\Delta E/kT), \tag{4}$$

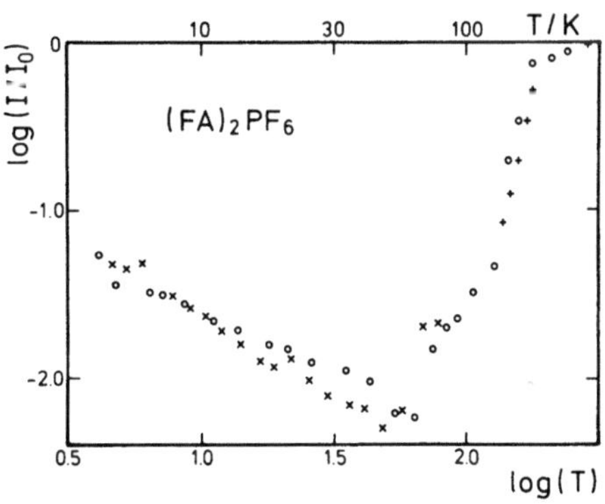

Fig. 4 Temperature dependence of the integrated ESR signal intensity I, normalized to the room temperature value I_o. Different symbols indicate different samples

where $\Delta E = (0.11 \pm 0.02)$ eV. This value qualitatively is in agreement with the activation energy for the dc conductivity and thus indicates that the structural phase transition produces a gap, which is responsible for both charge carrier and spin excitation energies. In this model the crystal has an antiferromagnetic ground state.

For $T<70$ K a slight Curie-like increase of the signal intensity with decreasing temperature is observed: $\chi \propto 1/T$. The corresponding Curie spin concentration is 10^{-3} spins per FA_2. It seems not unreasonable therefore to assign the signal in this low-temperature range to paramagnetic impurities.

On the other hand the low-temperature signal also could result from finite antiferromagnetic chains of different length, i.e. from chain defects[12].

For $T>180$ K the signal intensity only slightly increases with increasing temperature. In the limit of very high Fermi temperature T_F, the Pauli susceptibility is

$$\chi_{1d} = \mu_B^2 \, N/2kT_F \tag{5}$$

for a one-dimensional metal and

$$\chi_{3d} = 3 \, \mu_B^2 \, N/2kT_F \tag{6}$$

for a three-dimensional metal, respectively. For a conduction electron concentration of 1 per FA_2 we evaluate from (3), (5) and (6) experimental values for the Fermi temperatures T_F for the one-dimensional and three-dimensional models, respectively:

$$T_{F,1d}^{exp} = (2000 \pm 400) \text{ K and} \tag{7}$$

$$T_{F,3d}^{exp} = (6000 \pm 1200) \text{ K.} \tag{8}$$

The value for the 1d model is in excellent agreement with the theoretical value (1), calculated for the 1d free electron model. But the same is also true for the 3d case, where the theoretical value in the high-temperature limit is

$$T_F = \frac{\hbar^2}{2mk}\left(\frac{3\pi^2 N}{V}\right)^{2/3} = 5900\ \text{K}. \tag{9}$$

An indication whether the 1d or the 3d model is more appropriate is given by the temperature dependence of the Pauli susceptibility. The origin of this temperature dependence is the deviation from the high-temperature limit: the ratio of the Fermi temperature T_F and the temperature T is not infinite. Therefore the Fermi surface is not sharp. The following expressions for the susceptibilities have been calculated by Mott[14]:

$$\text{1d:} \qquad \chi(T) = \chi(o)(1+AT^2) \tag{10}$$

$$\text{3d:} \qquad \chi(T) = \chi(o)(1-BT^2) \tag{11}$$

(A,B>0). Comparison with the experimental values shows that the 1d model at least qualitatively is the correct one.

The experiments described above thus strongly indicate that we are observing the conduction electrons in the electron spin resonance signal if T>70 K.

4.2 Spin Relaxation

The spin relaxation experiments to be described below have been performed with the electron spin echo (ESE) spectrometer in the Huygens Laboratory. Its main components are a 400 mW klystron, a pulsed 1 kW microwave amplifier and a slotted tube cavity (Q = 1000). The signal was preamplified by a low noise TWT amplifier (Varian VTX 4276 J1), detected with a double balanced mixer and amplified with a 500 MHz broad band amplifier. It was recorded by a boxcar integrator and a transient recorder. The $\pi/2$ pulse time was $\leq$60 ns and the spectrometer's recovery time, which was mainly determined by the cavity's ringing time, was <0.5 μs. In order to avoid artifacts in the ESE intensities and thus in the transverse relaxation function, the low noise TWT tube had to be protected from saturation.

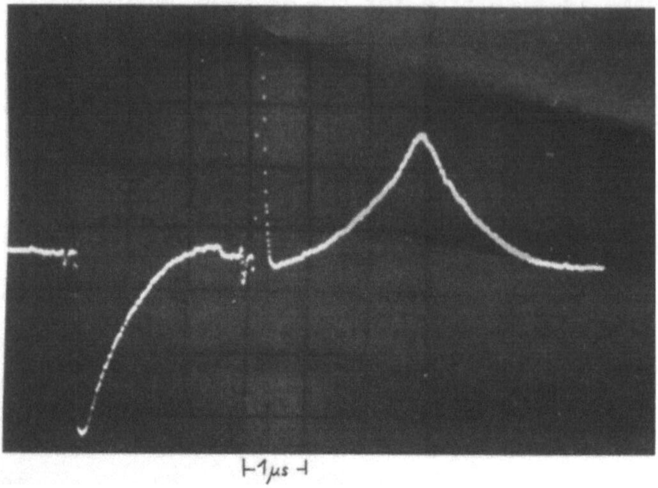

Fig. 5 FID and ESE during a $\pi/2-\tau-\pi-\tau$ echo pulse sequence. The microwave pulses themselves were suppressed almost completely by a pindiode switch gating the microwave amplifier in the recording system

It was gated by a pindiode switch, which switched off the microwave power at the TWT input during the pulses.

Again, all experiments were performed with a $(FA_2)PF_6$ single crystal.

Fig. 5 shows the result of a $\pi/2-\tau-\pi-\tau-$ Hahn echo sequence. The field inhomogeneity was large enough in order to supress the free induction decay (FID) signal early enough. The microwave pulses themselves are almost entirely suppressed, and the echo is strong, symmetrical and on a flat baseline.

Fig. 6 shows the echo amplitude versus the delay time τ in both a linear and logarithmic scale. The echo amplitude and thus the transversal magnetization M is decaying perfectly exponentially:

$$M(\tau) = M_o e^{-2\tau/T_2} \quad (12)$$

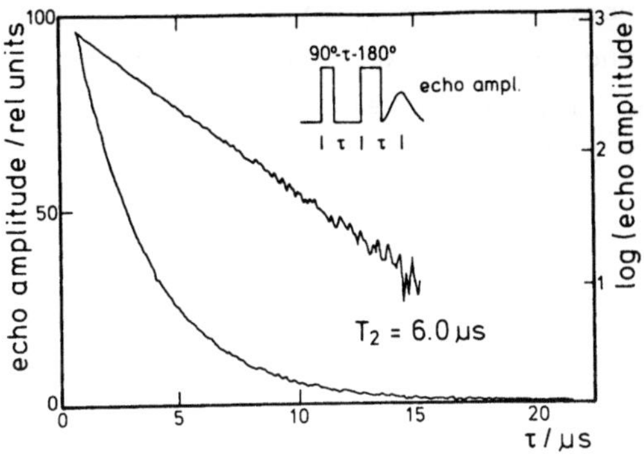

Fig. 6 Decay of the ESE amplitude after a $\pi/2-\tau-\pi-\tau$ echo sequence in a $(FA_2)PF_6$ single crystal. Linear and logarithmic scales, respectively

with $T_2=(6{,}2\pm0.5)$ μs at room temperature. The exponential decay is in perfect agreement with the result of Dormann and Sachs[11], who observed pure exponential decay at the much lower Larmor frequency of 84 MHz. But it is in strong disagreement with the results of Sigg et al.[15], who observed an $\exp(-(t/T)^{3/2})$ decay for a polycrystalline powder of $(FA_2)\,[(PF_6)_x\,(AsF_6)_{1-x}]$ mixed crystals which are believed to behave very similarly to the $(FA_2)PF_6$ single crystals.

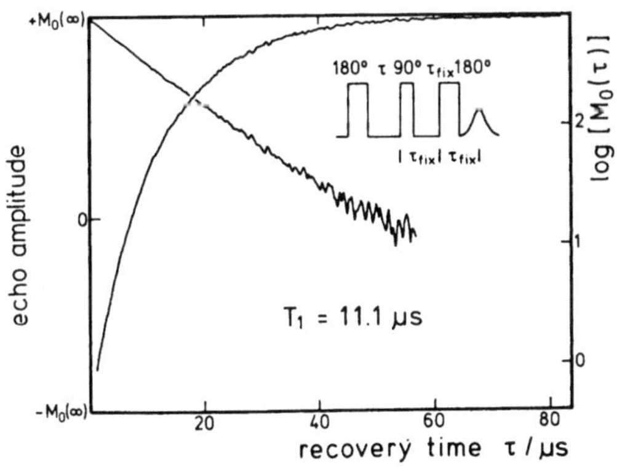

Fig. 7 Decay of the echo amplitude after a $\pi-\tau-\pi/2-\tau_{fix}-\pi-\tau_{fix}$ pulse sequence in a $(FA)_2PF_6$ single crystal. Linear and logarithmic scales, respectively

We did not observe any anisotropy of T_2 by rotating the external field with respect to the crystal axes. But we did observe a slight decrease of the transverse relaxation time to $T_2 = (5.4 \pm 0,5)$ µs by strongly increasing the inhomogeneity of the external field.

Fig. 7 shows the echo amplitude for the measurement of the longitudinal relaxation function via a $\pi-\tau-\pi/2-\tau_{fix}-\pi-\tau_{fix}$ echo sequence. This amplitude is proportional to the longitudinal magnetization M_z. Fig. 7 shows that M_z again is decaying exponentially to the steady state value M_o:

$$M_z(\tau) = M_o(1 - 2\exp(-\tau/T_1)). \qquad (12)$$

T_1 shows a slight anisotropy with respect to the stack axis: $T_1(\parallel) = (9.5 \pm 0.3)$ µs and $T_1(\perp) = (11.4 \pm 0.3)$ µs.

The observations of almost equal relaxation times T_1 and T_2, respectively, at microwave frequency and of equal T_1 and T_2 at radio frequencies[11] again indicate that we are observing the spin resonance of conduction electrons in a quasi metal. For a classical metal $T_1 = T_2$. There remains the question why other typical observables of one-dimensional spin dynamics do not show up in the experiments described above. For a classical one-dimensional spin system, for example for TMMC, the transversal relaxation function is not exponential and shows both strong angular dependence with a minimum of the linewidth at the magic angle and also a strong frequency dependence.

The decrease of T_2 with increasing field gradient could be an indication of spatial spin diffusion.

4.3 Overhauser shift

A further strong indication for the observation of conduction electron spin resonance is the Overhauser shift[16]. If the conduction electron density is restricted mainly to the organic part of the crystal, a shift of the ESR line should be observed

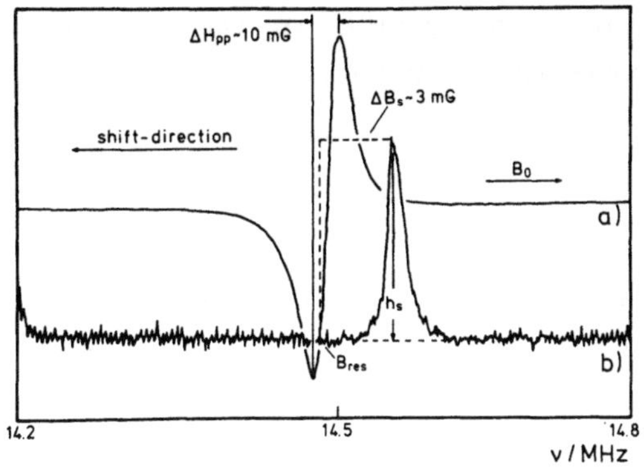

Fig. 8 Overhauser shift of the ESR line of conduction electrons in a $(FA_2)PF_6$ single crystal at room temperature. (a): ordinary ESR line for fixed microwave frequency and for sweep of the external field B_o. (b): variation of the ESR signal for fixed microwave frequency and for fixed external field $B_o=B_{res}$ by sweeping a radio frequency across the proton NMR

upon saturation of the proton NMR and no shift should be observed upon saturation of the fluorine NMR.

Fig. 8 shows the conventional ESR line for a good single crystal (a), and the variation of this ESR signal by sweeping an rf field through the proton NMR (b); the external field was locked at B_{res}. The maximum amplitude h_s corresponds with a shift of about 2.5 mG, and occurs at the radio frequency $\nu_s = \nu_p + (6{,}3\pm2)$ kHz (ν_p = free proton frequency). The width of the Overhauser shift signal is equal to the width of the proton NMR[17]. For radio frequencies in the fluorine NMR range no signal was observed.

Thus we are convinced by the experiments described above that we observed conduction electron spin resonance and that the conduction electrons move almost entirely within the organic part of the $(FA_2)PF_6$ crystals. This latter result is in complete agreement with the NMR results published by Höptner et al.[10] Further experiments and calculations should be able to reveal a complete quantitative description of the electronic structure of these new organic conductors.

5. Acknowledgements

We gratefully acknowledge the valuable help of V. Enkelmann and G. Wegner in the crystal structure analysis, which was performed by one of us (Bernd Bail) in Freiburg. We thank Brigitte Kraus for her assistance in crystal preparation and E.Dormann for many valuable discussions. This work is supported by Stiftung Volkswagenwerk.

References

1. Ch. Kröhnke, V. Enkelmann and G. Wegner, Angew.Chemie 92, 941 (1980)

2. H.J. Keller, D. Nöthe, H. Pritzkow, D. Wehe, M. Werner, P. Koch and D. Schweitzer, Mol.Liq.Cryst. 62, 181 (1980)

3. G. Wegner, Angew.Chemie 93, 352 (1981)

4. V. Enkelmann, B.S. Morra, Ch. Kröhnke and G. Wegner, Chem.Phys. 66, 303 (1982)

5. V. Enkelmann, Habilitationsschrift, Freiburg (1983)

6. H. Eichele, M. Schwoerer, Ch. Kröhnke and G. Wegner, Chem.Phys.Lett. 77, 311 (1981)

7. G. Sachs, W. Stöcklein, B. Bail, E. Dormann and M. Schwoerer, Chem.Phys.Lett. 89, 179 (1982)

8. T.C. Chiang, A.H. Reddoch and D.F. Williams, J.Chem.Phys. 54, 2051 (1971)

9. H.P. Fritz, H. Gebauer, P. Friedrich, P. Ecker, R. Artes and U. Schubert, Z.Naturforsch. 83b, 498 (1978)

10. W. Höptner, M. Mehring, J.U. von Schütz, H.C. Wolf, B.S. Morra, V. Enkelmann and G. Wegner, Chem. Phys. 73, 253 (1982)

11. E. Dormann, Physik. Blätter 39 (1983)

12. Schumacher-Slichter Verfahren, siehe z.B. A. Abragam, "The Principles of Nuclear Magnetism" Claredon Press, Oxford 1961

13. L.C. Tippie, Phys.Rev. B23, 5846 (1981), Phys.Rev. B23, 5854 (1981)

14. N.F. Mott, H. Jones, "The Theory of the Properties of Metals and Alloys", Oxford University Press, 186 (1936)

15. J. Sigg, H. Prisner, K.P. Dinse, H. Brunner, D. Schweitzer and K.H. Hausser, Phys.Rev. B27, 5366 (1983)

16. A. Overhauser, Phys.Rev. 92, 41 (1953)

17. Dormann, priv. comm.

Conducting Polymers Derived from Pyrrole

G.B. Street, R.H. Geiss, S.E. Lindsey, A. Nazzal, and P. Pfluger*
IBM Research Laboratory, San Jose, CA 95193, USA

Abstract

Using polymers to reduce the structural disorder it has been possible to interpret the results of electron diffraction data and to propose an idealized structure for neutral polypyrrole which is consistent in many respects with the structures of the dimer and trimer of pyrrole. XPS data have been used to define some of the deviations from this ideal structure that are seen in actual films and the origin of some of these defects is related to the mechanism of polymerization. In the case of the β,β'-dimethyl pyrrole polymer it has been possible to establish that the chain length corresponds to 100-1000 pyrrole units. The planarity of the chain structure is shown to be important for high conductivity in pyrrole polymers the mechanism for which appears to involve spinless carriers perhaps associated with bipolarons.

Introduction

Pyrrole blacks have been known since 1916 [1] when they were first obtained as powders by chemical oxidation of pyrrole monomer. These insoluble powders have not been characterized in great detail though they are known to consist of pyrrole units bonded predominantly via the α,α'-carbons. It was not until 1968 [2] that films of these materials were first made by the electrochemical oxidation of pyrrole. In 1979 [3], an improved electrochemical technique was used to prepare free standing films with sufficiently good mechanical properties to make practical the investigation of this system as a conducting polymer which might compete with polymers made conducting by carbon or metal loading. More recently it has been shown that the mechanical properties of these polymers can be comparable with standard polymers [5]. An attractive feature of the oxidized conducting form of polypyrrole is its stability in air. The conductivity of films which have been exposed to air for a period of one year decreases only by about 20%. Polypyrrole has been particularly difficult to characterize because of its insolubility and extremely poor crystallinity. The limited structural data that has been available has been obtained using the selective area capabilities of electron diffraction. A further problem with this polymer is its stoichiometry which is dependent on the method of preparation and was initially poorly defined probably because of the incorporation of water and oxygen during the preparation. Although the polymer was shown to be predominantly bonded via the α,α'-carbons of the pyrrole rings, the presence of other types of linking contributes to the disorder which dominates the transport properties of the polymer and also tends to

*Now at Brown Boveri Research Center, CH-5405 Baden, Switzerland.

confuse the interpretation of much of the characterization data [5]. Some of these problems have been resolved by the synthesis of polymers in which both the β positions of the pyrrole ring are blocked to ensure exclusive α,α' bonding of the polymer chain. In the case of the β,β'-dimethyl pyrrole, this leads to an increased chain order which in turn results in significant improvement in the crystallinity of the polymer, enabling a model for the structure of regular polypyrrole to be described in this paper. By tritiation of this monomer it has been possible to directly measure the chain length of the β,β'-dimethyl pyrrole polymer. The role of chain planarity and the hydrogen on the nitrogen in determining the conductivity will be discussed using data derived from planar and non-planar oligmers of pyrrole. Although all the polypyrrole films used for the detailed characterization discussed in this work were prepared using the dry box techniques described in our previous paper [5], we will also mention some of the properties of films grown from aqueous solutions.

Mechanism of the Electrochemical Polymerization of Pyrrole

The first step in the electrochemical polymerization of pyrrole is the generation of the radical cation at the anode as shown in Fig. 1. Chain propagation may then proceed by reaction of two radical cations, pairing the spins and eliminating two protons to produce the neutral dimer. The elimination of the protons is confirmed by the slow rise in the pH of the electrolyte solution during film growth. At the potentials necessary to oxidize the monomer, presumably the dimer and higher oligomers are also oxidized to the corresponding radical cation. Chain propagation continues by reaction of the oligomer radical cation primarily with the radical cation of the monomer, which is present in high concentration in the region of the anode. We have shown that the presence of the radical cation of the monomer is not essential to the polymerization process. In fact films can be grown in the absence of the monomer by the electrochemical oxidation of pyrrole dimer or trimer. As the chain grows the pyrrole oligomer becomes insoluble and eventually precipitates out on the electrode, where the chain can continue to grow until the oligomer radical cation becomes too unreactive or until it becomes prevented from reacting by steric reasons. From elemental analysis the final composition of the films corresponds to a polypyrrole cation with one positive charge for every 3-4 pyrrole units and the appropriate amount of an anion derived from the electrolyte solution. In the case of both polypyrrole and poly β,β'-dimethylpyrrole (PβDMP) grown under drybox conditions from silver perchlorate, acetonitrile solutions, chemical analysis indicates the presence of one perchlorate anion for every three pyrrole rings [5]. However, these analyses also reveal that, on the average, for every three pyrrole rings there is one more hydrogen than the idealized structure would indicate. Although the amounts of hydrogen and perchlorate are almost identical, spectroscopic data indicate that these species are not present in the form of perchloric acid.

Preparation of Polypyrrole from Aqueous Solution

In Dall'olio's original electrochemical preparation of polypyrrole he used an aqueous electrolyte solution of pyrrole acidified with sulphuric acid [2]. Although the conductivity of the resulting films was quite good (~7 ohm^{-1}-cm^{-1}) their mechanical properties were poor compared to those prepared from acetonitrile by DIAZ *et al.* [3]. We have now found that films with good mechanical properties can be prepared from aqueous solution. If pyrrole is electrochemically polymerized from aqueous solutions of copper salts,*e.g.*, $CuSO_4$, films of polypyrrole, mechanically similar to those grown in the absence of water, can be obtained. The $CuSO_4$ provides a facile reduction reaction depositing metallic

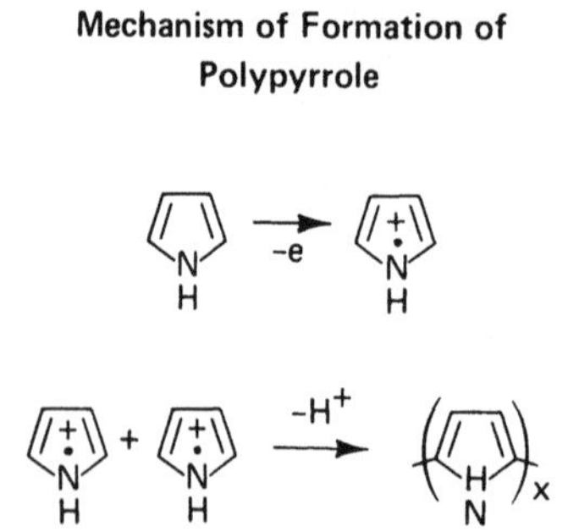

Figure 1. Formation and polymerization of the pyrrole radical cation

copper at the cathode. The electrical conductivities of the films are similar to Dall'Olio's. The conductivity shows a moderate increase as the concentration of pyrrole or the electrolyte salt in the electrolyte solution increases. Films have also been successfully grown using $CuCl_2$, $Cu(NO_3)_2$, $Cu(ClO_4)_2$ and $Cu_3(PO_4)_2$. During the polymerization, the pH of the solution decreased from ~4 to < 2, consistent with the elimination of protons described earlier. In contrast with the films prepared in the dry box the stoichiometries of these aqueously grown films are poorly defined. Although the carbon: nitrogen ratio is usually close to 4:1, the films contain large amounts of excess oxygen. In view of this poor stoichiometry these films were not subjected to the extensive characterization described here and elsewhere for the films grown under drybox conditions. Though there is obviously a serious difficulty in scientifically characterizing these films as opposed to those grown under dry-box conditions, the potential technological importance of conducting polymers grown from aqueous solutions is clear.

β,β'-Substituted Pyrrole Polymers

The β,β'-dimethyl pyrrole monomer can be polymerized to give films of very similar appearance to regular polypyrrole. The oxidation potential of the monomer is +1.0 relative to SSC electrode as shown in Table 1. This oxidation potential is lower than that of pyrrole monomer itself, presumably due to the effect of the methyl groups. The conductivity of the films of PβDMP is about 10 ohm^{-1}-cm^{-1}, a factor of four less than that of polypyrrole itself. Thus the conductivity of the PβDMP is similar to that of the mono β methyl pyrrole polymer [6]. By contrast the conductivity of the β,β'-diphenyl pyrrole is some four orders of magnitude less, 10^{-3} ohm^{-1}-cm^{-1},which is very similar to the conductivity of poly-N-methylpyrrole [7]. Models indicate that the three highly conducting pyrrole polymers in Table 1 can all be planar whereas neither of the low conductivity polymers can be planar. Some confirmation of the non-planarity of the N-methyl pyrrole polymer has been obtained from the crystal structure of the tetramer of N-methyl pyrrole which is grossly non-planar [8]. Thus it does appear that polymer chain planarity is important for high conductivity which is consistent with the basic premise of a delocalized π system. It is also interesting to note that the redox potentials of the three highly conducting, planar polymers are all of a similar negative potential while the two poorly conducting nonplanar systems have positive redox potentials of 0.5 volts *versus* SSC. It has frequently been suggested that the hydrogen on the nitrogen was somehow

Table 1. Conductivity and electrochemistry of pyrrole polymers

	Polypyrrole		Monomer
	Conductivity Ω^{-1} cm^{-1}	Redox Potential Volts	Oxidation Potential Volts
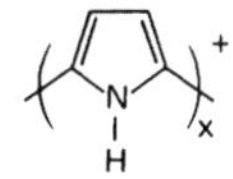	40	-0.2	+1.2
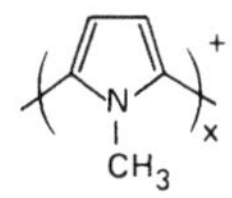	10^{-3}	+0.5	+1.2
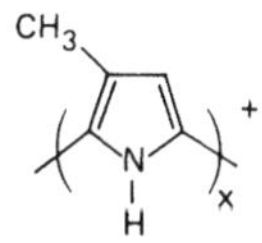	4	-0.3	
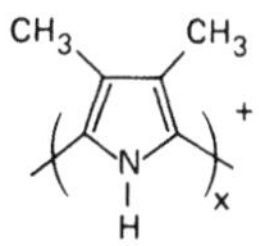	10	-0.1	+1.0
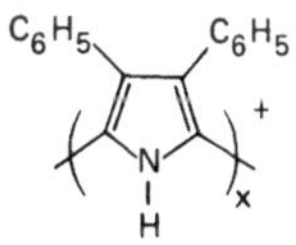	10^{-3}	+0.5	+1.2

important in the conductivity process [9]. These data clearly show that the presence of such a hydrogen does not convey conductivity on a polymer system which is not planar as seen from the case of the diphenyl substituted polymer. On the other hand, in the absence of a hydrogen on the nitrogen, the polymer cannot be planar and therefore cannot be highly conducting. However, the question of whether the hydrogen plays a more active role remains to be answered, though XPS data [10,11] certainly have been interpreted in terms of a labile NH group in polypyrrole. Similar conclusions as to the importance of planarity for high conductivity can also be obtained from a study of substituted polythiophenes which of course have no hydrogen attached to the heteroatom. Table 2 shows the conductivity of several polythiophenes and the corresponding pyrroles. It should be pointed out that the pyrrole polymers contain more anions, 1 per 3 rings, than the thiophenes, 1 per 7 rings [12]. However, the anion was bivalent (SO_4^{2-}) in the latter case so the degree of oxidation of both polymers is similar. These differences may partly explain the 3 orders of magnitude lower conductivity of polythiophene relative to the polypyrrole. However, the demonstrated sensitivity of the conductivity of chemically polymerized thiophenes to small amounts of non-α,α'-linkages [13] suggests that such linkages may also be important in electrochemically prepared polymer. This would be consistent with the higher conductivity measured for the mono β methyl thiophene polymer which [14] should contain fewer of these linkages. Eliminating these linkages

Table 2. Conductivity of pyrrole and thiophene polymers

Heterocyclic Polymers			
Polypyrroles	Conductivity $\Omega^{-1}\ cm^{-1}$	Polythiophenes	Conductivity $\Omega^{-1}\ cm^{-1}$
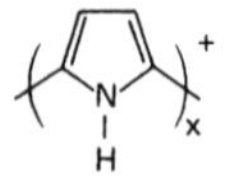	40	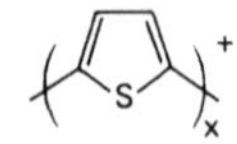	10^{-2}
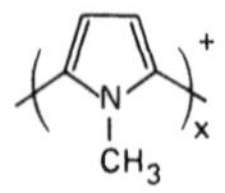	10^{-3}		
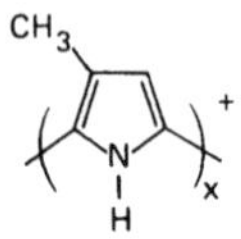	4	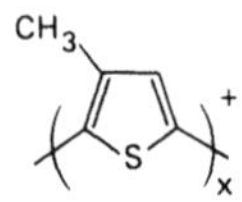	1
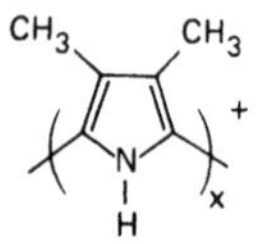	10	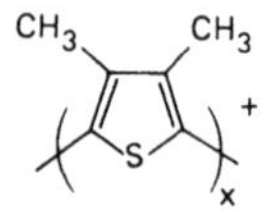	10^{-3}

entirely, by going to the β,β'-dimethyl thiophene,ought to lead to an even higher conductivity. However, as shown in Table 2, the reverse is actually observed and blocking the beta positions leads to a polymer of lower conductivity. The reasons for this are made very clear from molecular models which show that the dimethyl substituted thiophene polymer cannot be planar.

<u>Chain Length of β,β'-Dimethylpyrrole Polymers</u>

Although no way has yet been found to measure the chain length of polypyrrole itself, that of the PβDMP can be conveniently measured using the α,α'-tritiated monomer [15]. Assuming that all the tritium in the final polymer is located at the chain ends as suggested in the equations below, then the average chain length can be determined by comparing the activity of the monomer with the final polymer. It was demonstrated in separate experiments that under the pH conditions of the experiment the tritium or the α-carbons of the polymer and the monomer did not exchange with the solvent or with the hydrogen on the nitrogen. The average molecular weight determined in this fashion varied between 100-1000 pyrrole units depending on the preparation conditions. If all of the protons are not eliminated in the mechanism of the polymerization proposed earlier, then this number of 100-1000 pyrrole units represents the average separation along the chain between pyrrole rings containing a tritium atom and is the upper limit of the average conjugation length and the lower limit for the average molecular weight. Obviously the chain length of

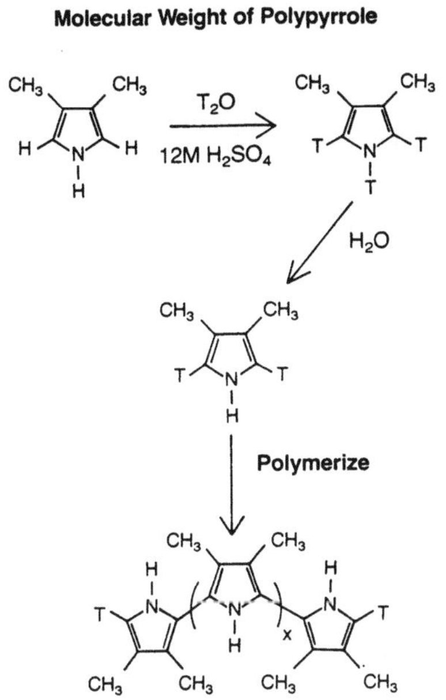

polypyrrole itself could be very different, particularly in view of the involvement of the β carbons in the chain linking in this case. From a study of the optical properties of polypyrrole at intermediate stages of oxidation between the fully oxidized and the neutralized states, YAKUSHI *et al.* [16] have concluded that as-grown polypyrrole has a spread of molecular weights including some very short chain oligomers.

Structural Model for Pyrrole Polymers

It is usually assumed in an ideal polypyrrole, ignoring the presence of excess hydrogen and non-α,α'-bonding, that the pyrrole rings alternate as shown in Fig. 2. This arrangement is required if the chain is to be linear. Some confirmation of the validity of this assumption has now been obtained at least in the case of the dimer and trimer of pyrrole both of which have been shown to have this kind of planar alternating structure [8]. It is also possible that pyrrole could form either a planar polymeric ring of ~10 pyrrole units with nitrogen of each pyrrole pointing to the center of the ring or even a spiral arrangement. Although we do not believe that either of these structures plays an important role in polypyrrole it does seem likely that the failure of adjacent rings to alternate could be an important source of defects. These defects would give rise to bends in the chain. Such defects are much more sterically hindered in the case of PβDMP. This, coupled with the fact that PβDMP is exclusively α,α'-bonded, no doubt plays a part in enhancing the order of this polymer relative to polypyrrole itself. This increased order is reflected in ESR [17], NMR [18], XPS [10,12] and electron diffraction studies [19]. Models of PβDMP suggest that it is possible to form a flat linear chain. Electron diffraction patterns [19] from planar specimens show a few diffuse maxima in the case of neutral polypyrrole while patterns from planar specimens of PβDMP contain more maxima and with some displaying texture suggesting a preferred orientation of the chains parallel to the plane of the film. It

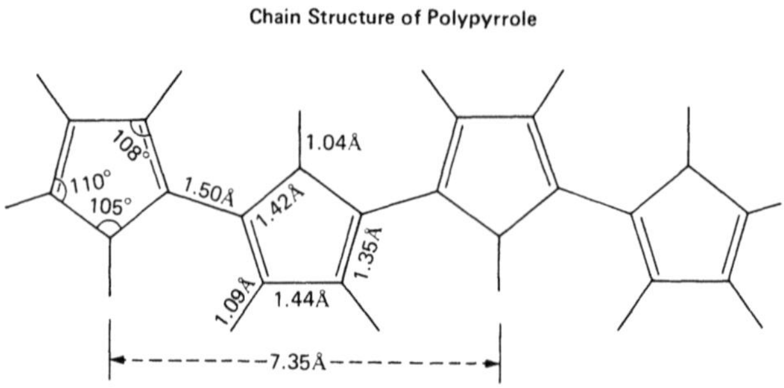

<u>Figure 2.</u> Chain structure of neutral polypyrrole

is believed that in both cases the polypyrrole chains lie with the pyrrole rings coplanar with the film surface. From the diffuseness of the diffraction patterns it is clear that both materials are very poorly crystalline. The observed diffraction maxima come from small crystalline regions separated by much larger amorphous regions. In the case of regular polypyrrole, these crystalline regions are randomly oriented with respect to each other, whereas the arcing of the pattern for the PβDMP suggests that the crystalline regions are aligned with one another. From diffraction patterns of cross sections of films it is possible to identify the spacings corresponding to the separation between the planes of chains. In polypyrrole this distance is 3.41Å, very close to the basal spacing in graphite, indicating that the separation between chains is determined by the π,π interaction of the pyrrole rings. In PβDMP the separation between chains is affected by the methyl groups and the interplanar spacing increases to 3.65Å. If the ideal polypyrrole chain is indeed linear and planar, and if the dimensions of the pyrrole ring do not change significantly from those of the monomer, it is possible to determine the chain geometry by assuming the value of 1.5Å for the ring, ring bond. These assumptions give rise to the geometry shown below with the translation vector along the chain of 7.35Å. Using this chain geometry it has been possible to model a structure which is consistent with all the observed spacings from the diffraction studies. For neutral polypyrrole the proposed packing consists of planes of chains arranged as shown in Fig. 3 and stacked on top of one another in an abab.... sequence to give a monoclinic structure with lattice constants a=8.2Å, b=7.35Å, c=6.82Å, $\alpha=\beta=90°$, and $\gamma=117°$. In the oxidized conducting form of the polymer the anions do not intercalate between the planes of the chains but instead appear to

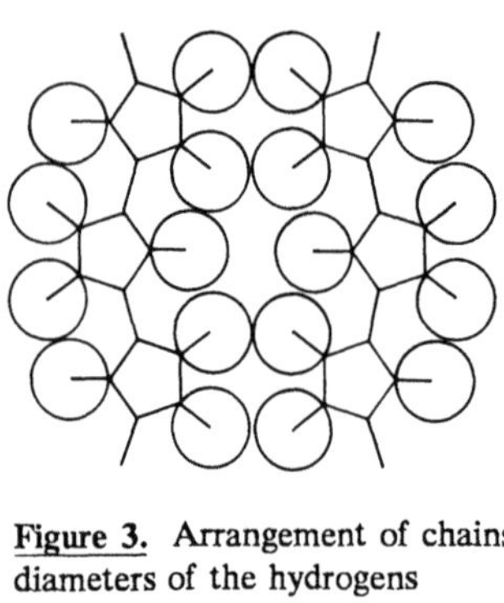

<u>Figure 3.</u> Arrangement of chains in neutral polypyrrole. The circles represent the VDW diameters of the hydrogens

intercalate between the chains within the plane. This arrangement has some important consequences, because for large anions the structure will be determined by the packing of the anions and the resulting structures will probably have very little relation to the structure proposed above. Large anions intercalating in this fashion will also tend to isolate the polymer chains leading to reduced interchain coupling and adversely affecting the conductivity. Preliminary attempts to grow polypyrrole with tetraphenyl borate or stearate anions led to very low conductivity ($<10^{-3}ohm^{-1}$-cm^{-1}) films.

Structural Disorder in Polypyrrole

In the previous section, though mention was made of structural disorder, an ideal chain geometry was assumed to model the polymer crystal structure. In actual films, there are gross deviations from this ideal chain structure, revealed by many techniques including C^{13}NMR studies and more specifically XPS studies. The C1s spectrum for polypyrrole perchlorate is shown in Fig. 4. The spectrum consists of one broad peak centered at 284.2 eV with a FWHM of 2.2 eV. Though the peak is broad and asymmetric the α and β carbons are not resolved as they are in the monomer. Lineshape analysis shows that this asymmetric peak can be fitted by 3 peaks. The two lowest energy peaks are centered at 283.6 eV and 284.5 eV. These two peaks are believed to be associated with the α and β carbons and their splitting is 0.9 eV as observed in the monomer. The third peak

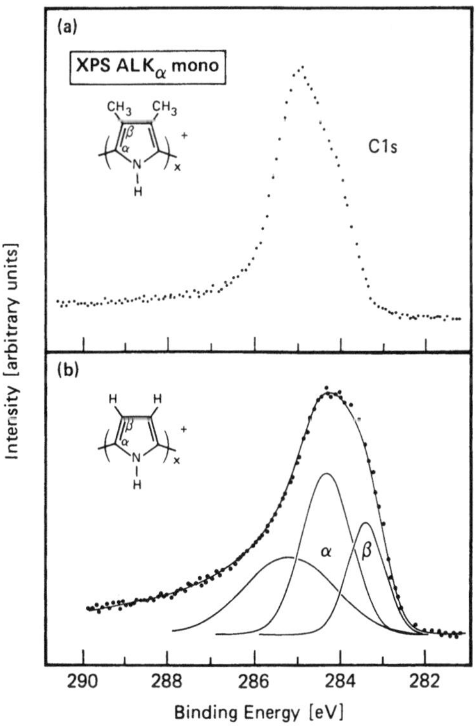

Figure 4. C1s XPS core level spectrum of(a) β,β'-dimethyl polypyrrole perchlorate and (b) polypyrrole perchlorate. The lower lines show the contributions from α, β, and disorder type carbons

centered at 285.4 eV is very broad with a FWHM of 2.4 eV and is believed to result from the non-ideal α carbons, *e.g.*, those associated with chain ends, those which are non-α,α'-bonded, and those in partially saturated rings. The intensity of the α and β peaks should be the same. However, the α peak and the disorder peak probably gain some intensity from hydrocarbon impurities present in almost any XPS spectrum in the region ~284.5-285 eV. Even after reducing the α and the disorder peaks by an appropriate factor, such as to equalize the intensity of the α and β peaks, the remaining intensity in the disorder peak implies that as many as 1/3 of the pyrrole rings are affected by structural disorder. The enhanced order of the PβDMP films is clearly shown in the XPS data by the linewidths of its C1s signal which is reduced to a FWHM of 1.7 eV compared to 2.2 eV for polypyrrole itself. In addition, there is much less asymmetry on the high binding energy side consistent with a reduction in the disorder peak.

The Transport Mechanism in Conducting Polypyrrole

The mechanism of electrical transport in polypyrrole is still an active area of study. However, it has become apparent that our earlier description of polypyrrole as a metal is unsatisfactory. Although the thermoelectric power is consistent with p-type metallic conduction the Hall coefficient is anomalously small and negative. The temperature dependence of the conductivity suggests a 3-dimensional range hopping mechanism. The structural disorder described in the previous section is certainly sufficient to cause extensive localization of the electronic wavefunctions. ESR studies show that the free spins seen in the polymer are not associated with the charge carriers. Indeed by appropriate electrochemical cycling these free spins can be removed and yet the film remains electrically conducting. The absence of paramagnetism associated with the current carrier can be explained in terms of bipolarons [17,20] which are doubly charged but carry no spin. It is important to appreciate that solitons cannot be mobile in polyprrole because the ground state is not degenerate. For pyrrole monomer the energy to switch the double and single bonds is ~10 Kcals/mole [21]. Recently Bredas has calculated that the corresponding energy in polypyrrole is ~14 Kcal/mole [20]. The possible importance of bipolarons in polypyrrole was mentioned previously by BRAZOVSKII and KIROVA [22]. Such a mechanism has also been suggested for polyparaphenylene [23] which in common with polypyrrole also has a nondegenerate ground state.

Acknolwedgments

We thank the Office of Naval Research for partial support of this work.

References

1. A. Angeli, L. Alessandri: Gazzetta Chimica Italiana **46**, 283 (1916).
2. A. Dall'olio, Y. Dascola, V. Varacca, V. C. R. Bocchi: Acad. Sci. Ser. C **267**, 433 (1968).
3. A. F. Diaz, K. K. Kanazawa, G. P. Gardini: J. Chem. Soc., Chem. Comm., 635 (1979).
4. A. F. Diaz, B. Hall: IBM J. Res. Dev. **27**, 342 (1983).

5. G. B. Street, T. C. Clarke, M. T. Krounbi, K. K. Kanazawa, V. Y. Lee, P. Pfluger, J. C. Scott, G. Weiser: Molec. Cryst. Liq. Cryst. **83**, 253 (1982).
6. M. Salmon, A. F. Diaz, A. J. Logan, M. T. Krounbi, J. Bargon, Mol. Cryst. Liq. Cryst. **83**, 265 (1982).
7. K. K. Kanazawa, A. F. Diaz, M. T. Krounbi, G. B. Street: Synth. Met. **4**, 119 (1981).
8. G. B. Street, A. Nazzal: To be published.
9. A. J. Diaz: Chem. Scripta **17**, 145 (1981).
10. P. Pfluger, M. T. Krounbi, G. B. Street, G. Weiser: J. Chem. Phys. **78**, 3212 (1983).
11. R. Erlandsson, I. Lundström: J. de Physique Coll., in press.
12. P. Pfluger, G. B. Street: J. Chem. Phys., to be published.
13. K. Sanechika, T. Yamamoto, A. Yamamoto: J. Polym. Sci., Polym. Lett. **20**, 365 (1982).
14. R. J. Waltman, J. Bargon, A. F. Diaz: J. Chem. Phys., to be published.
15. G. B. Street, A. Nazzal: To be published
16. K. Yakushi, L. T. Laughlan, T. C. Clarke, G. B. Street: J. Chem. Phys., in press.
17. J. C. Scott, P. Pfluger, M. T. Krounbi, G. B. Street, Phys. Rev. B, in press.
18. T. C. Clarke, J. C. Scott, G. B. Street: IBM J. Res. Dev. **27**, 313 (1983).
19. R. Geiss, G. B. Street, W. Volksen, J. Economy: IBM J. Res. Dev. **27**, 321 (1983).
20. J. L. Bredas: **Handbook on Conducting Polymers**, ed. by T. J. Skotheim (Dekker, New York), in press.
21. R. A. Jones, G. P. Beam: **The Chemistry of Pyrroles** (Academic Press, London, 1977).
22. S. A. Brazovskii, N. N. Kirova: JETP Lett. **33**, 4 (1981).
23. J. L. Bredas, R. R. Chance, R. Silbey: Mol. Cryst. Liq. Cryst. **77**, 319 (1981).

Organic Superconductors: Quasi One-Dimensional Conductors, Anomalous Superconductors

D. Jérome

Laboratoire de Physique des Solides, Université Paris-Sud
F-91405 Orsay, France

I. Introduction

The discovery of superconductivity in organic substances [1] is the achievement of an intense investigation of the electronic properties of organic molecular conductors. These organic conductors are somewhat special systems among the vast ensemble of organic semiconductors already known since the theory of electron energy bands in solids seems to account fairly well for their conduction properties. The research on organic conductors started at the beginning of the seventies with the preparation of the first stable highly conducting organic solids, namely those belonging to the TTF-TCNQ series [2,3]. It was quickly recognized that the unusual properties of these conductors - existence of phase transitions towards lattice periodically distorted non-magnetic state (Peierls state) [4] and precursor signs of these one-dimensional transitions observed at temperature as high as about three times the actual phase transition temperature, i.e.($\sim$3 x53 K) in TTF-TCNQ - required the use of One-Dimensional Physics [5,6] as a starting point of the interpretation.

The One-Dimensional (1-D) concept is relevant to organic conductors as a result of the planar shape of the constituent molecules and of their particular arrangement in the crystal, namely a packing of the molecules in stacks along a preferred direction, allowing strong and weak overlap between the π orbitals of near-neighbour molecules belonging to the same stacks and to adjacent stacks, respectively.

It should be clear at this point that the 1-D character of transport properties in organic conductors known up to now was not desired at the beginning. It is the result of the strong directionality of the electronic bonds of planar heterocylic molecules. In other words, One Dimensionality is the price to pay to achieve a good overlap between electron clouds of neighbour molecules. On the one hand the 1-D character of these materials provides an almost ideal possibility to improve our basic and theoretical understanding of 1-D Physics, but on the other hand (on a more pragmatic level), 1-D character prevents us from observing a superconducting state in the known materials at reasonably high temperature [5]. This lecture intends to deal with both aspects. The Peierls-Fröhlich nature of the phase transition at 53 K in TTF-TCNQ [4] is only one of the particular theoretical features exhibited by a 1-D electron gas which is predicted to be unstable at low temperature.

The nature of the instability depends on the interactions between the electrons within a given stack of molecules: direct coulombic interactions, indirect interactions mediated by phonons or possibly by the electric polarization of the highly polarizable organic molecules, and finally interactions between electrons and lattice modes.

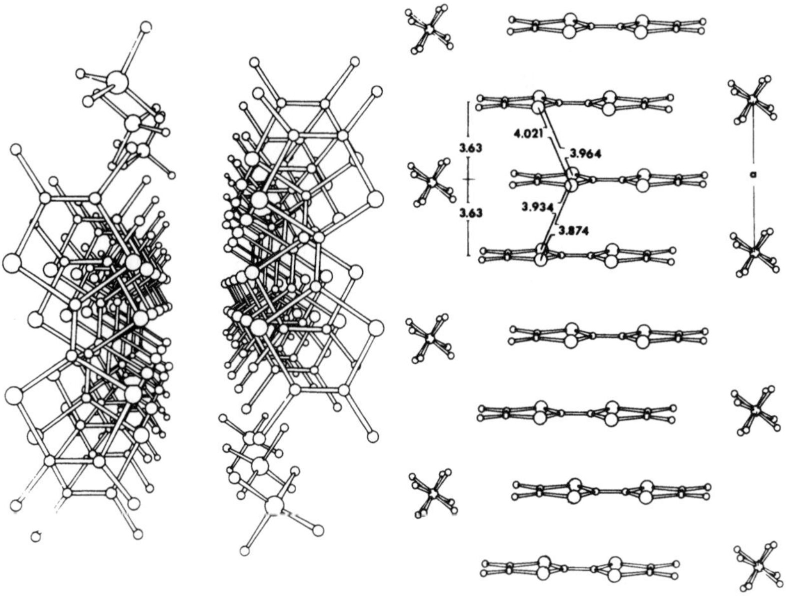

Figure 1. View of $(TMTSF)_2ReO_4$ along the stacking axis(a axis) and side-view of $(TMTSF)_2ClO_4$(right). The ReO_4^- ions show alternative order along the a axis (T<180 K) leading to a 2 a potential

In the presence of interactions a 1-D electron gas is unstable at low temperature against the formation of charge (or spin) density waves (CDW or SDW) [5]. Both instabilities compete with superconductivity which arises in the 1-D electron gas as well as in the usual 3D gas.

The interplay between spin (charge) density waves and superconducting instabilities of a 1-D electron gas becomes clear if the electron-hole (CDW-SDW channel) or the electron-electron (Cooper channel) response functions are calculated pertubatively. In the expansion of the superconducting response of a 1-D electron gas, divergent contributions from density fluctuations occur and vice versa,i.e.the two types of instability become intimately coupled.

Another problem raised by 1-D conductors is the easy excitation of large amplitude fluctuations of the order parameter at either finite temperature (thermal fluctuations) or even zero temperature (zero point motion or quantum fluctuations). The fluctuations prevent the establishment of long-range order. The onset of long-range order occurs only at zero temperature (or does not occur at all even at zero temperature if quantum fluctuations are taken into consideration) in a purely 1-D conductor. However, actual quasi 1-D conductors undergo phase transitions at a finite temperature but the transition temperature may still be much smaller than the value expected from the knowledge of intrachain electron-electron interactions: the so-called mean field approximation is inappropriate for the derivation of the transition temperature of a strongly anisotropic conductor.

The only molecular conductors exhibiting superconductivity so far belong either to the $(TMTSF)_2X$ [6] or to the $(BEDTTF)_2X$ [7] organic salts. A great deal of work has been devoted to the $(TMTSF)_2X$ selenium series in which su-

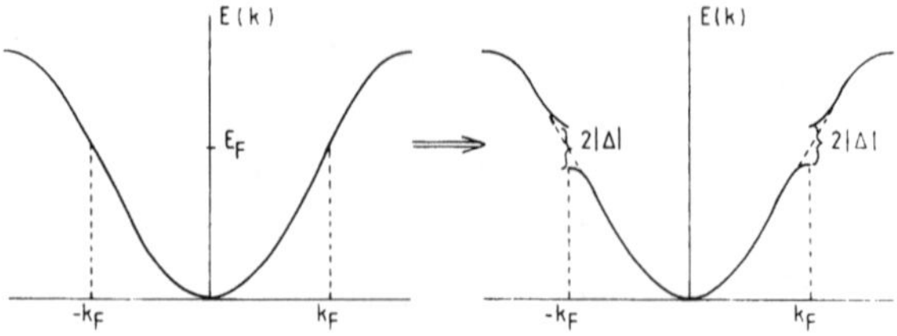

Figure 2. One-dimensional half-filled band in a uniform conducting state (left), in the presence of a distortion (Peierls or SDW) opening a gap at the Fermi level (right)

perconductivity has been unambiguously established for the six compounds $X = ClO_4, ReO_4, PF_6, AsF_6, SbF_6, TaF_6$ [8]. We shall not discuss the properties of the only salt $X = ReO_4$ of the sulfur $(BEDTTF)_2X$ series [7] since no remarkable difference from the $(TMTSF)_2X$ family has been estalished so far.

When the molecule tetramethyltetraselenafulvalene (TMTSF) is dissolved in an appropriate solvent containing a salt of a monovalent anion X and oxidized on a platinum anode in an electrochemical cell, high purity crystals of composition $(TMTSF)_2X$ separate on the electrode according to the reaction

$$2n\ \mathrm{TMTSF} + nX^- \xrightarrow{(-ne^-)} [(\mathrm{TMTSF})_2X]_n \ .$$

TMTSF are quasi-planar molecules stacked in a zig-zag pattern along the a axis of a triclinic crystal (fig.1). X is an inorganic anion such as PF_6^-, ClO_4^-,... which plays an essential role in allowing a partial filling of the outer shell of the organic molecule (or the valence band of the crystal) but otherwise does not contribute directly to electron transport. The anions do, however, in some cases affect the electronic properties of the $(TMTSF)_2X$ series, governing the stability of the ground state via complete or incomplete order at low temperature. The conducting band derived from the overlapping π-type molecular orbitals is half-filled rather than quarter-filled as inferred from the 2.1 stoichiometry of the salt (fig.2). This feature occurs due to the triclinicity and to a slight dimerization of the TMTSF units (Fig.1) which leads to a slight alternation of the band overlap integral $t_{//}$ along the a axis. The half-filling of the band (periodicity a) is expected to affect predominantly the low temperature behaviour, for the small dimerization gap can probably be neglected at room temperature.

After this very short introduction to the $(TMTSF)_2X$ series we wish to focus the rest of this lecture on the following question: can we use our knowledge of ordinary metals to understand both the phase transitions and the conducting state of organic superconductors such as $(TMTSF)_2X$ or alternatively do we need the development of new interpretations which take into account the particular low-dimensional properties of these new conductors? In section II we review the electronic properties of the $(TMTSF)_2X$ series (ground states and conducting state) and we try to see how they can be understood in terms of a Drude-like electron gas [9] and a mean-field treatment of the phase transition. Section III will be devoted to a very brief outline of a point of view which we feel has the merit to account for both "regular" and "anomalous" electronic properties. This model claims the existence of superconducting precursors in a temperature domain far above the onset of long-range order.

II. Electronic Properties

II.1. Anisotropy

According to Fig.1 which shows the typical structure of $(TMTSF)_2X$ salts, the stacking direction (a axis) is the direction along which the overlap between near-neighbour molecular orbitals is the strongest. However, because of short Se...Se contacts intermolecular overlap is also significant along the transverse b direction. The transverse c direction along which stacks of TMTSF molecules are separated from each other by X anions represents a direction of weak interchain coupling. When the conductivity proceeds via a coherent motion along the chain and a diffusive process from chain to chain (this is probably the situation at room temperature) the anisotropy of conductivity is given by the relation [10]

$$\sigma_a/\sigma_b = (at_a/bt_b)^2 \tag{1}$$

where $a \approx 3.6$ Å and $b \approx 7.7$ Å are the lattice parameters of $(TMTSF)_2X$ neglecting the on-stack dimerization. There have been several reports on σ_a/σ_b anisotropy data ranging from 25 (the smallest) in $(TMTSF)_2ClO_4$ [11] up to $(2\text{-}4)\times 10^2$ in other salts [12,13] . Hence, from the relation (1) the most conservative estimate of the band structure anisotropy is $t_a/t_b > 10$. The anisotropy is also very significant in the b-c plane since $t_b/t_c \approx (15$ to $30)$ according to conductivity data. Furthermore anisotropy data are in general agreement with semi-empirical molecular-orbital calculations [14] . In the rest of this lecture we shall take for the ratios $t_a:t_b:t_c$ the typical sequence 10:1:1/15 with $t_a \approx 0.12$eV. Considering the above determination of the band parameters, the Fermi surface of the conducting phase of $(TMTSF)_2X$ salts is warped but remains open and therefore does not display closed orbits (this would require a much smaller anisotropy $t_{//}/t_\perp \approx 2$ or so). However, closed orbits can still be obtained if (and only if) additional folding of the surface is provided by a doubling of the lattice periodicity along a (onset of spin density waves or anion ordering) as discussed in Sect II.2 of this lecture. An open Fermi surface does not necessarily mean that electron transport is diffusive along a transverse direction. Several experiments (optics and NMR relaxation for example) do show that the transverse motion becomes coherent below 100 K along the b direction in $(TMTSF)_2X$ salts. The observation of a plasma excitation in a Q-1-D conductor with the E field polarized perpendicular

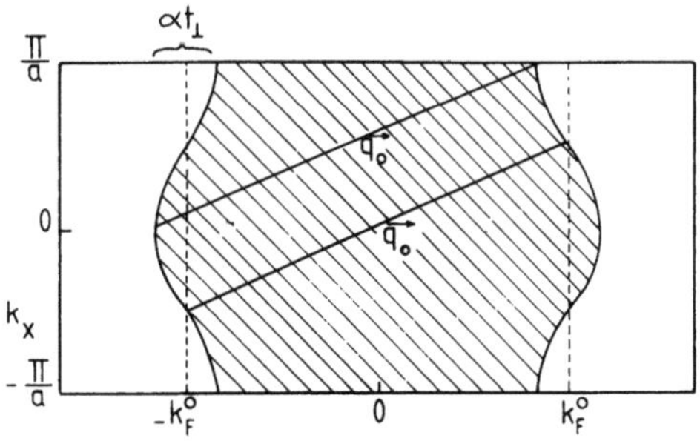

Figure 3. Warped Fermi surface of a Quasi One-Dimensional Conductor. The amplitude of the warping along the transverse direction is proportional to the interchain coupling $t_\perp$. The vector $(Q_o = 2k_F, 1/2)$ becomes the nesting vector of the warped surface instead of $(2k_F, 0)$ for a planar surface

to the chains [15] requires a corrugation of the Fermi surface due to interchain overlap $\sim 4t_\perp$ larger than the thermal broadening $\hbar/\tau_{//}$, due to the finite intra-stack electron scattering time $\tau_{//}$ [10] Fig.3, namely

$$\hbar/\tau_{//} \lesssim 2\pi\, t_\perp \ . \tag{2}$$

Condition (2) may be fulfilled in $(TMTSF)_2X$ salts if $t_\perp^b \approx 10$-12 meV at low temperature since a Drude-like fit of the longitudinal plasma edge of $(TMTSF)_2PF_6$ [15] or $(TMTSF)_2ClO_4$ [16] at 25 K leads to $\tau_{//} \approx 10^{-14}$s($\hbar/\tau_{//} \approx 80$meV). In case of open Fermi surface the conversion from plasma frequencies into bandwidth is linear [17,18]. The anisotropy of plasma frequencies leads again to the satisfying ratio $t_a/t_b^* \sim 10$, ($\omega_{pa} \sim 1.40$ eV). The c-axis transfer integral can also be derived by the field dependence of the nuclear spin-lattice relaxation rate. It is already known after extensive studies in TTF-TCNQ that the hyperfine spin-lattice relaxation is field dependent in a low-dimensional conductor [10] . If the electron Larmor frequency ω_e becomes larger than the electron hopping rate along particular directions in the crystal $1/T_1$ is no longer field independent as for regular isotropic conductors. The hopping rate can be made either very small along all transverse directions in a 1-D conductor or alternatively small only along a given direction perpendicular to a plane in which transport is coherent. The latter situation probably applies to $(TMTSF)_2X$ with a small hopping rate $1/\tau_\perp^{c^*}$ along the c* direction perpendicular to the a-b plane. The relaxation rate $1/T_1$ is field dependent ($\sim \log \omega_e$) or independent for $\omega_e \tau_\perp^{c^*} > 1$ and $\omega_e \tau_\perp^{c^*} < 1$ respectively. Given the experimental hopping rate $1/\tau_\perp^{c^*} = 2.10^{11}\,s^{-1}$ [19] and $\tau_{//} \approx 10^{-14}$s, the theory

$$\frac{1}{\tau_\perp^{c^*}} = 2\pi\, t_c^2\, \tau_{//} \tag{3}$$

leads to $t_c \approx 1$ meV. This value of t_c is in fair agreement with the derivation of the band structure anisotropy from transport properties or from molecular orbital calculation [14] . Another property depending on the band structure anisotropy is the critical field H_{c_2} of the superconducting state which [13] for an open Fermi surface superconductor is given within the Landau-Ginzburg formalism [13]by

$$H_{c_2}^i / H_{c_2}^j = d_i t_i / d_j t_j = (\sigma_i/\sigma_j)^{1/2} \ . \tag{4}$$

As mentioned earlier, the second part of (4) is justified only at high temperature when the transverse conductivity is diffusive. The existing data of critical fields in $(TMTSF)_2ClO_4$ or $(TMTSF)_2AsF_6$ show that transverse critical fields obey fairly well, order of magnitude wise, a relation such as (4). However a ratio $H_{c_2}^a / H_{c_2}^b \approx 2$ has been reported for a sample of $(TMTSF)_2ClO_4$ [11] leading to $t_a/t_b \approx 5$ through (4) instead of ≈ 10 from conductivity data. This discrepancy has been taken as evidence for the suppression of the superconducting state when the field is parallel to the chain axis by the pair-breaking effect of the magnetic field on the Cooper pairs [11] : the so-called Pauli limitation with

$$2\mu_B H_P = 2\Delta \tag{5}$$

where H_P is the Pauli-limited critical field and 2Δ is the binding energy of a Cooper pair. We do not believe, however, that experimental evidence in favour of a Pauli limitation along a are still very convincing (too small a critical temperature $T_c \sim 0.86$ K showing that the sample of $(TMTSF)_2ClO_4$ was not slowly cooled, ambiguous definition of H_{c_2}, possible misalignment of the crystal in the magnetic-field, linear T-dependent H_c^a in the neighbourhood of T_c instead of the $(1-T/T_c)^{1/2}$ dependence predicted by (5). Furthermore a first-order phase transition is expected at the Pauli critical field with a latent heat since the derivative of the energy versus temperature displays a discontinuity. To summarize, we have seen in this section how a large number of electronic properties of $(TMTSF)_2X$ (conductivity anisotropy, anisotropy of optical reflectance, NMR, and superconducting critical fields) can be understood within the framework of a rather simple model based on a very anisotropic electron energy dispersion.

II.2. Ground states

A 1-D electron gas tends to be unstable at low temperature. In the simplest theoretical treatment (mean-field approximation), the nature of the low-temperature divergent response depends on the sign and spatial shape of the electron-electron interactions. Approximating the Fourier transform of the interaction by two constants g_1 and g_2 (related to scattering at wave vectors $q = 2k_F$ and $q = 0$ respectively) allows a derivation of a phase diagram in the g_1-g_2 plane [15], Fig.(4). Treatments beyond mean-field approximation lead to results not significantly different from the mean-field approach. First-order renormalization group theory gives a power law for the

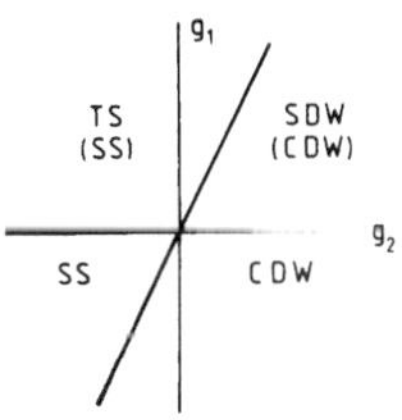

Figure 4. Most divergent response functions in a 1-D electron gas within the g_1-g_2 model. Divergences in brackets are weaker

temperature dependence of the $(q=\omega=0)$ correlation functions close to zero temperature. When the inequality

$$g_1 < 2g_2 \tag{6a}$$

is satisfied, electron-hole instabilities are dominant (spin or charge density waves according to the sign of g_1). The superconducting channel is dominant when

$$g_1 > 2g_2 , \tag{6b}$$

triplet pairing(TS) being more divergent than singlet pairing if $g_1 > 0$. The key question raised by present low-dimensional conductors is how do we understand phase transitions occurring in $(TMTSF)_2X$ at low temperature, in particular superconductivity? Can we expect the critical temperature to be linearly related to the interaction giving rise to the instability (mean-field approximation) or do we have to use another approach: exact 1-D treatment of the single chain problem followed by a mean-field approximation for the interchain coupling? In the latter case the critical temperature is given by:

$$1 - J\,\chi_{1D}(0)\Big|_{T=T_c} = 0 \tag{7}$$

where $\chi_{1D}(q)$ is the order parameter susceptibility of the single chain and J the transverse coupling. For an order parameter with two components (superconductivity or incommensurate CDW) the classical derivation leads to:

$$T_c \sim (J\,\alpha)^{1/2} \qquad (8)$$

in the weak interchain coupling limit (when the interchain coupling J is much weaker than the intrachain interaction α). Eq(8) provides two important consequences:

(i) The onset of long-range order in a Q-1-D conductor can occur at a temperature T_c much lower than the transition temperature T_c^{MF} which would be calculated assuming no fluctuation of the order parameter below T_c^{MF}, provided $J/\alpha \ll 1$.

(ii) The nature of the channel which is most divergent in the 1-D regime ($T>T_c$) is not necessarily the same as the ground state of the electron gas. The following may also be encountered: two competing instabilities A and B for which $\alpha^A>\alpha^B$ but $J^A<J^B$. The instability B may appear below a 3-D ordering temperature $T_c \gtrsim (J^B\alpha^B)^{1/2}$, but at temperatures well above T_c the divergence of the A instability will dominate the behaviour of electronic properties. This situation we think is relevant for several members of the $(TMTSF)_2X$ which display a magnetic insulating ground (SDW) state below $T_c \sim 12$-18 K. But above 12-18 K the 1-D superconducting divergence is dominant.

Various ground states observed in the $(TMTSF)_2X$ series may be classified as follows:

(i) Magnetic insulator

The typical example is the SDW state of $(TMTSF)_2PF_6$ or $(TMTSF)_2AsF_6$ firmly established by magnetic measurements: anisotropic magnetic susceptibility [20] , ^1H or ^{77}Se NMR [21,22] , and antiferromagnetic resonance [23] . If a spin structure doubling occurs at low temperature the periodicity of the electron potential becomes 2a in the SDW state and an exchange gap opens on the whole (or most of the) quasi planar Fermi surface. This process drives the conductor into a semiconducting ground state, Fig.(2).

(ii) Anion-ordered insulator

The ordering of non-centrosymmetric anions such as ReO_4 or FSO_3 can also lead to a doubling of the longitudinal lattice periodicity with the occurrence of an insulating state when the ordering alternates along the a axis. This situation is encountered in $(TMTSF)_2ReO_4$ [24] and $(TMTSF)_2FSO_3$ [25] below 180 K and 86 K respectively. We wish to emphasize that this kind of conducting to insulating phase transition is a by-product of the Q-1-D nature of the electronic spectrum. However the electrostatic interactions between anions leading to ordering are likely to be more isotropic than pairing or exchange interactions. The g_1g_2 model sketched previously has no relevance whatsoever to anion-ordered ground state. $(TMTSF)_2ClO_4$ is a very interesting system, since despite the non-centrosymmetric nature of the ClO_4 anions which order at 24 K, this system does not undergo any conductor-to-insulator phase transition down to 1.2 K where the superconducting state is stabilized. X-ray investigations [26] have shown that below 24 K a 3 D superlattice structure is observed with no change of the periodicity along the a and c axes and a doubling along b. This kind of uniform ordering along a does not affect the density of states at the Fermi level. While slow cooling results in quasi-perfect uniform ordering at low temperature (called the relaxed R state) [27,28], rapid cooling below 30 K leads to a partially ordered (quenched Q state) exhibiting an excess of entropy in the superconducting state with respect to the R state [29] . This excess of entropy has been attributed to frozen-in lattice (ClO_4 ions) disorder. We shall show later how this small disorder introduced by quenching governs the balance between ground states of different natures in $(TMTSF)_2ClO_4$.

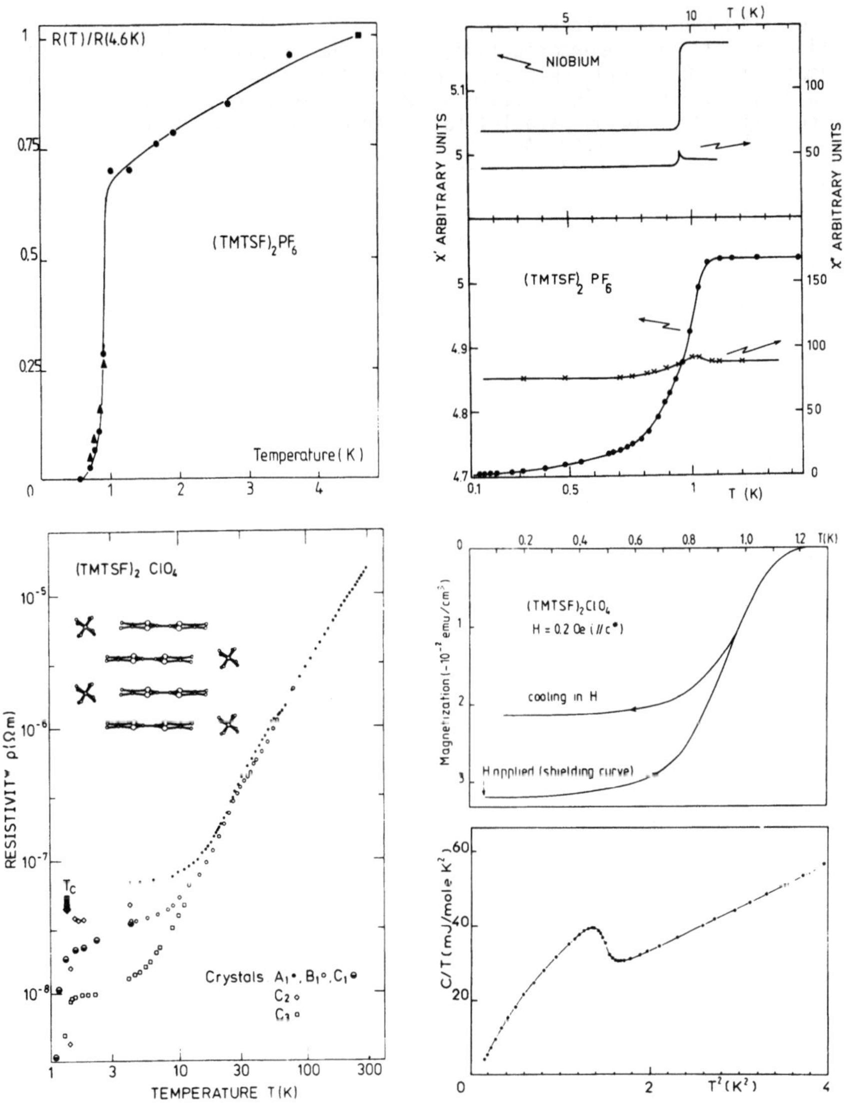

Figure 5. Observation of superconductivity in $(TMTSF)_2PF_6$ under 12 kbar, resistive (top left) and inductive (top right) methods with Nb superconducting signal shown for comparison, $(TMTSF)_2ClO_4$ at ambient pressure, resistive (bottom left), Specific heat and Meissner signals (bottom right)

(iii) Superconducting state

Superconductivity is observed below 1.2 K in the six salts $x = ClO_4$, ReO_4, PF_6, AsF_6, TaF_6, SbF_6 of the $(TMTSF)_2X$ series [8]. For all systems of the series except $(TMTSF)_2ClO_4$ superconductivity is observed provided the insulating ground state (magnetic or non-magnetic in $(TMTSF)_2ReO_4$) is removed by the application of a high pressure. At first glance the superconducting transition of organic superconductors does not look that different from the phase transition in metallic superconductors, Fig.5. The transition is rather sharp, for example in $(TMTSF)_2ClO_4$ (R state) the resistive transition is less than 0.15 K wide [30]. The Meissner flux expulsion is large enough (80% of the shielding signal for fields parallel to c* direction) to guarantee the bulk nature of the superconducting state [31] . As expected, the anisotropic character of the overlap parameters influences the Meissner effect as only a 1% flux expulsion is observed when the field is put along the a axis in agreement with smallness of the c-axis coupling. Additional interesting information related to the superconducting state are given by the specific heat data in $(TMTSF)_2ClO_4$ [32]. As shown in Fig.(6) the specific heat anomaly looks rather traditional. The jump of electronic specific heat, $\Delta C/C=1.67$, is only slightly bigger than the BCS weak coupling value of 1.43. The condensation energy and the thermodynamical critical field are 2×10^{16}eV/mole and 44 Oe respectively. An approximate determination of the gap parameter leads to a ratio $2\delta/T_c \simeq 4$ which is again not far from the value which is expected in an isotropic superconductor.

So far we have reviewed some properties of ground states which could be considered as regular in an anisotropic conductor by comparison with anomalous behaviours which we will review now. Non-magnetic impurity potentials have usually very little effect on superconductivity of metals. However, they are amazingly important with respect to the stability of long-range superconductivity of the $(TMTSF)_2X$ series. Weak and non-magnetic disorder can be introduced into $(TMTSF)_2ClO_4$ by various means:

(i) Alloying $(TMTSF)_2ClO_4$ with $(TMTSF)_2ReO_4$, which also becomes superconducting under pressure, allows the preparation of the solid solution $(TMTSF)_2ClO_{4(1-x)}ReO_{4x}$ [33]. Superconductivity is suppressed sharply with increasing x. No superconductivity is observed at x>3%, Fig.(7). Instead, the ground state of the solid solution shows all features of a magnetic insulator occurring below 5 K. Preliminary data indicate that the same effect of disorder is seen at the other end, namely as x is close to unity a very tiny amount of ClO_4 ions in $(TMTSF)_2ReO_4$ is sufficient to suppress the superconducting state (under pressure of 10 kbar).

(ii) The superconducting state of $(TMTSF)_2ClO_4$ is also suppressed by disorder of the conducting chain in the $[(TMTSF)_{1-x}(TMTTF)_x]_2ClO_4$ solution with

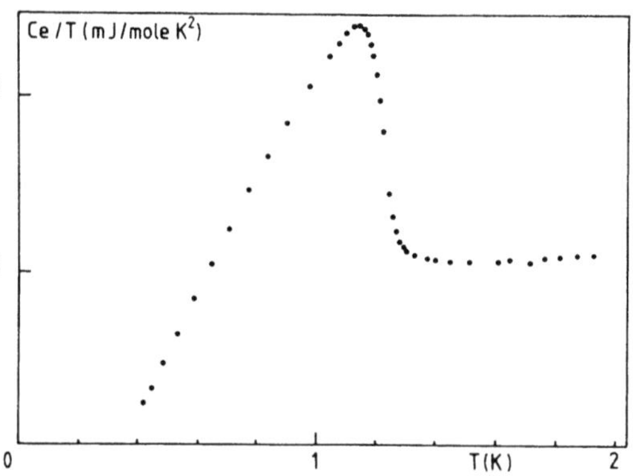

Figure 6. Ratio of the electronic specific heat to the temperature versus temperature in $(TMTSF)_2ClO_4$. The jump of specific heat at the transition is $\Delta C/C \simeq 1.6$

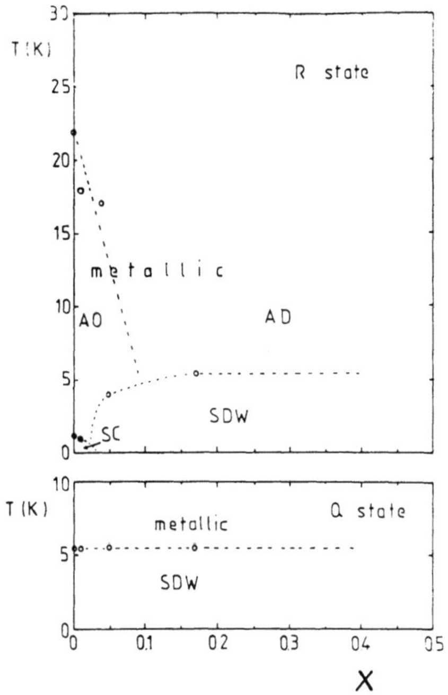

Figure 7. Phase diagram of the solid solution $(TMTSF)_2ClO_4{}_{(1-x)}\ ReO_4{}_x$. R state (top) and Q state (bottom)

x>2%, where TMTTF is the sulfur analog molecule of TMTSF [34]. Once more, after suppression of the superconducting state by disorder on the cation stack a magnetic ground state becomes stable below $\approx$10 K.

(iii) Another effect of lattice disorder on the nature of the ground state is provided by the R-state and Q-state studies of $(TMTSF)_2ClO_4$. A very quickly cooled $(TMTSF)_2ClO_4$ sample below 30 K (cooling rate >4 K/s) leads to the establishment of an insulating magnetic state below 3.5 K instead of the superconducting state observed at 1.2 K in an R sample [28]. All kinds of intermediate situations can be obtained between Q and R states where T_c (superconductivity) is lower than 1.2 K. For instance, for a cooling rate of $\approx$0.2 K/s, T_c drops to 0.9 K but surprisingly enough in spite of the significant lowering of T_c in the partial Q state no marked change of $N(E_F)$ can be detected via C_v measurements in the 1.2-2 K temperature domain [32].

We emphasize that the influence of non-magnetic impurities on the superconducting ground state is highly anomalous. Within the framework of a mean-field treatment of the superconducting transition the pair-breaking effect of such a small concentration of impurities should be negligible [35]. Furthermore, the restoration of a magnetic ground state at high concentration of impurities shows that their influence on the smearing of the Fermi surface is weak. In case of a transition involving a good nesting of the Fermi surface (SDW transition), the transition is destroyed when the Fermi surface broadening becomes larger than the transition temperature in the pure system.

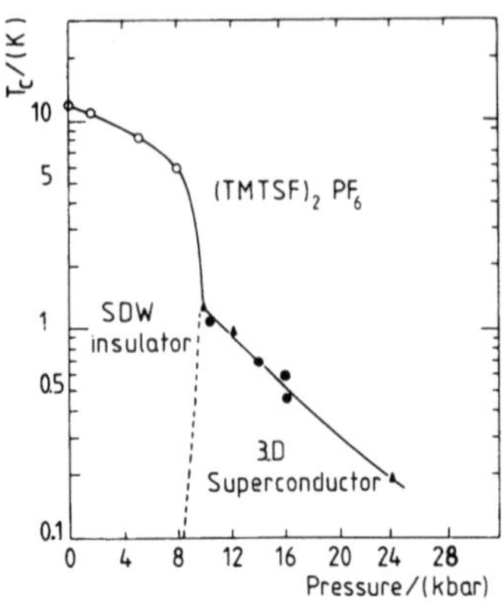

Figure 8. Phase diagram of $(TMTSF)_2PF_6$ under pressure

II.3. Magnetism versus Superconductivity

Interplay between magnetism and superconductivity is a constant experimental feature of the low-temperature properties of the $(TMTSF)_2X$ series. Pressure governs the balance between an itinerant antiferromagnetic ground state and superconductivity in all $(TMTSF)_2X$ salts when X is an anion of octahedral symmetry such as PF_6. Above 8 Kbar or so superconductivity becomes stable below 1.2 K as shown in Fig.(8). Close to the critical pressure the phase diagram shows a re-entrance of the superconducting ground state below the SDW phase [36,37] . Another remarkable feature is the strong pressure dependence of superconductivity, T_c decreasing at a rate of 10% per kbar and 30% per kbar in $(TMTSF)_2PF_6$ [38,39] and $(TMTSF)_2ClO_4$ [40] respectively.

The interplay between superconductivity and magnetism is also illustrated by the behaviour of $(TMTSF)_2ClO_4$ (R state) under magnetic field, Fig.(9). If a large magnetic field is applied along the c^* direction a magnetic state is restored above a "turn on" field H_{to} which depends on temperature and sample orientation ($H_{to} \approx 53$ kOe // c^* at T = 1 K). Furthermore, as shown in Fig.(9), a sequence of phase transitions between various magnetic states has been detected by Hall effect experiments [41] . These field-induced magnetic states exhibit a very small $N(E_F)$ at very low temperature [32] and a semimetallic character of the Fermi surface [42,43] . The phase diagram in Fig. 9 suggests the possible existence of a triple point at $T \approx 0$ K and $H = H_{c2}$ between superconductivity, highly conducting state and the SDW semimetal. More thorough experiments below 0.3 K are needed to establish fully this aspect of the phase diagram. We have already mentioned that non-magnetic disorder can induce a magnetic ground state (which can be further removed by pressure but without reappearance of superconductivity [33](Fig.7). Therefore experiments show very clearly how non-magnetic defects prevent very efficiently the establishment of superconducting long-range order to the benefit of a magnetic state. For all experimental data showing interplay between SDW and superconducting ground state a common phase border between the two states seems to be established (the field-induced SDW state of $(TMTSF)_2ClO_4$ requir s more data). There is so far no experimental evidence for the coexistence at a given temperature of magnetism due to the well-nested regions of the Fermi surface and superconductivity due to the weakly nested parts [44,45]. The specific heat jump at T_c shows that all electron states at the Fermi surface undergo the superconducting instability. Finally it has been argued that the SDW state of $(TMTSF)_2PF_6$ is suppressed by an enhancement of the interchain coupling under pressure [38]. This

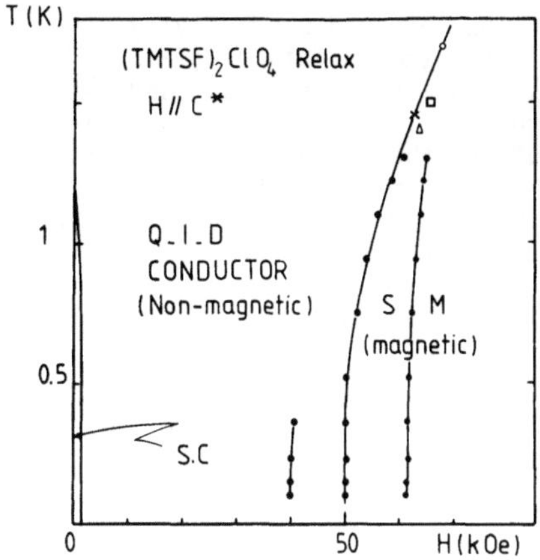

Figure 9. Phase diagram of $(TMTSF)_2ClO_4$ under magnetic field. The low field conducting state is non-magnetic. A sequence of transitions towards magnetic states are observed at high fields

possibility can be ruled out by the restoration of the SDW in $(TMTSF)_2PF_6$ at constant pressure by a large magnetic field [46] or the field-induced SDW state of $(TMTSF)_2ClO_4$. Both experiments leave the interchain coupling unaffected.

To summarize the presentation of the ground-state properties we wish to emphasize that while some properties of the phase transitions (specific heat data) resemble apparently the common behaviour of ordinary conductors, other properties of the superconducting state (large sensitivity to non-magnetic disorder, interplay between superconductivity and magnetism) require the development of new theoretical models especially adapted to organic superconductors.

II.4. The Conducting State

The electron gas of $(TMTSF)_2X$ salts has been studied fairly extensively via optical reflectance, DC conductivity, thermopower and Hall effect. The salient results are that the density of carriers (from Hall constant data at helium temperature) which amounts to 1.4×10^{21} cm^{-3}, i.e. one carrier per $(TMTSF)_2X$ unit, the optical relaxation time, $\tau \sim 3 \times 10^{-15}$s at room temperature, and an effective mass of unity lead to a room temperature conductivity ($\sigma \sim 1000 (\Omega cm)^{-1}$) within a factor 2 from the measured DC conductivity ($\sigma_{DC} \sim 500$-600 $(\Omega cm)^{-1}$ in the $(TMTSF)_2X$ series). However, this close agreement between optical and DC measurements no longer remains good upon cooling below nitrogen temperature (77 K) or so. For example, the optical relaxation time reaches $\sim 10^{-14}$s at low temperature in $(TMTSF)_2PF_6$ (T=25K) or $(TMTSF)_2ClO_4$ with very little temperature dependence [16] whereas the relaxation time derived from DC conductivity data should be larger than 10^{-13}s at 4.2 K to account for the spectacular increase of the DC conductivity which is observed below 10 K (1). At 1.5 K $\sigma_{DC} \sim 3 \times 10^{5} (\Omega cm)^{-1}$ in most conducting states of $(TMTSF)_2X$, a value reaching $10^6 (\Omega cm)^{-1}$ has even been reported in $(TMTSF)_2ClO_4$, [47] Fig.(5). This dramatic increase of the

DC conductivity is by no means reproduced by the infrared conductivity since $\sigma(\omega)$ is only $2500(\Omega cm)^{-1}$ at $\omega \simeq 200\ cm^{-1}$ for the R state of $(TMTSF)_2ClO_4$ at 2 K [48]. After the previous discussion of the electrical conductivity which has been shown to raise some problems in terms of a Drude-like single-particle model, we shall briefly outline the anomalous behaviour of the thermal conduction. The study of the thermal conductivity of $(TMTSF)_2ClO_4$ or $(TMTSF)_2PF_6$ (P=12 kbar) shows for both cases a very striking behaviour at low temperature. Below 60 K the thermal conductivity drops continuously down to 4.2 K, [49,50], Fig.(10). For both samples there is no typical increase of thermal conductivity at low temperature with a pronounced maximum around 10-20 K as it is the case for the lattice thermal conductivity of most pure metals and dielectric solids at about 1/20 of the Debye temperature [51]. This drop of κ is even more surprising if one considers that much heat should be carried by electrons at low temperature since the DC electrical conductivity overcomes $10^5\ (\Omega cm)^{-1}$ at helium temperature. A linear temperature dependence of the electronic thermal conductivity κ_e is expected at low temperature when the mean-free path of electrons is limited by impurities and as long as the collisions are elastic (Wiedemann-Franz law $\kappa_e/\sigma = LT$, where L is the Lorenz number). However, on the one hand within the Wiedeman-Franz formalism a maximum of κ_e should be observed around $\theta/10$, i.e. 20 K at about 30 $mWcm^{-1}K^{-1}$ if the DC conductivity is attributed to single particles with a very large scattering time. On the other hand the Wiedeman-Franz law seems to be correctly followed at room temperature as the ratio κ/σ remains pressure independent in $(TMTSF)_2PF_6$ up to 12 kbar [50]. Transport properties in the conducting state at low temperature are even more anomalous when the magnetic field response is studied. An extremely large, positive, non-saturating and orientation-dependent transverse magnetoresistance is observed at low temperature (T<30 K) in either $(TMTSF)_2ClO_4$, R state [52] or $(TMTSF)_2PF_6$ under pressure [53]. The effect of the magnetic field is largest for $H//c^*$ but practically negligible for H//a. The large magnetoresistance is admittedly hard to understand with the absence of closed orbits of the open Fermi surface. In such a situation the mere existence of any magnetoresistance is already an unexpected phenomenon for systems in which Hall effect experiments have shown that only carriers of one kind (very likely holes) are present in the conducting state. The only way large and non-saturating magnetoresistance in high fields could be expected (assuming a Fermi surface with electron and hole orbits!!) is the geometry where both E and H fields are applied in a plane perpendicular to the a axis [54]. The experimental conditions on $(TMTSF)_2X$ do not fulfill this requirement since the E field is parallel to a. The anomalous magnetoresistance (and the magnetothermopower not discussed in this lecture) have been attributed by Chaikin and co-workers [52] to an electron scattering which is strongly dependent on the location on the Fermi surface. Therefore, under magnetic field electrons at the Fermi surface are swept through regions of the surface where the scattering time could be anomalously short. Hence the average scattering time is decreased and the electrical resistance increased. If the above interpretation of the magnetoresistance is valid, similar arguments must be applied to the low-temperature thermal conductivity which is also remarkably field dependent in $(TMTSF)_2ClO_4$ [49] or $(TMTSF)_2PF_6$ under pressure,[50],Fig.(10). However, following the experimental data showing an anomalously large magnetoresistance and the single-particle variable scattering rate model, this is an increase of the thermal resistance which should be observed under magnetic field (the lattice thermal conduction is not sensitive to the magnetic field but the electronic contribution is suppressed following the electronic conductivity). The experimental data, Figs. (10,11), show exactly the opposite effect of a magnetic field. Below $T \simeq 25$ K, κ is greatly enhanced by the application of a magnetic field. Similar to orientation effects on the electrical conduction, the effect on the thermal conductivity is largest for $H//c^*$ and smallest when H is appro-

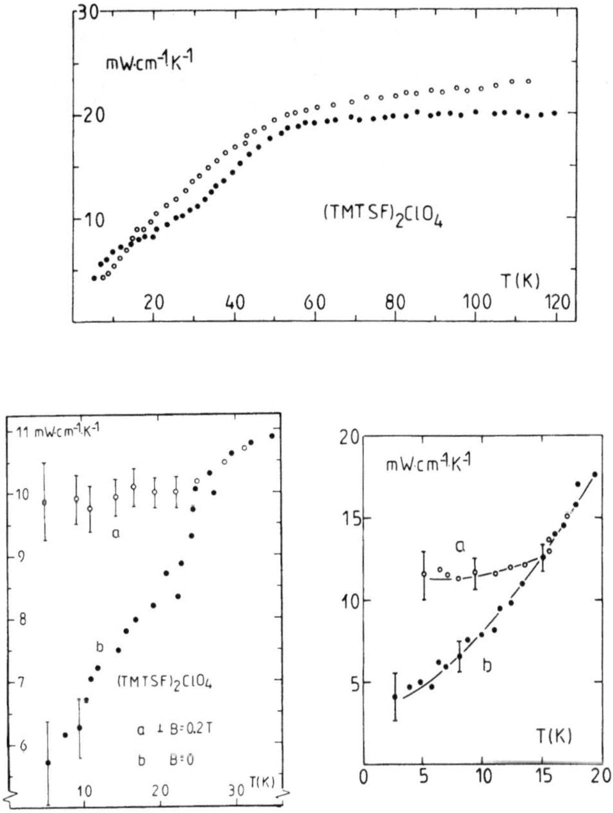

Figure 10. Temperature dependence of the thermal conductivity of $(TMTSF)_2ClO_4$ (R state), (top). Magnetic field dependence (H⊥a) (bottom left) and similar experiment performed in $(TMTSF)_2PF_6$ under 12 kbar (bottom right)

ximately parallel to the a axis. Magnetothermal transport data illustrate that firstly the drop of κ at low temperature must be attributed to a modification of the electronic structure (typically a decrease of $N(E_F)$ at low temperature) which can be removed under magnetic field, and secondly single-particle models of the electronic properties appear to be unable to explain simultaneously electrical and thermal conductance properties at low temperature. It has also been recognized that the Kohler's rule, namely $\Delta R/R_o$ = universal function of (H/R_o), is not obeyed in the R state of $(TMTSF)_2ClO_4$ [55], suggesting a change of mechanism for the conductivity below $\sim$30 K [50]. At low temperature (T<3K) the specific heat of $(TMTSF)_2ClO_4$ is well described by the typical law $C_v = \gamma T + \beta T^3$ where γ and β are related to $N(E_F)$ and the phonon spectrum respectively [32]. However, an important field dependence of $N(E_F)$ is observed at low temperature. For example at T = 0.6 K, $N(E_F)$ doubles in a field of 20 kOe (// c^*) after the suppression of the superconducting state which occurs at $H_{c2}^{c^*}$<1 kOe [56]. The sensitivity of $N(E_F)$ to a magnetic field agrees qualitatively with the field dependence of the thermal conductivity. The depression of the density of states at low temperature occurs in an energy width which can be determined by various spectroscopic studies. Far infrared reflectance of $(TMTSF)_2ClO_4$ (R state) reveals the existence of a gap at 30 cm^{-1} in the frequency dependence of the conductivity [48], (Fig.11). This low-energy gap ($\sim$ 3.8 meV) is most clearly seen at 2 K but remains present although shifting to lower energy value up to 30 K or so. The 30 cm^{-1} absorption threshold is also shifted to a lower frequency with a modest magnetic field of only a few kOe [48,57] . A similar pseudo-gap in N(E) over a width of $\sim$ 3.6 meV around the Fermi energy is revealed by Schottky tunneling of electrons between $(TMTSF)_2X$ and a layer of N-doped evaporated GaSb [58].

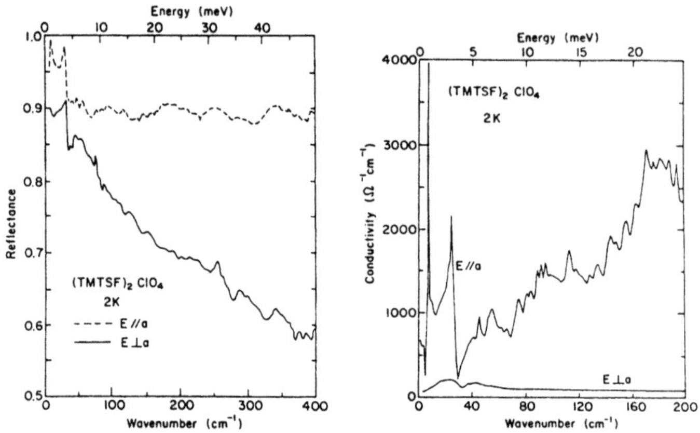

Figure 11. Far infrared reflectance of $(TMTSF)_2ClO_4$ (R state) (left) showing a drop at 30 cm^{-1}. Frequency dependence of the conductivity derived from a Kramers-Kronig analysis between 5 and 400 cm^{-1} (right), (According to reference [48])

III. A New Interpretation

The previous section has shown how our knowledge of electronic properties of metals and phase transitions in usual 3 D conductors is unable to explain a vast amount of important experimental features: interplay between magnetism and superconductivity, sensitivity of the superconducting ground state to non-magnetic defects, anomalous magnetotransport effects, large discrepancy between FIR and DC conductivity data. The Orsay group has proposed that 1-D theory and incipient superconductivity are reasonable interpretations [1,13] . However, how can one reconcile a precursor domain extending up to $\sim$ 30 K with a transition which behaves very much like a mean-field phase transition? The small critical width of the superconducting transition detected by specific heat ($\frac{\Delta T}{T_c} < 10\%$), Fig.6, can by no means support the existence of fluctuations up to $T \sim 10\ T_c$ in the framework of an anisotropic 3-D theory. Therefore the point of view of phase transitions in Q-1-D conductors has been suggested [13]. In such a case, for a two degrees of freedom order parameter (amplitude and phase) some decoupling of the two components occurs at $T \gg T_c$. The spatial correlation function of the order parameter is given by:

$$< \Delta (x)\, \Delta (0) > = |\Delta|^2 \exp - x/\xi(T) \qquad (9)$$

where $\xi(T) = \alpha/\kappa T$. Consequently, at T>0, the short-range order which arises with a coherence length $\xi(T)$ digs a pseudo-gap of width $2\Delta \sim \alpha$ at the Fermi level. The amplitude of the order parameter reaches a significant value ($\sim \alpha$) already below the 1-D mean-field temperature, $T_c^{MF} \sim \alpha$, whereas the phase remains free to take any value until T_c is reached. At T_c inter-chain locking of the phases occurs. If the zero point motion of the phase (quantum fluctuations) is taken into account a true gap develops below T_c , but its value 2δ is much smaller than the pseudo-gap 2Δ . For strong quantum fluctuations Schulz and Bourbonnais [59] have derived 2δ (T=0) = 3.5 kT_c; i.e. the mean-field behaviour is recovered for the phase-locking transition (see Sect II. 2 and Fig. 6). Moreover, with reasonable values of the intrachain coupling ($g_1-2g_2 \sim 0.6$) and band structure anisotropy ($t_{//}/t_\perp \sim 10$) the same authors have calculated that the superconducting precursor region will extend up to about 30 K when T_c=1.2K.

In conclusion organic conductors are newcomers among conductors. Given the very special nature of the lattice unit (molecules instead of atoms) it is not surprising that they exhibit properties different from those of metallic conductors. We have tried to show how new models are required for their understanding. Our main conclusion is that the Drude-like, Boltzmann formalism is inadequate especially below 40 K and several features of a Q-1-D electron gas are observed in this temperature regime. More thorough discussions of the fluctuation model are presented elsewhere [60,63] .

References

1) D. Jérome, A. Mazaud, M. Ribault and K. Bechgaard, J. Physique Lett. 41, L-95 (1980).

2) J. Ferraris, D.O. Cowan, V. Walatka and J.H. Perlstein, J. Am. Chem. Soc. 95, 948 (1973).

3) L.B. Coleman, M.J. Cohen, D.J. Sandman, F.G. Yamagishi, A.F. Garito, A.J. Heeger, Solid State Comm. 12, 1125 (1973).

4) F. Denoyer, R. Comès, A.F. Garito, A. Heeger, Phys. Rev. Lett. 35, 445 (1975).

5) D. Jérome and H.J. Schulz, Adv. in Phys. 31, 299 (1982).

6) K. Bechgaard, C.S. Jacobsen, K. Mortensen, H.J. Pedersen, N. Thorup, Solid State Comm. 33, 1119 (1980).

7) S.S.P. Parkin, E.M. Engler, R.R. Schumaker, R. Lagier, V.Y. Lee, J.C. Scott R.L. Greene, Phys. Rev. Lett. 50, 270 (1983).

8) S.S.P. Parkin, M. Ribault, D. Jérome and K. Bechgaard, J. Phys. C 14, 5305 (1981).

9) J. Ziman, Principles of the theory of solids, Cambridge University Press (1965).

10) G. Soda, D. Jérome, M. Weger, J. Alizon, J. Gallice, H. Robert, J.M. Fabre and L. Giral, J. Physique 38, 931 (1977).

11) R.L. Greene, P. Haen, S.Z. Huang, E.M. Engler, M.Y. Choi, P.M. Chaikin, Mol. Cryst. Liq. Cryst. 79, 183 (1982).

12) C.S. Jacobsen, K. Mortensen, M. Weger, K. Bechgaard, Solid State Comm. 38, 423 (1981).

13) H.J. Schulz, D. Jérome, A. Mazaud, M. Ribault, K. Bechgaard, J. Physique, 42, 991 (1981).

14) P.M. Grant, Phys. Rev. B26, 6888 (1982).

15) K. Kikuchi, I. Ikemoto, K. Yakushi, H. Kuroda and K. Kobayashi, Solid State Comm. 42, 433 (1982).

16) C.S. Jacobsen, D.B. Tanner, K. Bechgaard, Phys. Rev. Lett. 46, 1142 (1981).

17) K. Yamaji, J. Phys. Soc. Japan 51, 2787 (1982).

18) J.F. Kwak, Phys. Rev. B26, 4789 (1982).

19) L.J. Azevedo, J.E. Schirber and J.C. Scott, Phys. Rev. Lett. 49, 123 (1982).

20) K. Mortensen, Y. Tomkiewicz, K. Bechgaard, Phys. Rev. B25, 3319 (1982).

21) A. Andrieux, D. Jérome and K. Bechgaard, J. Physique Lett. 42, L-871 (1981).

22) J.C. Scott, H.J. Pedersen, K. Bechgaard, Phys. Rev. B24, 475 (1981).

23) J.B. Torrance, H.J. Pedersen, K. Bechgaard, Phys. Rev. Lett. 49, 881 (1982).

24) J.P. Pouget, R. Moret, R. Comès, K. Bechgaard, J. Physique Lett. 42, 543 (1981).

25) R. Moret, J.P. Pouget, R. Comès, K. Bechgaard, Proceedings of Int. Conf. (Les Arcs, 1982), J. Physique (1983).

26) J.P. Pouget, G. Shirane, K. Bechgaard, J.M. Fabre, Phys. Rev. B27, 5203 (1983).

27) T. Takahashi, D. Jérome, K. Bechgaard, J. Physique Lett. 43, L-565 (1982).

28) S. Tomic, D. Jérome, P. Monod, K. Bechgaard, J. Physique Lett. 43, L-839 (1982).

29) P. Garoche, R. Brusetti and K. Bechgaard, Phys. Rev. Lett. 49, 1346 (1982).

30) M. Ribault, Proceedings Int. Conf. Low Dim. Synth. Metals, Les Arcs 1982 J. Physique (1983).

31) D. Mailly, M. Ribault and K. Bechgaard ; J.M. Fabre and L. Giral, J. Physique Lett. 43, L-711 (1982).

32) P. Garoche, R. Brusetti, D. Jérome, K. Bechgaard, J. Physique Lett. 43, L-147 (1982).

33) S. Tomic, D. Jérome, D. Mailly and K. Bechgaard. Proceedings Int.Conf. (Les Arcs 1982) J.Physique 1983.

34) C. Coulon, P. Delhaes, J. Amiell, J.P. Manceau, J.M. Fabre, L. Giral, J. Physique 43, 1721 (1982).

35) P.G. de Gennes, Superconductivity in Metals and Alloys, W.A. Benjamin, New-York (1966).

36) R. Brusetti, M. Ribault, D. Jérome, K. Bechgaard, J. Physique, 43, 801 (1982).

37) R.L. Greene and E.M. Engler, Phys. Rev. Lett. 45, 1587 (1980).

38) H.J. Schulz, D. Jérome, M. Ribault, A. Mazaud and K. Bechgaard, J. Physique Lett. 42, L-51 (1981).

39) K. Andres, F. Wudl, D.B. Mc Whan, G.A. Thomas, D. Nalewajek and A.L. Stevens, Phys. Rev. Lett. 45, 1449 (1980).

40) D. Mailly and M. Ribault, Private Communication.

41) M. Ribault, D. Jérome, C. Weyl and K. Bechgaard, J. Physique Lett., to be published.

42) J.F. Kwak, J.E. Schirber, R.L. Greene and E.M. Engler, Phys. Rev. Lett. 46, 1296 (1981).

43) K. Kajimura, H. Tokumoto, M. Tokumoto, K. Murata, T. Ukachi, H. Anzai, T. Ishiguro, G. Saito, Solid State Comm. 44, 1573 (1982).

44) K. Machida, J. Phys. Soc. Japan, 50, 2195 (1981).

45) K. Machida and T. Matsubara, J. Phys. Soc. Japan, 50, 3231 (1981).

46) T. Takahashi, D. Jérome, K. Bechgaard, Proceedings of Int. Conf. (Les Arcs 1982), J. Physique (1983).

47) K. Bechgaard, K. Carneiro, M. Olsen, F. Rasmussen, C.S. Jacobsen, Phys. Rev. Lett. 46, 852 (1981).

48) H.K. Ng, T. Timusk and K. Bechgaard, Proceedings Int. Conf. (Les Arcs 1982), J. Physique (1983).

49) D. Djurek, M. Prester, D. Jérome, K. Bechgaard, J. Phys. C Solid State 15, L-669 (1982).

50) D. Djurek, D. Jérome, K. Bechgaard, J. Phys. C, submitted.

51) R. Berman, Thermal Conduction in Solids, Oxford University Press (1978).

52) P.M. Chaikin, M.Y. Choi, R.L. Greene, Proceedings Int. Conf. (Les Arcs 1982), J. Physique 1983.

53) D. Jérome, Chemica Scripta 17, 13 (1981).

54) A.A. Abrikosov, Solid State Physics, Suppl. 12, Academic Press, New-York (1972).

55) L. Forro, K. Biljakovic and J.R. Cooper, preprint (1983).

56) R. Brusetti, P. Garoche and K. Bechgaard, J. Phys. C.16, 3535, (1983).

57) H.K. Ng, T. Timusk, J.M. Delrieu, D. Jérome, K. Bechgaard, J.M. Fabre, J. Physique Lett. 43, L-513 (1982).

58) C. More, G. Roger, J.P. Sorbier, D. Jérome, M. Ribault and K. Bechgaard J. Physique Lett. 42, L-313 (1981).

59) H.J. Schulz and C. Bourbonnais, Phys. Rev. B27, 5856 (1983).

60) H.J. Schulz, D. Jérome, K. Bechgaard, Phys. Rev. B, submitted.

61) H.J. Schulz, Proceedings Int. Conf. (Les Arcs 1982), J. Physique (1983).

62) C. Bourbonnais, Proceedings Int. Conf. (Les Arcs 1982), J. Physique (1983).

63) D. Jérome, Proceedings of NATO Summer School Cambridge 1983, Plenum Press.

Part IX

Electronic Excitations in Photosynthetic Systems

Optical Investigations of Photosynthetic Systems

G. Paillotin and J. Breton

CEA-CEN/Saclay, Departement de Biologie, F-91191 Gif-sur-Yvette, France

1 - Introduction

The biological properties of plants and photosynthetic bacteria are intimately related to their optical properties. As a matter of fact, the photosynthetic organisms are able to convert light energy into stable chemical energy. This conversion comprises four main steps:

i) *Photon Absorption*: it is achieved by a large aggregate of pigment molecules, essentially chlorophyll (Chl) molecules (chlorophyll a and b (Chla, Chlb) for plants, bacteriochlorophyll (BChl) for photosynthetic bacteria). This Chl aggregate is named the "light harvesting antenna". When absorbed, a photon creates within the antenna a nearly localized exciton. Because of excitation energy transfer between Chl molecules, this exciton migrates, practically at random, in the antenna [1].

ii) *Electron-hole Pair Formation*: it occurs at the level of "molecular photocells" named "reaction centers" (RCs). Each RC contains an exciton trap (P). P is a Chl molecule or a dimer of Chl molecules. It can be excited by light absorption or by exciton transfer from the antenna. This second event is the most probable since the concentration of P is very small as compared to the total Chl concentration (less than 1%). Following its electronic excitation P acts as an electron donor to an acceptor A included in the RC:

$$PA + \text{exciton} \longrightarrow P^{*} A \longrightarrow P^{+} A^{-} . \qquad (1)$$

Then the positive and negative charges migrate to accessory electron donors and acceptors in order to stabilize the electron-hole pair [2].

iii) *Electron Transport and Storage*: Once created and stabilized the electric charges are transported and accumulated at the sites where they can be used for chemical storage of energy.

The efficiency of these reactions comes from the organization of the photochemical apparatus. The photon capture, the charge separation and the main steps of charge transport occur within a lipid and protein membrane (photosynthetic membrane) of low dielectric constant which has a thickness around 50 [Å]. The exciton and electron carrier molecules are fixed on polypeptides (intrinsic proteins) which determine at small distances (20 to 50 [Å]) a rigorous local order. It is a more flexible organization which appears at larger distances. It is governed by the spatial distribution of the intrinsic proteins within the photosynthetic membrane. This distribution is submitted to changes which are important for the biological regulation of photosynthesis [3]. The structure of the photosynthetic membrane is neither random, nor crystalline.

2 - Generalities about the Spectroscopic Properties of the Photosynthetic Membrane

The photosynthetic membrane contains a large quantity of absorbing molecules in the visible part of the spectrum. Thus optical investigations are very useful for the experimental study of both the structural and biological properties of photosynthetic organisms. Chl molecules are the major components of the photosynthetic organisms. The visible absorption spectrum of Chl reveals the contribution of four electronic transitions. The intense red absorption band corresponds to the transition to the first singlet excited state (named Qy transition). A second electronic transition appears around 575 nm for Chla (Qx transition). The directions of polarization of these two transitions are nearly perpendicular. The intense absorption band located in the blue region corresponds to at least two other electronic transitions.

Each purely electronic transition is coupled with the nuclei motion. For such a large molecule as Chl this motion is practically "classical". It explains the nearly Gaussian shape of the Chl absorption bands and justifies the validity of the HEMENGER formula when considering Chl-Chl interactions [4]. It remains that the photosynthetic apparatus is very complex. The absorbing molecules are fixed in intrinsic proteins embedded in a lipidic layer, itself surrounded by a liquid phase containing a large concentration of soluble proteins. Thus, from the experimental point of view, one must use several methods based on different means of optical selection:

i) *Resonance Selection*: At suitable wavelengths, light is essentially absorbed by Chl molecules. Of course it is the basis of any optical investigation but it gives interesting possibilities to the resonance Raman spectroscopy [5].

ii) *Geometrical Selection*: It is possible to orient a solution of photosynthetic membranes (by means of a magnetic or electric field, or by different mechanical ways [6]). Then, because of the polarization of the electronic transitions, it is possible to select with linearly polarized light the properly oriented molecules. This is the basis of the linear dichroism measurements.

iii) *Double Selection*: Fluorescence emission and absorption changes make possible double resonance experiments. An exciting light beam is used for a resonance or a geometrical selection and analyzing light probes the sample in a selective way (polarization of fluorescence, photoselection). This kind of experiment can be extrapolated to EPR measurements (magnetophotoselection). If fluorescence is a common process for light absorbing molecules, a more interesting absorption change process is induced by the creation of electrical charges at the level of the RCs. For instance,the formation of P^+ (see equation 1) induces a bleaching of the P absorption spectrum [2] .

3 - Organization of the Photosynthetic Pigments

3-1) *The Pigment-Protein Complexes*

a) Composition

Over the past ten years it has been recognized that, in vivo, the photosynthetic pigments (Chl and the accessory pigments as carotenoids) were specifically bound to hydrophobic proteins. This conclusion came from the observation that upon dissolution of the photosynthetic membrane with detergents, most of the pigments remain physically -but not covalently- bound to proteins [7]. From a given photosynthetic membrane several of these pigment-protein complexes can usually be isolated and, depending upon their functional roles in the native membrane, they are classified as either antenna or RC complexes. Many of these complexes in which the native state of the pigments seems to be preserved have been obtained in a purified form. They have been extensively characterized in terms of their composition both in polypeptides and in pigments. Although always greater than one, the number of Chl molecules per complex varies largely depending on the biological sample. Usually membranes from photosynthetic bacteria yield Chl-protein complexes which contain a rather small number of pigment molecules: six or seven for RCs, two to four in the antenna although large aggregates of the latter are usually obtained. In contrast the complexes which can be obtained from plants and algae are characterized by a large number (20 to 100) of Chl molecules per complex.

b) The Pigment Binding Sites

Optical spectroscopy has proven to be of great value in the investigation of the nature and specificity of the pigment binding sites. Low temperature absorption measurements have allowed the resolution of several forms of Chla in vivo corresponding to different local environments of the same chemical species [8].

The orientation of the various pigments has been extensively investigated by using absorption of linearly polarized light on oriented membranes and isolated complexes [9]. The specific orientation which has been detected for each of the various pigment forms is taken to indicate that there is a well-defined anchoring of these molecules within their protein hosts.

Most of our knowledge on the nature of the interactions between Chl molecules and the host protein derives from resonance Raman spectroscopy [5]. This technique is especially well adapted to probe the network of bonding interactions holding the various Chl molecules to their biological sites. In resonance Raman spectra of photosynthetic material from plants and algae, at least five different kinds of interaction are distinguished among the ketone carbonyls of Chla and probably two for the ketone carbonyls of Chlb while at least two are observed for the formyl carbonyls of Chlb. Furthermore most of the Chl molecules have their Mg central atom five-coordinated while the remaining present a six-coordinated Mg atom. Although the frequencies of some components of the cluster of the carbonyl stretching modes could match with frequencies observed for oligomers of Chla and b, detailed comparison between the other regions of the in vivo spectrum with the spectra of monomers, oligomers and hydrated polymers led LUTZ [5] to conclude that no sizable amount of antenna chlorophylls could be present as oligomer either bridged or unbridged by water. A similar picture has also emerged for BChla in vivo.

Furthermore a detailed picture of a BChla protein complex has been recently obtained by X-ray crystallography [10]. Although this complex, extracted from the green bacteria Prosthecochloris aestuarii, is quite atypical in being a water-soluble protein, it constitutes a reference model useful for interpreting the Raman spectra of the other Chl-protein complexes. More precisely the pattern observed in the Raman spectrum of this complex for the various carbonyl and Mg modes can be accounted for by the specific interactions between the peptidic residues of the protein backbone and the BChla molecules which are deduced from the X-ray data.

c) Antenna Complexes

In the previous section we have indicated that in the photosynthetic membrane the antenna Chl and BChl are mostly monomeric and are locked into position by interacting, at least by their conjugated carbonyls and their Mg, with specific protein groups. Furthermore, the striking similarity between the resonance frequencies of the carbonyls which can be noticed for the antenna pigments in a large variety of organisms (plants, algae, bacteria) suggests that the binding sites of the pigments are similar in all these cases. These observations further indicate that the possible hydrogen-bond donor sites of the proteins are not randomly involved in the pigment binding, but that only certain specific ones actually assume this function [5]. Finally the existence of discrete

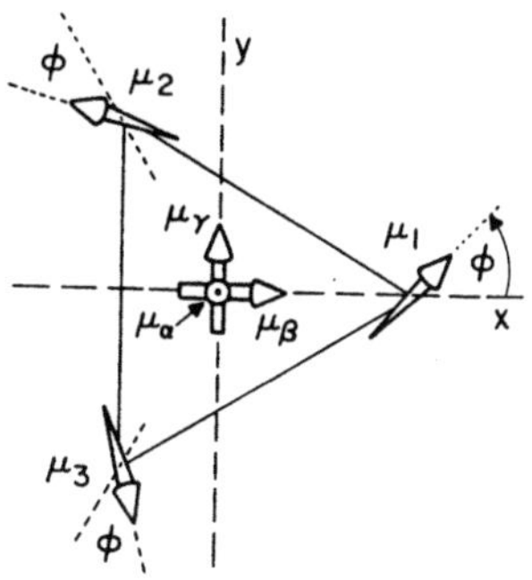

Fig. 1: VAN METTER's postulated arrangement of chlorophyll b transition moments having C_3 symmetry. Moments 1 through 3 are those of individual chromophores, tilted away from the z direction by an angle θ (not shown). Moments α, β and γ are those of the exciton eigenstates, α being a state not necessarily degenerate with the degenerate states β and γ. μ_α is parallel to z. Reproduced from SHEPAŃSKI and KNOX (1981) Israel J. Chem. 21, 325

spectral forms in the in vivo low-temperature absorption spectra of Chla [8] as well as the assignement of a specific orientation for the transition moments associated with these spectroscopic forms [11], [12] also suggest that these pigments are anchored to the protein at a limited number of well-defined and ubiquitous binding sites.

Although it seems clearly established that most of the antenna Chl molecules cannot be considered as dimers or oligomers in a chemical sense (large interaction in the ground state) this obviously does not mean that they are not coupled in the singlet excited state. The in vivo Chla concentration is about 0.1 M and in the antenna complex from P. aestuarii the BChl concentration reaches 1 M. Under such conditions electrical interaction between neighboring molecules can become of importance and eventually lead to an excitonic coupling of these molecules.

Even if great care must be exercised in interpreting the circular dichroism (CD) spectra of pigments in vivo because of Chl-protein interactions, there is some evidence in favour of such excitonic coupling. VAN METTER [13] described in such a way the CD spectra of the Chl-protein complex CP 2. This complex, obtained with the anionic detergent sodium dodecyl sulfate, represents a solubilized form of the major antenna complex of plants. It contains three Chla and three Chlb per complex. By analyzing the absorption, the CD and the fluorescence polarization properties of this complex it has been concluded that a coupling exists between the three Chlb but not between the three Chla molecules. A model with C_3 symmetry was introduced [13] to describe the arrangement of the Qy transition moments of the three interacting molecules (fig. 1). There are three collective excited states: two (β,γ) absorb at 652 nm, well separated from the state (α) absorbing at 665 nm. The parameters calculated from a simultaneous deconvolution of the absorption and CD spectra led these authors to the following model: 88% of the oscillator strength resides in the 652 transition and 12% in the 665 nm transition, φ is comprised between 60 and 90° and the tilt angle θ of the transition moments of the monomers out of the plane of the centers of the pigments (fig. 1) is estimated at 70°.

While investigating the low-temperature linear dichroism spectra of intact membranes and isolated Chl-protein complexes from plants we have noticed several unusual features in the Chlb absorption region which seem to be relevant to the model described above. The antenna complex which we have mostly investigated is isolated by using the non-ionic detergent Triton X-100. It is named the light-harvesting complex (LHC) and represents a more native state of the antenna pigments than CP 2. The low-temperature linear dichroism spectrum of LHC oriented by the polyacrylamide gel technique is presented in fig. 2a together with the corresponding absorption spectrum and has been described in detail [11].

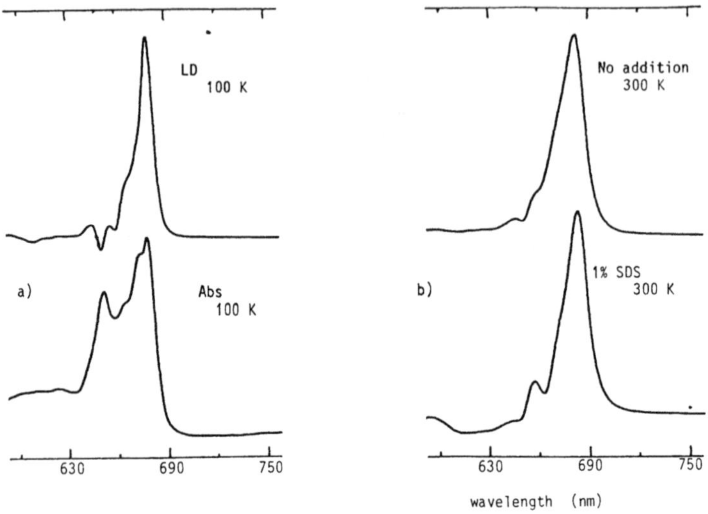

Fig. 2a: Linear dichroism (top) and Absorption (bottom) spectra of the light-harvesting complex from green plants. Temperature is 100 K

Fig. 2b: Linear dichroism spectra of the light-harvesting complex either in the native state (top) or after treatment with the detergent Sodium dodecyl sulfate. Temperature is 300 K

In both fig. 2a and 2b the vertical scale is in arbitrary absorption units.

Focussing only on the Chlb signals, a complex set of linear dichroism bands is observed between 640 and 652 nm. Furthermore a shoulder at 664 nm does have an associated linear dichroism signal. All these features can also be recognized in the linear dichroism spectrum of intact membranes and have been considered as a "fingerprint" for LHC *in vivo* [11, 12]. The extreme sharpness of the negative component at 648 nm for which we have estimated a full width at half maximum of 5 ± 1 nm indicates that it might represent the sum of two close absorption bands exhibiting dichroisms of opposite signs. We suggest that these bands, together with the shoulder at 664 nm, can be related to the excitonic components described by SHEPANSKI and KNOX [14]. Several observations, which will be detailed below, lead us to propose that in the isolated LHC as well as *in vivo* the two excitonic components (β, γ) are not totally degenerate as they are in CP 2 but absorb at slightly different wavelengths. A splitting of the 650 nm band into its two components has been reported in the CD spectra of isolated LHC [15,16] and chloroplasts [17,18]. Furthermore the evolution of the linear dichroism spectrum of LHC upon its transformation into a CP 2-like complex by gradual addition of sodium dodecyl sulfate has been investigated. The changes observed in the 640-665 nm region during this treatment are taken to indicate a redistribution of intensity between the three excitonic components leading to the gradual evolution of the 648 nm S-shaped signal into a single positive band centered at 652 nm at ambiant temperature (fig. 2b). This last feature is also present in the linear dichroism spectrum of CP 2 prepared directly (J. BRETON, unpublished results). These observations point towards a coalescence of the β and γ excitonic components of Chlb when going from a native state to the CP 2 complex. In addition we note that the existence of a Chlb population which is not present *in vivo* but is unique to the CP 2 complex has been reported [5]. By analyzing its resonance Raman spectra it has been concluded that this population has lost the native interaction between its formyl group and an unknown binding site as well as its sixth external ligand on the Mg atom. Taken together, all these observations indicate that the three interacting Chlb molecules present either *in vivo* or in the isolated LHC can be characterized by two nearly degenerate excited levels (absorption around 650 nm) and a third level at 665 nm. Sodium dodecyl sulfate treatment triggers a change in the interactions between the Chlb molecules which induces a coalescence of the two quasi-degenerate levels.

d) Reaction Centers

So far only the photosynthetic bacteria have yielded small functional RC particles free of antenna pigments. In the case of plants a large fraction (50 to 100 Chl molecules) of specific antenna Chl remains bound to the isolated particles which contain the RC. The high ratio of antenna to photoactive pigments in these last systems precludes easy investigation of their absorption properties. As isolated from purple bacteria the RC contains usually four bacteriochlorophyll (BChl) and two bacteriopheophytin (BPheo) molecules, one carotenoid and one quinone complexed with an iron atom. These molecules are bound to a 75 KDaltons, hydrophobic protein constituted by three distinct peptides. The nature of this binding has been investigated by resonance Raman spectroscopy. For the four BChl molecules it has been concluded that the type of bonding interactions were very similar in RCs and in antenna complexes [5] in spite of the fact that the RCs and antenna pigments exhibit very different kinds of coupling in the excited states. Very likely the geometrical arrangement of the pigments in both the antenna and the RCs is mainly determined by the pigment-protein interactions and not by the pigment-pigment ones. In purple bacteria the trap or the primary donor P (see introduction) consists of a dimer of BChl molecules named P_{870} according to its spectrum. The Raman spectroscopy indicates that the "special pair" model in which the two interacting BChl of P_{870} are linked by small bifunctional ligands has no real grounds. For the two BPheo molecules the Raman scattering reveals an unusual environment for these pigments which might indicate either some excitonic coupling of the two molecules or a perturbation by very close π conjugated systems.

Although the RC is the object of considerable interest due to its important functional role, the precise geometry of the pigments (distances, relative orientations) as well as the nature of the electronic interactions between them is still debated. One of the techniques which has been successfully applied to gain information on the relative geometry between the pigments is referred to as photoselection. In this elegant method outlined by ALBRECHT [19], an excitation beam of linearly polarized light is used to create an anisotropy in a randomly oriented sample. This anisotropy, which corresponds to a known spatial distribution of excited molecules, can be probed by means of different physical properties: light emission (polarization of fluorescence), absorption changes (photodichroism) or EPR (magnetophotoselection). These techniques can give precise information on the relative positions of the transition moments involved in the absorption and the detection provided that several conditions are fulfilled: the rotational motion of the object containing the pigments must be much slower than the relaxation of the photo induced excited state, the angle between the two investigated transition moments should not vary and finally the intensity of the exciting beam must be kept far below the level of saturation of the observed phenomenon.

By using RC prepared from the bacteria Rhodopseudomonas sphaeroides and immobilized in a glycerol glass at 150 [K], VERMEGLIO et al. [20] have investigated the polarization of the photobleaching of the Qy transition moment of the primary donor P_{870} upon excitation at each wavelength in the

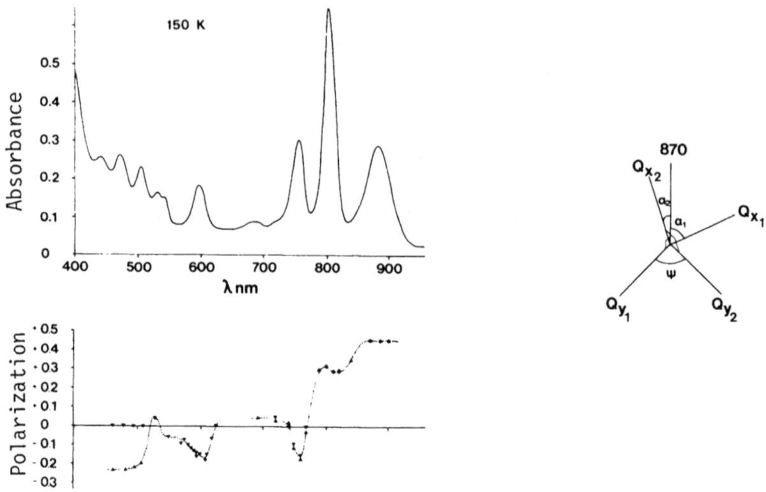

Fig. 3: *Left*: Absorption spectrum (top) of reaction centers from the photosynthetic bacteria Rhodopseudomonas sphaeroides and polarization spectrum (bottom) of the bleaching observed at 870 nm. The value of the polarization can vary between + 0.5 for a transition moment parallel to the 870 nm transition and - 0.33 for a transition perpendicular to the 870 nm transition. Temperature is 150 K. *Right*: Schematic representation of the arrangement of some of the transition moments within the reaction center. 870 represents the long-wavelength transition moment of the primary donor. Qx_1, Qy_1 and Qx_2, Qy_2 are the transition moments of the pheophytin molecules absorbing at 546 and 530 nm respectively. $\Psi = 55°$ or $125°$. With the hypothesis that the two pheophytins are not in excitonic interaction, $\alpha_1 \cong 60°$ and $\alpha_2 \cong 50°$

absorption spectrum. Let us recall that this bleaching corresponds to the oxidation of P_{870} according to reaction 1 (see introduction). Figure 3 represents the result of such experiments together with the absorption spectrum of the RCs. Upon excitation and detection within the 870 nm band a constant polarization value $p = +0.45$ is observed demonstrating that this absorption band is due to either a single transition or to two parallel transitions. When exciting at 530 and 546 nm it is possible to discriminate between the Qx transitions of each of the two BPheo molecules. Upon excitation at 760 nm, where the two BPheo absorb, a value of 65 to 70° is calculated for the average angle of the two corresponding transition moments with the 870 nm oscillator. Additional information may also be obtained upon excitation within a single transition absorption band and observation of light-induced absorption changes -other than the bleaching of P 870- at other wavelengths. This is the case, for example, upon excitation of the two spectrally resolved Qx transitions of the BPheo molecules and observation of the shift of their absorption spectra induced by the appearance of a positive charge on P 870 (see reaction 1). From such experiments VERMEGLIO et al. [20] were able to propose a model for the arrangement of the four transition moments of the two BPheo molecules relative to the 870 nm oscillator. This model is presented in fig. 3.

Although such experiments are potentially very informative there are several points which lead to ambiguities in their interpretation. Among these the problem of the significance of the absorption in terms of the electronic interaction between the molecules is probably the most important. This problem will be approached in a later section.

3-2) *Supramolecular Organization of the Complexes in the Photosynthetic Membrane*

As isolated with detergents the pigment-protein complexes can be oriented by mechanical or electrical methods [9,11]. By comparing the orientation of the pigments both in the intact membrane and in the isolated complexes it is possible to determine the orientation of the complexes themselves with respect to the membrane [12]. Furthermore the orientation of the pigment molecules with respect to the plane of the membrane has been investigated. So far it has always been observed that the antenna Chl molecules absorbing at the long wavelength side of the spectrum have their Qy moment practically in the plane of the membrane. A similar situation has also been found for the Qy moments of all the primary donors both in plants and in bacteria [9]. This last information is obtained by measuring, on oriented membranes, the dichroism of absorption changes linked to the photooxidation of the primary donor.

Although the orientation of the Chl molecular plane cannot always be unambiguously determined there seems to be a general tendency for this plane not to lie parallel to the membrane plane. There are however some exceptions like one of the three BChla molecules belonging to an important class of antenna complexes in bacteria [21]. In the case of strongly interacting molecules, the orientation of the monomers cannot be easily determined and only the orientation of transition moments of the excitonic components can be obtained. For example, in the case of LHC, the two excitonic oscillators absorbing at 665 and 652 nm are parallel to the membrane, and the oscillator absorbing at 648 nm is perpendicular to the membrane plane [11,12].

Another aspect of the organization of the pigments in the photosynthetic membrane is linked to the type of interaction which takes place between adjacent complexes. These interactions may play an important role in the processes of energy transfer (interactions between antenna complexes), of energy trapping (interactions between antenna and RC) and of charge transfer (interactions between charge carriers). These interactions might also be involved in the regulation of the energy distribution between different regions of the photosynthetic apparatus [3]. A specific association between one class of bacterial antenna complex and the RC has been characterized. These antenna are presently thought to form a small core of pigments surrounding each RC. It is reasonable to assume that in green plants such a core also exists and is formed by the set of specific Chla molecules found in close association with the RCs. In both green plants and bacteria another pool of antenna (the light harvesting complexes) functions to connect these cores associated with each RC.

Although this last field of research is still in an early stage of development it seems to offer a great deal of potential. Very likely new spectroscopic tools like exciton annihilation processes, phase fluorimetry and time-dependent fluorescence depolarization measurements will help in gaining a better understanding of the supramolecular organization of the photosynthetic membrane.

4 - Charge Separation Process

In addition to structural information, optical characterization of photosynthetic membranes also provides information on the functional properties of these systems. Moreover these two kinds of studies are complementary. Of particular interest is the charge separation which occurs at the level of the RCs.

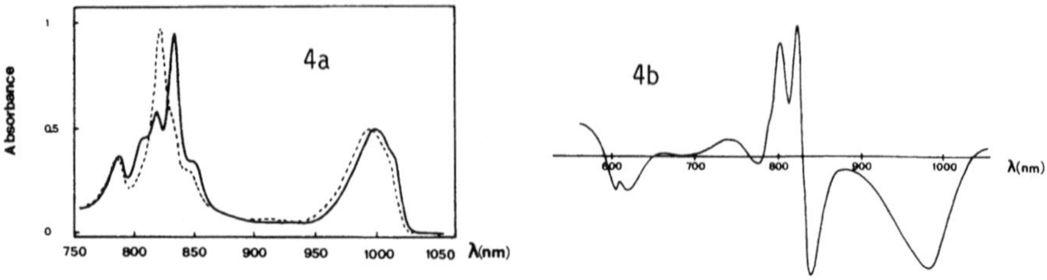

Fig. 4a: Absorption spectra of Rhodopseudomonas viridis, recorded at 4.2 K, for the two following states: (—) state (P), (---) state (I^-)
4b: Difference spectra of the two states (P^+) and (P) recorded at 160 K

Although this process is intensively studied in plants [2], photosynthetic bacteria offer the simplest model for optical investigations. This comes from the fact that the RCs of bacteria are easily solubilized and also because the various pigment molecules in the RCs do not absorb all at the same wavelength.

As indicated previously, the bacterial RC contains four BChl and two BPheo molecules. They serve as a "local" antenna which collects the excitons coming from the large antenna surrounding each RC. But these molecules also play important role in the electron-hole pair formation. The trap for excitons constituted by the primary donor P consists of a dimer of BChl molecules: |BChl-BChl|. Following its excitation this dimer can give an electron to a primary electron acceptor A which is again a BChl molecule. This primary electron transfer from A to a secondary electron acceptor I is performed in about 4 psec [23]. This acceptor is one of the two BPheo molecules included in the RC. Then the electron is transferred to a quinone molecule Q in about 200 psec. In the state $P^+ Q^-$ the electron-hole pair is stabilized. All the reactions which permit the charge separation may be summarized in the following scheme:

$$\text{exciton} + PAIQ \rightarrow P^*AIQ \xrightarrow{<1\ \text{psec}} P^+A^-IQ \xrightarrow{4\ \text{psec}} P^+AI^-Q \longrightarrow P^+AIQ^- \longrightarrow PAIQ^- . \qquad (2)$$

In the last step of the scheme (2) the positive charge located on P is transferred to a cytochrome.

On the other hand, when Q is previously reduced the electron transfer process is limited to I:

$$\text{exciton} + PAIQ^- \rightleftarrows P^*AIQ^- \rightleftarrows P^+A^-IQ^- \rightleftarrows P^+AI^-Q^- \rightleftarrows PAI^-Q^- . \qquad (3)$$

As indicated in the series of reactions (3) the back reactions play in this case an important part and induce a fast luminescence emission phenomenon.

All the states of the RC which appear in reactions (2) and (3) have optical properties which can be used to obtain information on the charge separation process. Only three of these states involving P, A and I are stable or nearly stable: PAI, P^+AI and PAI^-. For simplicity these states are respectively named (P), (P^+) and (I^-). Figure 4 represents the main features of the spectroscopic properties of these three states. These data were obtained by VERMEGLIO and PAILLOTIN [24] with R. viridis. Although the photosynthetic reactions of these bacteria are quite similar to those of R. sphaeroides, the spectrum of its RC is shifted towards the long wavelengths. For example, the absorption band corresponding to the transition from the ground state to the first singlet excited state of P is located around 970 nm at room temperature when it appears around 870 nm for R. sphaeroides. When studying the properties of the RCs it is usually of interest to investigate the RCs extracted from R. viridis as well as R. sphaeroides because, although their various chromophores have different characteristics, both undergo the same reactions. Beside the 970 nm absorption band the primary donor P exhibits a second absorption band around 850 nm. From photoselection experiments [24] it was shown that the oscillator corresponding to this transition is oriented perpendicularly to the 970 nm absorbing oscillator as it could be predicted for a dimer in the framework of the DAVYDOV theory (see also [25]).

Upon oxidation of P one observes the bleaching of both the 970 and 870 nm absorption bands (fig. 4). Correlatively the absorption bands of the two other BChl and the two BPheo molecules are shifted because of the electrical effect of the positive charge located on P. Furthermore a new absorption band appears around 808 nm. It corresponds to the absorption of the BChl molecule of P^+ which does not carry the positive charge. In first approximation these bleachings, shifts and absorption increases give a rather good description of the difference spectrum between the states (P^+) and (P) (fig. 4). On the other hand, in the state (I^-) one observes, as expected, the bleaching around 810 nm of the absorption band of the BPheo molecule which carries the negative charge. As previously observed in the (P^+) state the negative charge induces a shift of the absorption bands of the other components of the RC.

Nevertheless this first approximation does not solve some important problems: if the two excited states of the dimer P absorb at 970 and 850 nm it is surprising, according to the DAVYDOV theory, to observe the absorption of the monomeric form of the BChl forming P at 808 nm. At the same time it is difficult to explain why the bandwidth of the 970 nm band is much larger than the monomeric one when one expects a narrowing of this band. Furthermore the oscillatory strength of the 808 nm band is much smaller than expected (fig. 4).

VERMEGLIO and PAILLOTIN [24] propounded the following answers to these different problems:

i) As it can be observed on fig. 4, the 970 absorption band is due to two different transitions: the first one is the ordinary transition from the ground state to the first singlet excited state, but the second one is associated with a transition to a charge transfer state. This state can be the $P^+ A^-$ one or the inner charge transfer state $|BChl^+\text{-}BChl^-|$ of the primary donor P: $|BChl\text{-}BChl|$. They suggest that the first hypothesis is the most probable one and estimate the edge-to-edge distance between P and A equal to 3.5 [Å]. At the same time the participation of a charge transfer state in the absorption around 970 nm could explain the large broadening of this band as compared with the monomeric absorption band.

ii) The position and the intensity of the monomer absorption band around 808 nm comes from the effect of the positive charge located on P^+. Let us recall that the 808 nm band is assumed to correspond to the absorption of the $|BChl^+\text{-}BChl|$ species and essentially of the unoxidized BChl molecule. Because of the presence of an unpaired electron on the neighboring BChl, a mixing of the singlet and triplet excited states arises for the uncharged molecule. It explains the large shift towards the short wavelengths and the decrease in intensity of the absorption band of the BChl, which although included in P, does not carry the positive charge. Correlatively an extra-absorption band is expected around 1 300 nm. Such a band is actually observed in the (P^+) state. According to their model, VERMEGLIO and PAILLOTIN [24] were able to calculate an edge distance between the two BChl molecules of the dimer P equal to 2.8 [Å].

5 - Conclusion

In the previous sections we have tried to demonstrate that optical investigations offer unique possibilities for the experimental study of photosynthesis.

This comes from an obvious reason: the photosynthetic membrane contains many light-absorbing intrinsic probes. Thus light, which induces the photosynthetic reactions, principally interacts with pigment molecules the biological role of which is predominant. The resulting frequency selection justifies the use of resonance Raman spectroscopy which has significantly contributed to our knowledge of the binding sites of the pigments [5]. But this frequency selection is also underlying several of the experiments we have described: optical properties of the Chlb molecules which in plants absorb in a rather different region compared to the major component formed by Chla molecules, or in bacteria of the primary donor of the RC. In other respects the geometrical selection obtained by using linearly polarized light brings additional but essential structural information.

The possibility of isolating pigment-protein complexes -the building blocks of the photosynthetic membrane- [3] gives the opportunity of studying the local order of this membrane. Moreover this geometrical selection permits a better analysis of the functional role of different components of the photosynthetic apparatus. From this point of view, Raman spectroscopy performed with oriented samples may bring new and interesting data. The benefit which can derive from using both the frequency and the geometrical selections is also illustrated by the photoselection experiments.

Furthermore optical investigations offer two other advantages: the effect of light may be easily controlled and it is of special interest for time-resolution experiments. Most of our knowledge of the primary processes which govern the charge separation comes from very fast flash spectroscopy experiments [22,23]. Also of interest is the possibility of using very large exciting light intensities. Recently, original experiments were performed in order to analyze the exciton migration within the antenna. The elementary event in this process is a very fast one: the jump of an exciton from one Chl molecule to another takes about 1 psec. Thus it is very difficult to obtain a time resolution of this exciton motion. The experimental approach of this problem is grounded on the following idea: with very short (several psec) and intense light pulses it is possible to create, almost at the same time, several excitons within the antenna. When two of these excitons collide there is a probability that annihilation of one or even two of these excitons occurs. This induces a quenching of the fluorescence. The efficiency of this quenching is correlated to the dimension of the domain where the excitons can migrate freely. From the study of the "quenching curves" obtained with increasing pulse intensity it is possible to determine the dimension of the domains [26]. Thus, with the help of this non-linear optical investigation , topological information on the photosynthetic apparatus can be obtained.

We can conclude that optical spectroscopy is the most useful tool for the study of the photosynthetic apparatus. It is interesting to notice that the conceptual tools derived for such studies can also be very useful for the analysis of other biological processes.

REFERENCES

1 R.S. KNOX: in "Primary Process of Photosynthesis" - (Elsevier, Amsterdam, 1977) 55/97

2 P. MATHIS & G. PAILLOTIN: in "The Biochemistry of Plants", Vol. 8 (Academic, New York, 1981) 97/161

3 G. PAILLOTIN, A. VERMEGLIO & J. BRETON: in "Trends in Photobiology" (Plenum, New York, 1982) 539/547

4 R.P. HEMENGER: J. Chem. Phys. 67, 262 (1977)

5 M. LUTZ: in "Advances in Infrared and Raman spectroscopy", vol. 11 (Wiley, New York, 1983) in press

6 J. BRETON, M. MICHEL-VILLAZ & G. PAILLOTIN: Biochim. Biophys. Acta 314, 42 (1973)
7 J.P. MARKWELL, J.P. THORNBER & R.T. BOGGS: Proc. Natl. Acad. Sci. US, 76, 1233 (1979)
8 C.S. FRENCH: Proc. Natl. Acad. Sci. US, 68, 2895 (1971)
9 J. BRETON & A. VERMEGLIO: in "Photosynthesis: Energy Conversion by Plants and Bacteria" - (Academic, New York, 1982) 153/194
10 B.W. MATTHEWS, F.E. FENNA, M.C. BOLOGNESI, M.F. SCHMID & J.M. OLSON: J. Mol. Biol. 131, 259 (1979)
11 P. HAWORTH, P. TAPIE, C.J. ARNTZEN & J. BRETON: Biochim. Biophys. Acta 682, 152 (1982)
12 P. TAPIE, P. HAWORTH, G. HERVO & J. BRETON: Biochim. Biophys. Acta 682, 339 (1982)
13 R.L. VAN METTER: Biochim. Biophys. Acta 462, 642 (1977)
14 J.F. SCHEPANSKI & R.S. KNOX: Israel J. Chem. 21, 325 (1981)
15 R.P.F. GREGORY, S. DEMETER & A. FALUDI-DANIEL: Biochim. Biophys. Acta, 591, 356 (1980)
16 E. BRECHT, S. DEMETER, A. FALUDI-DANIEL Photobiochemistry, Photobiophysics 3, 153 (1981)
17 J. BRETON & M. HILAIRE: C.R. Acad. Sci. Paris, Ser. D, 274, 678 (1972)
18 O. CANAANI & K. SAUER: Biochim. Biophys. Acta 501, 545 (1978)
19 A.C. ALBRECHT: J. Mol. Spectrosc. 6, 84 (1961)
20 A. VERMEGLIO, J. BRETON, G. PAILLOTIN & R. COGDELL: Biochim. Biophys. Acta, 501, 514 (1978)
21 J. BRETON, A. VERMEGLIO, M. GARRIGOS & G. PAILLOTIN: in "Proc. Int. Congr. Photosynth. 5th", vol III (Balaban international Science Services, Philadelphia, 1981) 445/459
22 V.A. SHUVALOV, A.V. KLEVANIK, A.V. SHARKOV, Y.A. MATVEETZ & P.G. KRUKOV Febs. Lett. 91, 135 (1978)
23 D. HOLTEN, C. HOGANSON, M.W. WINDSOR, C.C. SCHENCK, W.W. PARSON, A. MIGUS, R.L. FORK a C.V. SHANK: Biochim. Biophys. Acta 592, 461 (1980)
24 A. VERMEGLIO & G. PAILLOTIN: Biochim. Biophys. Acta 681, 32 (1982)
25 A. VERMEGLIO & R.K. CLAYTON: Biochim. Biophys. Acta 449, 500 (1976)
26 G. PAILLOTIN, C.E. SWENBERG, J. BRETON & N.E. GEACINTOV: Biophys. J. 25, 513 (1979)

Index of Contributors

MIX
Papier aus verantwortungsvollen Quellen
Paper from responsible sources
FSC® C105338

If you have any concerns about our products,
you can contact us on
ProductSafety@springernature.com

In case Publisher is established outside the EU,
the EU authorized representative is:
Springer Nature Customer Service Center GmbH
Europaplatz 3, 69115 Heidelberg, Germany

Printed by Libri Plureos GmbH
in Hamburg, Germany